Beno Eckmann

Mathematical Survey Lectures 1943–2004

Beno Eckmann

Beno Eckmann

Mathematical Survey Lectures 1943–2004

Springer

Beno Eckmann
Institute for Mathematical Research
Swiss Federal Institute of Technology Zurich
8092 Zurich
Switzerland
e-mail: eckmann@math.ethz.ch

Library of Congress Control Number: 2006929803

Mathematics Subject Classification (2000): 55-XX, 53-XX, 16-XX, 20-XX

ISBN-10 3-540-33790-3   Springer Berlin Heidelberg New York
ISBN-13 978-3-540-33790-4   Springer Berlin Heidelberg New York

Springer is a part of Springer Science+Business Media

springer.com

© Springer-Verlag Berlin Heidelberg 2006
Printed in Germany

Production: LE-TEX Jelonek, Schmidt & Vöckler GbR, Leipzig
Cover design: Erich Kirchner, Heidelberg
Cover picture: Courtesy of Heinz Niederer

Printed on acid-free paper   41/3100/YL   5 4 3 2 1 0

TO DEAR DORIS
WE HAVE BEEN MARRIED SINCE 1942

Preface

In my long professional life as a mathematician I have had to deliver various survey lectures. They were of different character according to the occasion. A selected number are reproduced in this volume. They were, with few exceptions, printed in books or in periodicals, and are reproduced here without any modification.

I started as a student at the ETH Zurich in 1935 and have spent my working life there, with the exception of the period 1942–1948 (lecturer then associate professor at the University of Lausanne) and 1947 and 1951/52 (Institute for Advanced Study Princeton). In 1948 I was appointed as a professor at the ETH; I retired in 1984 and have since then been professor emeritus.

The surveys in this volume are lectures given at congresses and workshops, at international congresses of mathematicians (plenary or section lectures), and at special occasions like birthdays or commemorations. They all reflect the special interests of the respective times. Although they are in fields of topology, algebra and differential geometry they reveal the big changes mathematics has undergone over the years from 1943 to 2004. New thinking has been developed; it has created new problems, or simply has let old problems appear in a new light.

Several topics and methods appeared that were very new and then gradually became routine. They are of course present, in some way or other, in my later surveys. Let me mention here the categorical and functorial thinking, the exact sequences, the geometric ideas applied to groups, the role of the combinatorial and the analytic Laplace operator, the passage from vector space to Hilbert space and operator space methods, and so on. All the survey lectures are in the spirit of pure mathematics. I know that some of the results have had applications in various concrete fields of intellectual enterprise (I am convinced that all of pure mathematics has practical application sooner or later). These are not mentioned, except for the last item in this book. The second-to-last item is a personal account of some of the history of algebraic topology; it is a talk given at a special occasion, namely the eve of the 40-year jubilee of the Institute for Mathematical Research (FIM) which I founded in 1964 at the ETH in Zurich.

Zurich, April 2006   *Beno Eckmann*

Table of Contents

L'idée de dimension

Revue de Théologie et de Philosophie 127 (1943), 3–17

1. L'idée de dimension, bien qu'elle soit une des plus intuitives et anciennes de la géométrie, n'a été l'objet d'une théorie exacte et satisfaisante que depuis vingt ou trente ans, et ce n'est que ces derniers temps qu'elle a atteint une certaine perfection. La question, touchée déjà dans les *Eléments* d'Euclide, a été reprise autour de 1900 sous un nouvel aspect, entre autres par le célèbre mathématicien français Henri Poincaré, et ses idées sont à la base des recherches ultérieures qui ont donné lieu à une des plus belles théories géométriques.

Si je vais essayer de vous présenter quelques résultats et idées simples à ce sujet — dans la forme plus ou moins incomplète que le cadre de cette leçon m'impose et que mes collègues mathématiciens voudront bien excuser — c'est parce que je crois qu'il s'agit là d'une chose dont on parle assez souvent en disant que l'espace a trois dimensions, qu'une surface en a deux, que le temps a une dimension, en faisant allusion, avec un sous-entendu mystérieux, à la quatrième dimension — sans bien se rendre compte de ce qu'on entend par là, et sans savoir qu'il s'agit d'une question d'importance fondamentale pour la géométrie et pour toutes les sciences.

Toutefois, je ne considère tout ceci que comme un exemple, et le but de mon exposé sera atteint, si j'arrive à vous donner une idée des relations assez délicates qui existent entre l'intuition, l'expérience et l'abstraction, et qui sont caractéristiques de la façon de penser et de travailler en géométrie moderne.

2. Venons en à notre problème : Que notre espace soit à trois dimensions, qu'une courbe soit à une dimension, etc., qu'est-ce que cela veut dire ? Quelles sont les raisons et les conséquences de ce fait ?

Avant tout : Qu'est-ce que l'espace ? Il nous faut bien distinguer deux choses : l'espace de notre intuition et expérience, où nous vivons — je veux l'appeler dans ce qui suit *l'espace réel* (seulement à titre d'abréviation, car il y a plusieurs sortes ou gradations de réalité) — et *l'espace géométrique* qui est une création abstraite de l'esprit.

L'espace réel — je ne veux pas essayer d'en donner une définition fermée ; ses propriétés sont plus ou moins imprécises ; car les objets considérés ne sont pas des points, droites, etc., mais des arêtes d'un corps, des rayons lumineux, des réticules, des plans plus ou moins rugueux, des figures dessinées d'une manière un peu inexacte ; et si on essayait de rendre plus exactes les propriétés de ces objets par approximations successives, on n'y arriverait pas, même théoriquement ; comment arriver, par exemple, à une droite précise, la matière étant composée d'atomes, ou de molécules en mouvement thermique ? Et un rayon lumineux, dès qu'on essaie de le rendre suffisamment fin, commence à se disperser ! On peut même dire que les propriétés de tous ces objets changent de temps en temps — suivant nos moyens et nos possibilités expérimentales. Sans trop en discuter, nous voulons admettre que tout ce qu'on dit de l'espace réel est vrai dans un sens naïf et pas tout à fait définitif.

Dans *l'espace géométrique*, les choses sont bien différentes. Ses objets ont des propriétés tout à fait exactes : ce sont ou bien des axiomes qu'on ne démontre pas, ou bien des théorèmes qu'on démontre à l'aide des axiomes et de la déduction logique. Mais que sont ces objets ? On n'en dit rien ; ce ne sont en tout cas pas les objets inexacts de l'espace réel, mais des êtres abstraits qui ont seulement les propriétés qu'on leur a attribuées sous forme d'axiomes. Ces axiomes ne sont donc sûrement ni vrais ni faux ni évidents, mais tout simplement des postulats, des conventions qu'on impose à des êtres abstraits appelés points, droites, etc. Ces conventions sont naturellement inspirées par le réel ; elles idéalisent des choses qu'on a constatées dans le réel d'une manière assez grossière. Mais elles dépassent tout ce que le réel peut nous donner : si l'on dit, par exemple, que deux droites se coupent en un point, ou si l'on parle du comportement à l'infini, l'espace réel ne nous donne jamais de renseignements précis et directs là-dessus. Il y a donc une part d'arbitraire dans les axiomes ; l'espace géométrique serait-il ainsi une construction purement logique qui se base sur des conventions arbitraires ? Heureusement il est plus que cela, plus qu'un simple jeu logique : il est une image schématique de l'espace réel, extrêmement utile d'ailleurs ; on s'en sert à tout instant dans notre vie, dans la technique, dans les sciences.

En conclusion : l'espace géométrique est une construction logique dont la base est formée par les axiomes, c'est-à-dire des conventions arbitraires du point de vue logique, mais inspirées par le réel et en conséquence justifiées. L'espace géométrique n'est pas identique à l'espace réel, mais il en est — pour employer une expression due à M. Gonseth — un schéma simple et efficace.

On peut naturellement créer d'autres géométries, en choisissant comme base de la construction logique des axiomes un peu différents, et on l'a fait. C'est alors l'expérience qui nous amène à prendre l'une de ces géométries plutôt qu'une autre comme schéma de l'espace réel. La géométrie ordinaire ou euclidienne est considérée comme la plus simple et la plus efficace pour les besoins ordinaires ; mais il se peut que, pour des cas extraordinaires, en astronomie ou en atomistique, par exemple, on se voie contraint de préférer un autre schéma.

3. Maintenant que nous avons bien distingué deux choses, l'espace réel, d'une part, et l'espace géométrique, d'autre part, nous pouvons préciser notre question :

A quelles propriétés de l'espace géométrique abstrait fait-on appel en disant que l'espace réel est à 3 dimensions ?

Une méthode très simple pour y répondre est la suivante : Tout point de l'espace géométrique peut être caractérisé par trois nombres réels, appelés coordonnées ; remarquons que d'une manière analogue on fixe un point sur une droite par un nombre (on fait cela sur chaque règle) et un point dans le plan par deux nombres (on le fait sur toute carte topographique). Les trois coordonnées d'un point P dans l'espace sont par exemple les trois distances de P à trois plans perpendiculaires deux à deux, ou bien les deux coordonnées de la projection de P dans le plan horizontal et la hauteur au-dessus de ce plan (fig. 1).

Si on fait varier ces trois nombres indépendamment les uns des autres, on obtient tous les points de l'espace. Connaissant les coordonnées de deux points de l'espace, on peut calculer leur distance par une formule simple ; à l'aide des coordonnées, on peut calculer des angles, déduire des propriétés géométriques par de simples calculs, etc. — c'est la méthode bien connue sous le nom de *géométrie analytique*. Dans cette géométrie, un point c'est *trois nombres*, et l'espace c'est l'ensemble obtenu en faisant varier ces trois nombres indépendamment. C'est pourquoi l'on dit que l'espace géométrique

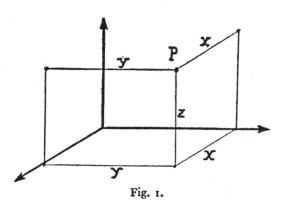

Fig. 1.

a trois dimensions, en entendant par nombre des dimensions le nombre des coordonnées variant indépendamment.

4. Oublions maintenant pour un instant la signification extérieure de l'espace géométrique, oublions qu'il est le schéma de l'espace réel. Alors on ne comprend plus le rôle extraordinaire du nombre 3 dans cette construction ; car il est clair que la même construction logique peut se faire avec 4 ou 5 ou un nombre quelconque n de coordonnées ; on obtient alors l'espace à n coordonnées ou à n dimensions.

Un point de cet *espace à n dimensions*, c'est n *nombres réels*, et on obtient tout l'espace, si on les fait varier indépendamment. On peut y faire de la *géométrie analytique* : la distance de deux points se calcule à partir des coordonnées à l'aide d'une formule analogue au cas des trois dimensions, etc.

Inutile de demander si cet espace à 4 ou à 5 ou à n dimensions existe ou non — il est simplement une construction logique qui ne prétend pas donner des renseignements sur quelque chose de réel — l'espace à 3 dimensions ne le fait pas non plus ; il peut servir comme schéma de l'espace réel, voilà son rôle particulier ! (D'ailleurs l'espace à deux dimensions est, d'une manière analogue, le schéma du plan, et celui à une dimension de la droite.)

On va se demander : Ne pourrait-on pas prendre comme schéma de l'espace réel l'espace à 4 ou 5 ou à un autre nombre de dimensions aussi bien que celui à 3 dimensions ? Cela nous donnerait une géométrie bien différente de la nôtre !

Mais parmi toutes ces possibilités, l'expérience nous en a fait choisir une : 3 dimensions sont, comme on l'a toujours constaté, juste ce qu'il faut pour décrire (d'une manière schématique, mais efficace) les points de notre espace réel. Et même si l'on a essayé,

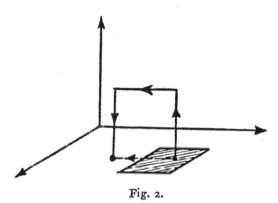

Fig. 2.

comme je l'ai déjà dit, de modifier un peu notre géométrie (par exemple en prenant une autre formule pour la distance), on n'a jamais été amené à changer ce nombre 3 des dimensions, des coordonnées variant indépendamment.

Mais il y a une autre possibilité qui reste plus ou moins ouverte : Il se pourrait que notre espace réel fît partie d'un espace réel à 4 dimensions (c'est-à-dire de quelque chose dont le schéma devrait être l'espace abstrait à 4 dimensions), comme un plan qui fait partie de l'espace ordinaire, qui est « plongé » dans cet espace. Certaines propriétés géométriques diffèrent essentiellement suivant qu'on reste dans l'espace ou qu'on en sort dans l'espace à 4 dimensions.

Pour mieux comprendre ceci, examinons la situation pour le plan. Comparons la géométrie dans le plan, selon qu'on reste dans le plan en faisant abstraction de l'espace qui l'entoure, ou qu'on n'y reste pas.

Considérons un rectangle, et un point à l'intérieur (fig. 2) ; dans le plan, il est impossible de faire sortir le point de l'intérieur du rectangle sans qu'il traverse un des côtés ; donc, s'il lui est défendu de les traverser, s'il est « enfermé », il ne peut pas sortir sans qu'on « n'ouvre une porte » ! Or, à travers l'espace, cela est bien possible : on élève le point dans la direction d'un troisième axe, perpendiculaire au plan, on le déplace parallèlement au plan, on le fait retomber dans le plan.

Considérons la situation analogue dans l'espace : si un objet est enfermé dans une armoire (dans un cube), il est impossible de l'en faire sortir sans « ouvrir la porte », sans traverser les faces, sans y percer un trou. Or, si notre espace était plongé dans un espace à 4 dimensions ou plus, cela serait bien possible. On peut le vérifier aisément et rigoureusement dans la géométrie analytique de l'espace

Fig. 3.

à 4 dimensions, en y donnant par formules le mouvement nécessaire : on déplace le point dans la direction d'un quatrième axe, on le transporte parallèlement à l'espace, et on le fait retomber dans l'espace, dans notre monde.

On peut indiquer d'autres phénomènes de ce genre qui pourraient se produire, si notre espace était plongé dans l'espace à 4 dimensions : On pourrait transformer par un simple mouvement un gant droit en un gant gauche, on pourrait résoudre un nœud fermé sans couper la ficelle (fig. 3), on pourrait séparer deux anneaux enlacés sans les ouvrir (fig. 3), et ainsi de suite.

Si de tels phénomènes se produisaient régulièrement et qu'ils fussent confirmés par des expériences physiques, le moyen le plus simple et clair pour s'en rendre compte et pour les formuler et expliquer serait le schéma d'un espace à 4 dimensions dans lequel se trouverait notre espace. Or, excepté quelques trucs de prestidigitation, ces phénomènes, désignés comme surnaturels, n'ont jamais été observés. C'est un résultat empirique (comme, par exemple, la non-existence du mouvement perpétuel de première ou de seconde espèce). Pour la description de notre espace et de ses phénomènes, l'hypothèse d'une quatrième dimension est superflue.

5. Ce qui, naturellement, ne nous empêche pas, nous autres mathématiciens, de parler de l'espace à quatre dimensions ou plus, considéré uniquement comme système logique, et d'examiner ses propriétés. On garde le même langage géométrique et on parle de points, de droites, de plans, d'angles, etc., bien qu'en général ces choses ne puissent pas être mises en correspondance avec des objets de l'espace réel ! Néanmoins cette construction est très importante et efficace, pour les raisons suivantes :

Elle permet au mathématicien de traduire en langage géométrique des faits et problèmes analytiques ou algébriques, ce qui simplifie parfois énormément la solution et suggère des méthodes et résultats ; on peut presque dire qu'en pratiquant cette géométrie, on arrive à une certaine intuition de l'espace à n dimensions — je ne sais pas si c'est la grande analogie avec l'espace à 3 dimensions (qui naturellement peut aussi nous tromper !) ou tout simplement l'habitude de penser à ces choses.

Plus importante encore est l'application que voici : il y a, soit dans le monde de l'expérience et de l'intuition, soit en physique, soit dans les différentes branches mathématiques, des objets et phénomènes pour lesquels il y a grand avantage à prendre l'espace à n dimensions comme « schéma », dans ce sens qu'ils peuvent être décrits par n nombres réels variant indépendamment comme les coordonnées dans l'espace à n dimensions. On appelle alors ceci *un continu à n dimensions ou à n degrés de liberté*.

On en trouve des exemples dans tous les domaines de la science :

a) *Une courbe est un continu à une dimension ;* on peut « numéroter » ses points par un nombre variable, à savoir par la longueur de l'arc de courbe à partir d'un certain point fixe. En d'autres termes, on peut faire correspondre aux points de la courbe les points d'une droite (espace à une dimension !) de telle sorte qu'à deux points distincts correspondent deux points distincts, et qu'à des points voisins correspondent des points voisins. On parle dans ce cas d'une correspondance bi-univoque (c'est-à-dire univoque dans les deux sens) et continue.

b) Autre exemple : *le temps* est un continu à une dimension, car on fixe les instants divers par *un* nombre.

c) *Un morceau de surface*, par exemple d'une sphère, est un continu à 2 dimensions : on peut én décrire les points par 2 nombres, les 2 coordonnées d'une carte topographique de la surface (soit la longitude et la latitude).

d) *Les mouvements dans le plan* (par exemple d'un segment) forment un continu à 3 dimensions : tout mouvement est donné par 2 déplacements indépendants et par une rotation. On dit qu'un corps possède pour les mouvements dans le plan 3 degrés de liberté, et on peut représenter ces mouvements par 3 nombres, donc par les points de l'espace à 3 dimensions.

e) D'une manière analogue, *les mouvements d'un corps dans l'espace* ont 6 degrés de liberté : 3 translations indépendantes et une rotation qui est donnée par trois nombres. Les mouvements dans l'espace, donnés par 6 nombres, peuvent donc être représentés par des points de l'espace à 6 dimensions. (Si on fixe un point du corps en mouvement, il n'a que 3 degrés de liberté, ceux de la rotation. Si on fixe 2 points, il ne reste qu'un degré de liberté, celui de la rotation autour de la droite qui joint les 2 points.)

Les physiciens se souviennent bien de l'importance toute particulière du nombre de ces degrés de liberté, par exemple dans la théorie de la chaleur spécifique.

f) Pour fixer le lieu et le temps d'un point qui se déplace, ou d'une observation, il nous faut 4 nombres : *le continu espace-temps* est à 4 dimensions, on peut le représenter par les points de l'espace à 4 dimensions, et cela est même d'une grande importance dans la théorie de la relativité.

g) *L'état d'une molécule* en mouvement peut être donné par 6 nombres : les 3 coordonnées du lieu, et les 3 composantes de sa vitesse. Dans la théorie cinétique des gaz, l'état d'un gaz constitué par N molécules sera donc donné par $6N$ nombres et les états divers du gaz peuvent être représentés par les points de l'espace à $6N$, disons 10^{24} dimensions. Cela semble assez drôle, mais c'est très pratique dans la théorie statistique des molécules.

Je ne veux pas parler des nombreux exemples que le mathématicien rencontre en géométrie algébrique et dans l'analyse.

6. En résumé :

A la question : « Pourquoi dit-on que notre espace a 3 dimensions », on répondra : Parce qu'il a pour schéma l'espace géométrique à 3 dimensions ou coordonnées.

Et à la question : « Pourquoi tel et tel continu a-t-il n dimensions ou *n* degrés de liberté » : Parce qu'il peut être décrit par *n* nombres variant indépendamment comme les *n* coordonnées dans l'espace à *n* dimensions, autrement dit : parce que les éléments de ce continu peuvent être représentés par les points de cet espace, ou *mis en correspondance avec les points de cet espace à* n *dimensions.*

Tout cela semble très simple, et c'est bien ce qu'on a longtemps regardé comme définition de la dimension, sans bien préciser ce qu'on entend par cette *correspondance* (en pensant peut-être surtout à des correspondances données par des formules simples). Mais c'est en ce moment que de graves *difficultés* s'élèvent, difficultés qui sont de caractère mathématique, c'est-à-dire interviennent dans la construction logique, et qu'on doit surmonter avant que la définition de la dimension ne soit justifiée.

7. Ces difficultés, les voici :

Si l'on arrive à décrire un continu, disons par 2 nombres, ne pourrait-on pas le faire aussi bien par 3 (ou 4 ou 1) — naturellement en procédant d'une tout autre manière ? Y a-t-il en somme une différence entre les espaces à différentes dimensions, par exemple entre celui à 2 et celui à 3 ? Ne pourrait-on pas établir une *correspondance* entre les points de ces deux espaces qui permettrait de remplacer tout simplement l'un par l'autre, surtout pour la description d'un continu ?

Vous le voyez : La notion de dimension qui semble si simple et qui est si importante pour maintes applications, on ne sait pas si elle a un sens !

Vous me répondrez : Ce qu'on peut décrire par 2 nombres variant indépendamment, on ne le peut pas par 3, car dans le plan il n'y a pas autant de points que dans l'espace !

Or, cela n'est pas vrai dans un certain sens (bien que, si on pense à un plan plongé dans l'espace, il y ait des points en dehors du plan !) En effet, deux faits extrêmement surprenants, découverts vers la fin du XIXe siècle, l'ont bien mis en évidence.

Premièrement, l'école de Cantor a trouvé des correspondances entre deux espaces à différentes dimensions (par exemple entre plan et espace, ou droite et espace à 4 dimensions, etc.), correspondances qui sont bi-univoques, c'est-à-dire qu'à deux points différents correspondent deux points différents. On pourrait dire que de cette manière les deux espaces contiennent « le même nombre de points », et on pourrait, en effet, remplacer l'un de ces espaces par l'autre !

Et maintenant ? Faut-il pour cela abandonner l'idée de dimen-

sion comme nous l'avons formulée ? Non. Car cette correspondance
de Cantor, bien qu'elle soit bi-univoque, *n'est pas continue*, c'est-à-
dire qu'à des points voisins correspondent des points éloignés dans
l'autre espace, ou dans le continu qu'on veut décrire, et dont la
connexion serait ainsi complètement détruite ! Or, pour définir la
dimension, on n'a sûrement pas envisagé de telles correspondances ;
elles sont à exclure.

Le second fait, c'est la célèbre *courbe de Péano*. C'est une courbe,
en ce sens que c'est l'ensemble des points parcourus par un point
qui se déplace dans le plan ; on peut numéroter ces points à l'aide
du temps, donc d'*un* nombre, c'est-à-dire on peut les mettre en correspondance avec les points de la droite. Or, cette courbe de Péano
a la propriété de remplir complètement un carré, donc un continu
à 2 dimensions, et elle établit ainsi une correspondance entre droite
et plan. Cette fois, la correspondance est continue — mais *elle n'est
pas bi-univoque*, car il y a des points du carré qui sont parcourus
plusieurs fois, et pour cela elle est aussi à exclure.

8. Si ces résultats vous semblent un peu étranges, n'oubliez pas
qu'on parle des espaces géométriques abstraits ; il ne s'agit donc pas
de l'intuition, mais de la construction logique qui, bien qu'elle soit
inspirée par le réel, dépasse — même dans son fondement axiomatique ! — les possibilités de notre intuition et expérience. Et comme
cette construction est à la base de l'idée de dimension exprimée dans
la définition, on voit maintenant ce qu'il nous faut pour *sauver*
cette idée de dimension : il faut démontrer qu'il est impossible
d'établir une correspondance qui soit en même temps bi-univoque
et continue entre deux espaces à différentes dimensions. En d'autres
termes : qu'il est impossible de construire une image bi-univoque et
continue de l'espace à n dimensions dans un espace à dimension infé-
rieure, par exemple d'un plan dans une droite.

On peut démontrer cela. Cette fois l'intuition ne nous trompe pas.
La première démonstration a été donnée en 1911 par Brouwer [1] ;
en 1913, il en a donné une autre, plus profonde que la première, en
précisant des idées énoncées en 1912 par Poincaré dans son célèbre
mémoire : « Pourquoi l'espace a trois dimensions » [1]. Les recherches
de Brouwer et l'importance de ses résultats ne furent guère connues
pendant plusieurs années. En 1922, les mêmes idées furent reprises

[1] Voir la note bibliographique à la fin de cette conférence.

indépendamment par Menger et Urysohn [1] et bien approfondies. C'est alors qu'une belle théorie a été créée avec la collaboration d'un grand nombre de mathématiciens. On peut dire que le point final de ce développement, bien que maints problèmes y restent ouverts, c'est le livre de Hurewicz et Wallmann, *Dimension Theory*, paru en 1941 à Princeton [1].

9. Un des principaux résultats de toutes ces recherches — mais non le seul ! — c'est donc que l'ancienne définition commune de la dimension, comme nous l'avons formulée, subsiste, qu'elle a un sens. Quelle est au fond la signification de ce fait ? A quel ordre d'idées appartient-il ?

Le point essentiel, nous l'avons vu, c'étaient les correspondances ou images bi-univoques et continues. Considérons, par exemple, un cercle, et un cercle mal dessiné, c'est-à-dire une *déformation* du premier. C'est bien une image bi-univoque et continue du cercle. On constate que presque toutes les propriétés géométriques se sont perdues, longueur, angles, directions, la courbe n'a plus de centre, etc. Il y a pourtant quelque chose qui subsiste : la courbe est toujours un continu à une dimension, et elle est fermée (non ouverte). Ces propriétés qui n'ont pas changé sont d'un autre ordre que celles qu'on considère en géométrie ordinaire, on dirait d'un ordre plus élémentaire. La discipline qui s'occupe de telles propriétés, s'appelle *analysis situs* ou *topologie*.

Par exemple, il n'y a pas de différence topologique entre une sphère et un œuf ou un ellipsoïde ou une surface fermée de ce genre, mais bien entre une sphère et un tore. Et c'est une propriété topologique de dire que deux courbes fermées dans l'espace soient enlacées (fig. 3) ou non. Le célèbre théorème d'Euler sur les polyèdres (la somme des sommets et des faces moins la somme des arêtes est égale à 2) est de caractère topologique ; car il faut seulement que le polyèdre ait la connexion, le genre d'une sphère ; la forme, les angles, les côtés des faces n'y jouent aucun rôle.

Et le sens de ce théorème de Brouwer qui était nécessaire pour sauver l'idée de dimension (qui dit qu'il est impossible d'altérer le nombre des dimensions par une simple déformation) — c'est que *la dimension d'un espace est une propriété topologique*, bien qu'on

---

[1] Voir la note bibliographique à la fin de cette conférence.

l'ait définie à l'aide des coordonnées, donc d'éléments non-topologiques, à savoir des angles, des longueurs, des droites, etc.

On peut dire — les exemples vous l'ont bien montré — que les propriétés topologiques sont suggérées par l'expérience et l'intuition purement géométriques, tandis qu'en géométrie ordinaire on fait appel à des notions arithmétiques et analytiques d'un tout autre ordre. Autrement dit : Pour arriver à la géométrie ordinaire, il faut restreindre le domaine des correspondances envisagées jusqu'ici ; par exemple si, au lieu de considérer des déformations bi-univoques et continues, on considère seulement les mouvements dans l'espace, alors les longueurs et les angles, etc., restent invariants, donc les choses, dont on s'occupe en géométrie élémentaire métrique. Ainsi les différents points de vue géométriques sont caractérisés par les correspondances admises ; le cas le plus général, c'est la topologie ; le cas le plus restreint, la géométrie métrique, et il y a des cas intermédiaires.

10. On a donc constaté que l'idée de dimension est d'un caractère topologique ; aussi devrait-on chercher une *définition* qui le mette en évidence, c'est-à-dire une définition n'utilisant que des notions topologiques.

Pour cela, laissons-nous inspirer tout simplement par l'espace réel. Dans les *Eléments* d'Euclide, on trouve l'idée suivante : L'extrémité d'une courbe, c'est un point, le bord d'une surface, c'est une courbe, la frontière d'un corps, c'est une surface. Dans le mémoire que j'ai cité, Poincaré propose de caractériser la dimension par de telles propriétés, ou des propriétés très voisines. La définition de Menger et Urysohn, admise aujourd'hui comme étant la meilleure, n'en diffère presque pas.

On prend un point du continu en question et un voisinage entier de ce point, et on essaie d'extraire le voisinage hors de ce continu (fig. 4). Pour cela, on est obligé de couper ou de déchirer le continu en certains points qu'on appelle points *frontières du voisinage*.

S'il s'agit d'un continu à *une* dimension — pensons à une courbe ou un fil de fer — il suffit de le couper en quelques points isolés (qui ne forment donc eux-mêmes aucun continu).

S'il s'agit d'un continu à 2 dimensions — pensons à une surface — il ne suffit pas de la couper en quelques points isolés, on doit la couper suivant une courbe (donc un continu à une dimension).

S'il s'agit d'un continu à 3 dimensions — pensons à l'espace — ni des points isolés ni des courbes ne suffisent ; la frontière d'un voisinage (par exemple d'une boule solide) et formée par une surface (donc un continu à 2 dimensions) — et ainsi de suite.

On dira donc *qu'un continu est à* n *dimensions, lorsque les points frontières d'un voisinage forment un continu à* n — 1 *dimensions.*

On peut alors démontrer que l'espace abstrait à *n* coordonnées jouit précisément de cette propriété ; ceci montre que l'ancienne et la nouvelle définitions sont équivalentes. Mais la nouvelle n'utilise ni longueurs, ni angles, ni droites, seulement les voisinages, qui sont de caractère purement topologique (leur forme ne joue aucun rôle !) C'est peut-être la meilleure méthode pour mettre en évidence ce fait

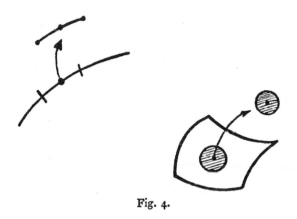

Fig. 4.

que la dimension, le nombre des degrés de liberté, exprime une propriété topologique d'un continu.

11. Je pourrais encore vous parler d'autres méthodes qui permettent de caractériser la dimension d'une manière topologique, par exemple celle proposée par Lebesgue utilisant la notion de recouvrement ; elle était très importante pour le développement de la théorie. Je pourrais aussi attirer votre attention sur les espaces à une infinité de dimensions, ou sur des espaces (abstraits) plus généraux encore que ceux donnés par des coordonnées ; ce sont les *espaces topologiques,* où on n'a presque qu'une seule notion géométrique, celle du voisinage ; mais comme nous l'avons vu, ces voisinages nous permettent de parler de la dimension, et on a même trouvé le résultat intéressant qu'en principe tous ces espaces généralisés sont des parties des espaces donnés à l'aide de coordonnées. Je pourrais encore vous parler des problèmes non résolus, des applications à

l'algèbre et à l'analyse — mais cela me conduirait trop loin, et il est temps de conclure.

12. Vous avez sûrement constaté que toutes ces méthodes géométriques sont caractérisées par une collaboration tout à fait particulière entre l'intuition, l'expérience et l'abstraction, où l'on distingue toujours strictement entre le réel et la théorie qui peut être son schéma. C'est d'ailleurs typique pour tout travail mathématique et même pour les sciences, surtout pour la physique théorique.

Cela ne fut pas toujours le cas. Autrefois on pensait que l'espace géométrique et l'espace réel, qu'une théorie et son objet, c'était la même chose. Et tout au fond, disons d'une manière inofficielle, on le pense peut-être aujourd'hui encore.

Si l'on fut obligé d'abandonner cette idée, si l'on dit sans ambiguïté que tout ce qu'une théorie énonce ne concerne pas la réalité, mais seulement son schéma abstrait — n'est-ce pas faire de nécessité vertu ? L'intelligence n'est-elle pas ainsi plus modeste qu'elle ne voudrait l'être ?

Peut-être devrait-on formuler la chose un peu différemment. Car ce qu'on appelle la réalité, on ne peut au fond la voir et décrire qu'à travers un certain schéma abstrait. Sans l'idée préexistante d'un continu, toute expérience serait impossible, elle se réduirait à des sensations isolées sans relations. Des concepts simples, considérés comme appartenant à l'expérience et à l'intuition — comme par exemple celui de la dimension — supposent au fond déjà une théorie abstraite. Souvent même les mathématiques, avec leurs constructions et formalismes abstraits, ont permis de concevoir de nouvelles idées intuitives ou de découvrir de nouveaux faits expérimentaux ; l'histoire des sciences nous en donne bien des exemples.

Vous me permettrez donc de voir dans les mathématiques non seulement un instrument très utile pour les sciences et la technique, non seulement le langage qui nous permet de mettre en relations les phénomènes, de formuler des lois et d'en tirer les conséquences, mais beaucoup plus : *J'y vois l'expression de notre façon de penser*. Et si le mathématicien, comme un géographe qui ne se contente pas de connaître la géographie de son village natal, semble s'éloigner de plus en plus des schémas ordinaires et ose aller d'une abstraction et généralisation à l'autre, il ne s'éloigne pas plus pour cela du réel ; au contraire, il crée de nouvelles possibilités de penser, et de voir et comprendre notre monde.

NOTE BIBLIOGRAPHIQUE

Nous n'avons pas indiqué les mémoires et livres qui traitent les questions mathématiques abordées dans cette conférence. On les trouve cités dans les deux livres suivants consacrés à la théorie de la dimension :

K. Menger, *Dimensionstheorie* (Leipzig, 1928).

W. Hurewicz et H. Wallmann, *Dimension Theory* (Princeton, 1941). Voir aussi le livre de P. Alexandroff et H. Hopf, *Topologie I* (Berlin, 1935), dont plusieurs chapitres sont consacrés à la notion de dimension

Ces questions sont traitées d'une manière plutôt générale dans

H. Poincaré, *Dernières pensées* (Paris, 1913), le chapitre intitulé « Pourquoi l'espace a trois dimensions », p. 57-97 ;

Krise und Neuaufbau in den exakten Wissenschaften (Leipzig und Wien, 1933), cinq conférences, dont en particulier :

H. Hahn, *Die Krise der Anschauung*, p. 42-64.

G. Nœbeling, *Die vierte Dimension und der krumme Raum*, p. 66-92.

Eidgenössische Technische Hochschule Zürich, ETHZ

Topologie und Algebra

Vierteljahrsschrift der Naturforschenden Gesellschaft in Zürich 89 (1944), 25–34

Antrittsvorlesung, gehalten am 22. Mai 1943 an der Eidgenössischen
Technischen Hochschule in Zürich

1. Es sei mir gestattet, an zwei einfache
Tatsachen anzuknüpfen. Die eine ist der
berühmte EULER'sche Polyedersatz: Beim
Würfel ist die Anzahl der Ecken und Flächen zusammen um zwei grösser als die
der Kanten. Und das ist nicht nur beim
Würfel so, sondern bei jedem Körper, ob
er regelmässig oder unregelmässig, einfach
oder kompliziert sei, wenn er nur, grob gesagt, etwa die Gestalt einer deformierten
Kugel hat. Das zweite Beispiel: Wenn man
auf der unendlichen Geraden nach beiden
Seiten Punkte in gleichen Abständen mar-
kiert und dann die Gerade auf einen Kreis
aufrollt, so liegen die markierten Punkte
auf dem Kreis überall vollkommen dicht
und gleichmässig verteilt — wenn nicht gerade zufällig der Abstand der Punkte ein
ganzzahliger Bruchteil des Kreisumfangs
ist (oder ein Vielfaches davon); in diesem
Falle kommen die markierten Punkte nur
auf vereinzelte, regelmässig verteilte Stellen des Kreises zu liegen.

Ich will vorläufig mit diesen Beispielen
nur andeuten, dass es überraschende
Beziehungen gibt zwischen einfachen Ge-
genständen der räumlichen Anschauung und
den Zahlen, dem Rechnen. Solche Beziehungen finden sich in allen Teilen der Mathematik, in den abstraktesten Theorien wie
in den Anwendungen in Physik und Technik.

R a u m u n d Z a h l — wenn sie auch fast
immer gleichzeitig auftreten, so ist man
doch geneigt, sie auf den ersten Blick als
sehr verschiedenartige Dinge anzusehen,
die zu ebenso verschiedenartigen Methoden
und Denkweisen Anlass gegeben haben;
man unterscheidet ausdrücklich zwischen
algebraischen und geometrischen Überlegungen, d. h. solchen, die auf dem Rechnen,
und solchen, die auf dem Raumbegriff beruhen. Bei näherem Zusehen zeigt es sich
aber bald, dass die beiden Denkweisen eng
miteinander verknüpft, ja ineinander ver-
flochten sind, so dass es gar nicht leicht
wäre, eine Grenze zwischen ihnen zu zie-
hen. Ihre eingehende Analyse und die Ver-
folgung ihrer Wechselbeziehungen, das ist
für das Verständnis vieler Teile der Mathematik besonders wertvoll und hat in neuerer Zeit zu ihrer Entwicklung entscheidend
beigetragen.

Wenn ich es im folgenden versuche, mit
einigen Andeutungen und an Hand einfach-
ster Beispiele auf solche Zusammenhänge
aufmerksam zu machen, so weiss ich wohl,
dass man damit das Wesen der Sache nur
sehr unvollkommen wiedergeben kann, und
dass ohne die wirkliche Durchführung der
Ideen in Theorie und Praxis das Bild immer verzerrt bleibt. So werde ich mich vor
allem bemühen, die gedankliche Atmosphäre fühlbar zu machen, in welcher sich

solche Beziehungen abspielen. Man wird es wohl entschuldigen, wenn ich im Rahmen dieses Vortrages manches nur sehr unvollständig formuliere, mit allzu allgemeinen Wendungen über die Schwierigkeiten hinweggehe und hier oder dort etwas vereinfache oder übertreibe; und auf der andern Seite wird man es mir nicht übel nehmen, wenn ich gelegentlich in abgelegenere Gebiete der Wissenschaft gerate und dabei nicht umhin kann, die dort gebräuchliche Geheimsprache zu verwenden.

2. Wenn man sich einzelne, besonders wichtige Züge eines etwas komplexen und unübersichtlichen Gegenstandes klar machen will, so löst man sie vom Ganzen los und untersucht sie selbständig; man macht sich ein vereinfachtes Bild, ein S c h e m a, des Gegenstandes, oder sogar mehrere verschiedene, indem man ihn von verschiedenen Seiten zu beleuchten sucht. In diesem Sinn hat man als Schema des Rechnens die abstrakte Algebra und als Schema der Raumbeschreibung die abstrakte Geometrie herausgeschält und verselbständigt.

Ein Schema — man denke etwa an dasjenige einer komplizierten Maschine — ist kein getreues Abbild der Wirklichkeit, sondern sehr unvollständig und meistens verzerrt. Aber das, worauf es ankommt, kann man ihm übersichtlicher entnehmen als der Wirklichkeit selbst: Es lässt Überflüssiges weg, und hebt dafür das Wesentliche um so besser hervor; es erhebt nicht den Anspruch darauf, die Dinge richtig wiederzugeben, sondern nur bestimmte Beziehungen zwischen ihnen. Deswegen umfasst es oft gleichzeitig verschiedene ähnliche Fälle und erleichtert das Verständnis der Grundidee, die gewöhnlich nicht nur hinter einer Fülle technischer Einzelheiten, sondern auch hinter einer gefälligen äussern Form verborgen ist. Und nicht viel anders ist es eigentlich in der Mathematik: selbst wenn man die mehr oder weniger verwickelten Einzelheiten blinder Rechnungen oder unübersichtlicher Konstruktionen verstanden hat, so empfindet man doch oft das Bedürfnis, den Gedankengang als Ganzes zu überblicken, ihn in allgemeine Zusammenhänge einzuordnen, das «Wesen der Methode» oder die «Idee des Beweises» zu kennen. Man möchte über die speziellen Anwendungen hinaus alle Möglichkeiten erfassen und

sucht selbst Einzelprobleme durch allgemeingültige Schlüsse zu meistern; man gibt sich nicht mit «zufälligen» Resultaten zufrieden, sondern fragt nach ihrem «tiefern» Grund.

So hat es sich als besonders lohnend erwiesen, aus der Fülle der Möglichkeiten die beiden nächstliegenden Gesichtspunkte herauszugreifen und sich über das Schema des Rechnens und dasjenige der Raumbeschreibung Rechenschaft zu geben.

3. Darf ich zunächst über das erste, die A l g e b r a, einige ganz kurz und allgemein gehaltene Bemerkungen machen.

In der abstrakten Algebra nimmt man vom Zahlbegriff nur eines, das Rechnen, sieht aber von der Natur der Zahlen selbst völlig ab; man ersetzt sie ja durch Buchstaben und führt mit diesen alle Operationen aus, Addition, Multiplikation und ihre Umkehrungen. Dabei ahmt man die Regeln des Zahlenrechnens mehr oder weniger vollständig nach; so soll z. B., weil $3+5=5+3$ ist, auch $a+b = b+a$ sein (Kommutativgesetz). Und das hat diesen Sinn: Formeln und Resultate bleiben richtig, wenn man für die Buchstaben irgendwelche Zahlen einsetzt. Natürlich bleibt auf diese Weise manches am Charakter der reellen Zahlen unausgenützt: etwa dass sie, auf der Zahlgeraden dargestellt, eine lückenlose Menge bilden. Denn rechnen kann man ja auch mit den ganzen Zahlen, die doch in mancher Hinsicht einen andern Charakter haben. Überhaupt gibt es noch viele andere Dinge als reelle oder ganze Zahlen, die man auch addieren kann (oder multiplizieren, oder beides), und für diese gilt genau derselbe Formalismus. So kann man z. B. die Winkel in einem Punkt addieren, oder die Drehungen um diesen Punkt; obschon dabei dieselben formalen Regeln gelten wie etwa für die Addition reeller Zahlen, so ist doch dieses System ganz anders beschaffen; durch wiederholtes Addieren kommt man wieder auf Null zurück. Systeme von Dingen, mit denen man eine Operation ausführen kann, nennt man G r u p p e n (genauer kommutative, wenn, wie in allen unsern Beispielen, das Kommutativgesetz gilt). Es gibt auch Gruppen, die nur aus endlich vielen Dingen bestehen; z. B. nehme man statt aller Winkel oder Drehungen nur die Achtelsdrehungen und ihre Vielfachen und numeriere sie mit den Zahlen 0 bis 7; dann ist $1 + 2 = 3$, $2 + 4 = 6$, aber $3 + 5 = 0$, $4 + 7 = 3$ usw., und man spricht von einer Gruppe endlicher Ordnung (hier 8). Die Gruppe der Ordnung 2 besteht nur aus zwei Dingen, etwa 0 und 1, und es ist $1 + 1 = 0$.

Die Theorie der Gruppen sucht die Beziehungen aufzufinden, die allen Systemen mit einer Operation gemeinsam sind, und alle überhaupt möglichen solchen Systeme zu bestimmen. Da man in den verschiedensten Gebieten auf solche Systeme stösst, vereinigt diese Theorie Gedankengänge, die in der Zahlentheorie, beim Auflösen von Gleichungen, in der Funktionentheorie, in der Atomphysik, in der Lehre von den Kristallen usw. vorkommen — sie bildet für sie alle ein gemeinsames Schema, das natürlich komplizierter ist, als es meine anspruchslosen Beispiele vermuten lassen.

In analoger Weise untersucht man die Systeme mit beiden Operationen (Addition und Multiplikation und ihre Umkehrungen); man nennt sie **Körper**. Vom Kommutativgesetz der Multiplikation ($a . b = b . a$) sieht man dabei oft ab, da es viele Zahlsysteme gibt, in welchen es nicht gilt, und die in mancher Hinsicht, sogar für Anwendungen in Physik und Technik, wichtig sind. Ich will darauf verzichten, über die Fülle der Probleme und Möglichkeiten, die dieser Körperbegriff in sich birgt, auch nur Andeutungen zu machen.

Der wesentliche Schritt der abstrakten Algebra besteht darin, dass man das Schema verselbständigt, jede Gruppe und jeden Körper gewissermassen als eine Welt für sich betrachtet, in der es zwischen den Dingen keine andern Beziehungen gibt als die reinen Rechenoperationen. Was man darin vornimmt, sind also endliche, Schritt für Schritt zu überblickende Vorgänge. Es entsteht so ein abstraktes Gebäude von seltener Geschlossenheit und Harmonie, die auch in allen Anwendungen — in der Mathematik oder in unserer Umwelt — aufs schönste zur Geltung kommt.

Die abstrakte Algebra ist keine alte Disziplin; in der zweiten Hälfte des neunzehnten Jahrhunderts unter den Händen von DEDEKIND und KRONECKER entstanden, ist sie erst in jüngerer Zeit von EMMY NOETHER und ihrem Kreis zur vollen Blüte gebracht worden.

4. Nun zur **Geometrie**, zum Schema des Raumes. Als einfachste Eigenschaften und Beziehungen, die man aus dem ziemlich komplexen Raumbegriff herausgreifen könnte, erscheinen einem zunächst sicher diejenigen, welche vom Verbinden von Punkten, vom Schneiden von Geraden, vom Aufspannen von Ebenen usw. handeln, also etwa das, was man als **projektive Geometrie** bezeichnet. Man wird hier wieder von der Natur der Dinge Punkt, Gerade, Ebene absehen und nur auf ihre gegenseitigen Beziehungen achten und erhält so eine Theorie, die sich in ihrer Einfachheit und Allgemeinheit sehr wohl der abstrakten Algebra gegenüberstellen lässt. Wenn man aber nun meint, dass man in dieser Weise streng zwischen algebraischen und geometrischen Überlegungen unterscheiden könne, dass man genau trennen könne, was im einen und was im andern Schema wurzelt oder sich beweisen lässt — dann ist das ein Irrtum.

Es handelt sich wohl um einen methodischen, aber nicht um einen inhaltlichen Unterschied. Es hat sich gezeigt, und das ist eines der Ergebnisse der berühmten Untersuchungen von HILBERT[1][1]) über die «Grundlagen der Geometrie», dass die beiden Theorien, so verschieden ihr anschaulicher Ursprung auch sein mag, genau dasselbe leisten (falls man die räumliche projektive Geometrie zugrunde legt, die ebene genügt nicht!). Das wird den, der mit der «analytischen Geometrie» vertraut ist, vielleicht wenig wundern; in dieser wird ja einfach ein Punkt der Ebene ersetzt durch zwei Zahlen, seine Koordinaten (wie das jedem bei topographischen Karten geläufig ist), ein Punkt im Raum durch drei Zahlen, und das geometrische Konstruieren wird ersetzt durch das Rechnen mit diesen Zahlen, und umgekehrt. So kann man z. B. ein Geschütz mit Hilfe von Koordinaten und Berechnungen richten — und das Ziel auch tatsächlich treffen!

Aber die Übereinstimmung ist tiefer; das liegt daran, dass den Grundbegriffen der projektiven Geometrie (d. h. dem Verbinden, Schneiden, Aufspannen) analytisch die einfachsten Rechenoperationen entsprechen, wie man sie nicht nur mit reellen

---

[1]) Solche Zahlen [1] usw. verweisen auf das Literaturverzeichns am Schluss dieses Vortrags.

Zahlen, sondern in jedem Körper ausführen kann. So bedeutet z. B. das Schneiden von Geraden in der Ebene oder von Ebenen im Raum nichts anderes als das Auflösen linearer Gleichungen. Die Verwandtschaft lässt sich bis in die Grundregeln verfolgen: der Regel a(bc) = (ab)c, die für jedes Zahlsystem unentbehrlich ist, entspricht die als «Satz von Desargues» bekannte ebene Schnittfigur (Abb. 1; die Ecken der

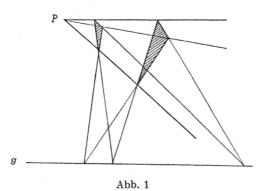

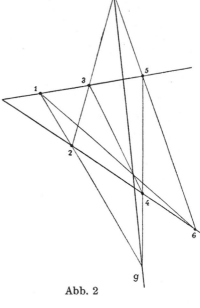

Abb. 1   Abb. 2

drei Dreiecke liegen auf den drei Strahlen durch *P*, und ihre verlängerten Seiten müssen sich immer auf einer Geraden *g* schneiden — falls wir uns in einer Ebene befinden, die Teil der räumlichen Geometrie bildet; in diesem Fall erkennt man die Richtigkeit sofort, wenn man die Abbildung als Photographie einer räumlichen betrachtet). Und die Rechenregel a.b = b.a entspricht dem «Satz von Pascal» (Abb. 2; man schreibt dem Winkel ein Sechseck 123456 ein und findet, dass die drei Schnittpunkte (12)(45), (23)(56), (34)(61) auf einer Geraden *g* liegen).

Dem abstrakten Standpunkt entsprechend wird man jedes System von Dingen, an welchem sich die projektiven Operationen des Schneidens von Geraden usw. ausführen lassen, einen projektiven Raum nennen, auch wenn dieses System mit unserem Erfahrungsraum nicht viel zu tun hat. Dann gibt es — ähnlich wie das Rechnen in den verschiedenen Zahlsystemen der abstrakten Algebra möglich ist — verschiedene projektive Geometrien; jede aber kann man als analytische Geometrie betreiben, wenn man nur für die Koordinaten ein geeignetes Zahlsystem wählt. Die Geometrie be-

stimmt die Eigenschaften des ihr angemessenen Zahlsystems (Körpers) selbst (z. B. gilt — im Gegensatz zum Satz von DESARGUES — der Satz von PASCAL nicht immer, was zur Folge hat, dass dann im zugehörigen Körper das Kommutativgesetz a.b = b.a nicht gilt). Zu jeder projektiven Geometrie gehört ein Zahlsystem und umgekehrt: man ist wohl berechtigt, hier mit HERMANN WEYL von einer «prästabilierten Harmonie zwischen Geometrie und Algebra» zu sprechen. Man sieht sich zur Auffassung der Griechen zurückgeführt, wonach ein Sachgebiet aus sich heraus einen ihm zukommenden Zahlbegriff bestimmt. Diese Auffassung hat lange Zeit einer wohl erklärlichen Vorrangstellung der (reellen) Zahlen weichen müssen, ist aber in den letzten Jahren in der Physik besonders eindrücklich wieder hervorgetreten. Es hat sich dort gezeigt, dass die reellen Zahlen wohl zur Wiedergabe des makroskopischen Geschehens und experimenteller Messungen, aber nicht zur Beschreibung der Systeme von Atomdimensionen geeignet sind; da treten andere Zahlsysteme an ihre Stelle, in denen vor allem das Kommutativgesetz a.b = b.a nicht gilt, und es lässt sich genau verfolgen, wie ihren algebraischen Eigenschaften gewisse physikalische (Ladung, Eigendrehimpuls usw.) entsprechen.

5. Man muss also feststellen, dass der als besonders einfach betrachtete Teil der Raumbeschreibung, die projektive Geometrie, uns genau auf die abstrakte Algebra zurückführt; und man fragt nach andern dem reinen Raumbegriff entspringenden Ideen, die man dem Algebraischen mit Recht als geometrisch gegenüberstellen darf. Dies besonders auch deshalb, weil man ja mit dem bisher Erwähnten vielen Eigenschaften von Raum und Zahl gewiss nicht gerecht worden ist. Das erhellt schon aus dem einfachen Beispiel der r e e l l e n Z a h l e n.

Wenn man diese, wie es der reine Algebraiker tut, nur addiert und multipliziert, so betrachtet man sie als isolierte Dinge, mit denen man gewissermassen nur sprunghaft umgeht; denn man behandelt sie ja nicht anders als die ganzen Zahlen oder endliche Gruppen. Man denkt nicht daran, dass sie gleichzeitig eine kontinuierliche, lückenlose Mannigfaltigkeit bilden — die Zahlgerade, den Maßstab —, in welcher man stetig von einer Zahl zur andern übergehen und jedes Intervall unbegrenzt beliebig fein unterteilen kann. Der algebraischen Auffassung, in welcher das Einzelwesen isoliert, ohne Rücksicht auf seine Nachbarschaft, betrachtet wird, steht eine andere gegenüber, die es auf Gedeih und Verderb mit seiner Umgebung verbindet und für die der stetige Zusammenhang des Ganzen ausschlaggebend ist: die topologische Auffassung.

In der T o p o l o g i e oder Analysis situs, wie sie früher genannt wurde, untersucht man solche Eigenschaften geometrischer Gebilde, die eben auf ihrem stetigen Zusammenhang, aber nicht auf Grösse, Form, Länge beruhen; d. h. solche Eigenschaften, die sich nicht ändern bei stetigen Deformationen, bei allen Übergängen, welche nichts zerreissen, was zusammengehört, und nichts zum Zusammenfallen bringen, was getrennt ist. Man denke etwa an einen Kreis, und einen, der von Hand schlecht gezeichnet wurde: zwar hat man vieles zerstört (Symmetrie, Länge, Form), aber er ist eine einfach geschlossene Linie geblieben; das wäre nicht mehr der Fall, wenn man ihn einschnürt oder zerreisst. Bei dem am

Anfang erwähnten EULER'schen Polyedersatz handelt es sich um etwas, das allen Flächen und Polyedern zukommt, die «von der Art der Kugel sind», d. h. sich in sie deformieren lassen; sie dürfen krumm sein, verschieden an Form und Grösse, immer ist die Zahl Ecken + Flächen — Kanten gleich 2. So wie man elementargeometrisch zwischen zwei kongruenten Figuren nicht zu unterscheiden braucht, so gibt es topologisch keinen Unterschied zwischen Kugel, Würfel, Eifläche usw., wohl aber zwischen Kugel und Ringfläche und Doppelringfläche (vgl. Abb. 3). Dass es nicht möglich ist, das

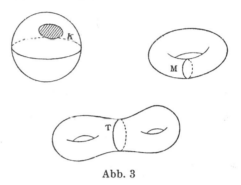

Abb. 3

eine stetig in das andere zu deformieren, ergibt sich schon aus der Betrachtung geschlossener Linien auf solchen Flächen; es gibt da verschiedene Möglichkeiten: Kurven, die sich auf einen Punkt zusammenziehen lassen, solche, die etwas beranden und solche, die es nicht tun (in Abb. 3 sind verschiedene Fälle eingezeichnet; ein Kreis K auf der Kugeloberfläche berandet — das Innere fällt heraus, wenn man längs des Kreises aufschneidet — und ist sogar zusammenziehbar; der «Taillenkreis» T, auf der Doppelringfläche berandet — etwa die linke Hälfte —, aber lässt sich nicht zusammenziehen; der «Meridiankreis» M auf der Ringfläche tut beides nicht.)

An den Unterschieden zwischen diesen Flächen äussern sich topologische Eigenschaften der G e s t a l t. Es gibt auch solche der L a g e: Zwei verschiedene Ringe lassen sich nicht trennen, ohne dass man sie aufbricht; ein Knoten lässt sich durch topologische Deformation des Raumes nicht in einen Kreis überführen (Abb. 4). Es handelt sich jeweils um dasselbe Gebilde, das

Abb. 4

aber auf verschiedene Arten im Raum liegen kann.

Ein topologischer Kern steckt in den verschiedensten mathematischen Theorien, und es ist oft sehr lohnend, ihn aus den Sätzen und Formeln herauszuschälen. So besonders in der Theorie der analytischen Funktionen, auf Grund von RIEMANN's Idee, die Gesamtheit der Funktionselemente einer analytischen Funktion als eine Fläche zu betrachten (vergleiche [3]); zwei wichtige Sätze handeln von Kurven auf dieser Fläche: der CAUCHY'sche Integralsatz, der aussagt, dass das Integral längs jeder geschlossenen Kurve = 0 ist, falls sie berandet, und der Monodromiesatz, der voraussetzt, dass die Kurve zusammenziehbar ist. Überhaupt sind die Grundgedanken der RIEMANN'schen Funktionentheorie, der Lehre von den algebraischen Funktionen und den Integralen auf RIEMANN'schen Flächen in hohem Masse topologischer Natur, und deshalb sind sie auch so übersichtlich und anschaulich.

6. Der allzu anschauliche Charakter solcher Betrachtungen ist manchmal geradezu gefährlich; man vergisst fast, dass etwas zu beweisen ist, und die Anschauung ist zwar sicher eine unerschöpfliche Quelle der Entdeckung, aber oft nur eine Fiktion oder eine Sache der Gewohnheit. Das hatte früher zur Folge, dass die Topologie als eine etwas ungenaue Disziplin angesehen und behandelt wurde, und ihre strenge Ausarbeitung ist auch wirklich erst sehr spät gekommen. Noch GAUSS sagte, dass man von ihr nicht viel mehr als nichts wisse. Es war wohl gerade die RIEMANN'sche Funktionentheorie, die zum erstenmal nicht nur den Anstoss zur Untersuchung der geschlossenen Flächen gab, sondern auch zeigte, dass topologische Gedanken sich durch ganze Theorien hindurchziehen. Die neuere strenge Begründung nahm ihren Ausgang von CANTOR (um 1880) und POINCARÉ (kurz vor 1900) und führte in den letzten Jahren zu einem ungeheuren Aufschwung dieses Kapitels der Geometrie. Es zeigte sich dabei immer mehr, dass es gar nicht so sehr

die Beweise sind, welche Schwierigkeiten bieten, als vielmehr die richtige Wahl der Grundbegriffe — in unserer Sprechweise: das geeignete Schema.

Der Begriff, den man hier aus dem komplexen Ganzen herausgelöst hat, so wie es in der abstrakten Algebra mit den Rechenoperationen geschieht, ist derjenige der Nachbarschaft oder U m g e b u n g ; er verknüpft die einzelnen Punkte mit benachbarten, eine Beziehung, die nicht zerstört werden darf (das nennt man Stetigkeit). Auf den Geraden oder auf Kurven sind die Umgebungen kleine Bögen, also von vereinzelten Punkten begrenzt, auf Flächen sind es von Kurven begrenzte Stücke, und im Raum von Flächen begrenzte Raumstücke. Diesen Begriffen, so klar und präzise sie sich fassen lassen, haftet im Vergleich zu den algebraischen Vorgängen etwas Transzendentes, Unergründliches an, das mit den unbegrenzten Möglichkeiten der Verfeinerung und Unterteilung zusammenhängt (und mathematisch zu den Begriffen Konvergenz, Häufungspunkt usw. führt).

Beim Ausbau des Schemas geht man nun wiederum, ähnlich wie man in der Algebra einen abstrakten Standpunkt einnehmen kann, zur abstrakten Raumkonstruktion über. Als geometrische Figuren gelten dann nicht nur Gebilde des gewöhnlichen Raumes; man kann als «Punkte» irgendwelche Dinge nehmen, deren individuelle Natur gleichgültig ist, wenn sie nur durch Beziehungen der Nachbarschaft verknüpft sind und infolgedessen ein Kontinuum, eine stetige Mannigfaltigkeit bilden, in der es stetige Übergänge und beliebig feine Unterschiede gibt. Der Begriff eines solchen Kontinuums darf wohl als eine der ursprünglichsten Ideen nicht nur der Geometrie, sondern auch aller Naturbeschreibung bezeichnet werden. Ihr steht das Algebraische gegenüber, in welcher es nur isolierte Dinge gibt, und zwischen ihnen nur sprunghafte Übergänge, die Rechenoperationen. Im Begriff der reellen Zahl ist beides vereinigt und erweist sich beides als unerlässlich.

7. So hat man die Dinge fein säuberlich getrennt und glaubt nun mit den zwei entgegengesetzten Prinzipien: T o p o l o g i e und A l g e b r a , Umgebung und Rechnen, transzendent und algebraisch, kontinuierlich und sprunghaft (oder «diskret») an die

Fragen herantreten zu können. Man möchte vielleicht schon darüber streiten, welchem der Vorrang zu geben sei, welches die «tieferen Gründe» an den Tag bringe, welches aufschlussreicher oder bequemer sei, und so fort. Aber im gleichen Augenblick melden sich auch schon die Z u s a m m e n h ä n g e. Sie sind verschiedener Art. Naheliegend, aber darum nicht weniger interessant sind diejenigen, die auf der gewöhnlichen analytischen Geometrie beruhen; da übersetzt man einfach Sätze des Raumes vermittelst der Koordination in solche der reellen Algebra — es handelt sich also hier um «konkrete» Topology and Algebra.   Als Beispiel sei folgender Satz erwähnt, den schon POINCARÉ bewiesen hat: Wenn auf der ganzen Kugeloberfläche eine Strömung gegeben ist, also ein Feld von Richtungen, die sich von Punkt zu Punkt stetig ändern, so müssen immer Wirbel oder Quellen oder andere Ausnahmepunkte (Unstetigkeiten) entstehen (versucht man es etwa mit einer Strömung längs der Meridiane oder der Breitenkreise, so bilden die Pole solche Unstetigkeiten). Algebraisch kann man diesen Satz dahin aussprechen, dass gewisse Gleichungssysteme mit drei Unbekannten reelle Lösungen besitzen. Man kann aber nur den Satz, nicht etwa den Beweis übertragen, und es ist übrigens bisher gar kein rein algebraischer Beweis für ihn bekannt; so löst also die Topologie gewisse Probleme der reellen Algebra, die diese selbst noch nicht gelöst hat. — Ähnliches gilt für viele andere Sätze über Richtungsfelder und Strömungen, und ihre algebraischen Konsequenzen sind gar nicht oder nur teilweise wirklich algebraisch bewiesen worden.

8. Diese Anwendungen sind deswegen nicht so überraschend, weil sie die Vermittlung der reellen Zahlen benützen, in welchen eben beide Gesichtspunkte schon drinstecken. Es gibt aber auch Anwendungen der abstrakten Algebra auf die Topologie, und das ist überhaupt die wichtigste Methode zur Untersuchung topologischer Eigenschaften.

Denken wir etwa an die Kugel, die Ringfläche oder ähnliche Flächen. Wir können sie mit einem Netz von Dreiecken überspannen; diese Dreiecke sind zwar krummlinig begrenzt, können aber in geradlinige deformiert werden. Dieses Dreiecksnetz ist eine Art Gerüst für die Fläche und ersetzt sie in grober Weise. Wir können nun jedes Dreieck — das liegt im Wesen des Kontinuums — beliebig fein unterteilen, nach einer festen Vorschrift, und so jeden Punkt der Fläche beliebig genau einschliessen. Aber dieses Verfeinern, durch welches die Fläche immer genauer beschrieben wird, ist ja durch das ursprüngliche Netz schon vollständig festgelegt, d. h. die Eigenschaften der Fläche müssen schon in diesem Netz enthalten sein, in der Verknüpfung der Eckpunkte zu Dreiecken; man muss nur wissen, welche drei Punkte jeweils ein Dreieck bilden, und welche Dreiecke aneinanderstossen. Damit wird aber die Untersuchung zu einem Kombinieren von endlich vielen diskreten Dingen (den Eckpunkten), und man verwendet dabei mit Vorteil den Formalismus der abstrakten Algebra; diese k o m b i n a t o r i s c h e o d e r a l g e b r a - i s c h e T o p o l o g i e (vgl. [2], [4]) ist geradezu ein Zweig der reinen Algebra, und vom Kontinuum ist darin nicht mehr die Rede. Dass es trotzdem möglich ist, damit geometrische Eigenschaften zu untersuchen, das beruht auf einer ebenso merkwürdigen wie wichtigen Tatsache, welche die Verbindung mit dem Kontinuierlichen herstellt: man kann im kombinatorischen Schema «topologische Invarianten» finden, d. h. algebraische Grössen, die nur von der topologischen Beschaffenheit des betreffenden geometrischen Gebildes, aber nicht von Grösse, Form usw., insbesondere nicht vom speziellen Eckpunktgerüst abhängen, an welchem man algebraische Topologie treibt. Z. B. ist die Zahl: Anzahl der Ecken —Anzahl der Kanten + Anzahl der Flächen bei einer Triangulation einer geschlossenen Fläche eine solche Grösse, sie ist für jede Triangulation einer Fläche und auch jeder anderen Fläche vom gleichen topologischen Typus gleich gross (was man ja etwa von der Anzahl der Ecken nicht behaupten kann); für die Kugel oder eine Fläche vom Typus der Kugel hat sie den Wert 2 — das ist der EULER'sche Polyedersatz — und für den Typus der Ringfläche den Wert 0.

Besonders die Theorie der Berandung, von der oben die Rede war, ist vollständig von der algebraischen Methode durchdrungen, während in der Theorie der Deformationen (Zusammenziehen einer Kurve u. a.) erst wichtige Ansätze dazu vorhanden sind. Alles, was ich bisher an Sätzen über Flächen — Berandung, Richtungsfelder usw. — oder über Kurven im Raum erwähnt habe, beweist man am besten mit dieser algebraischen Methode, und noch viel mehr; man kann direkt geometrische Eigenschaften «berechnen» aus der Zahl der Ecken, Kanten, Dreiecke usw. und aus ihrer Verknüpfung, und die Methode lässt sich nicht etwa nur auf Gebilde unseres Raumes oder seiner Verallgemeinerung auf mehr Dimensionen anwenden, sondern, nach den neuesten Untersuchungen, auch auf viel allgemeinere abstrakte Räume. Dagegen ist es bisher noch nicht gelungen, die geometrische Struktur vollständig durch die algebraische zu charakterisieren, d. h. zu ersetzen; es sind nur einzelne Eigenschaften eines geometrischen Gebildes, die man so bestimmen kann, und es bleibt in dieser Hinsicht noch manches Problem ungelöst.

9. So äussert sich in mannigfachen Anwendungen eine gewisse Verwandtschaft unserer beiden scheinbar wesensverschiedenen Schemata Algebra und Topologie. Besonders eindrücklich aber tritt sie zutage in ihrer S y n t h e s e : Man betrachtet Systeme, die gleichzeitig eine algebraische und eine topologische Struktur besitzen (d. h. Gesamtheiten von Elementen, mit denen man rechnen kann, und die gleichzeitig Punkte eines Umgebungsraumes bilden).

Das Beispiel, an dem ich versuchen möchte, die besondern Züge dieser Synthese anzudeuten, ist die Theorie der «kommutativen topologischen Gruppen», wie sie vor wenigen Jahren von PONTRJAGIN [5] entwickelt worden ist — nicht etwa, dass es das einzige Beispiel wäre, aber es ist besonders einfach und übersichtlich, und auch in mancher Hinsicht von grundlegendem Interesse. Ich darf wohl zuerst einige Gedanken und Resultate dieser Theorie in kurzen Worten skizzieren.

Der Kreis vereinigt topologische und algebraische Eigenschaften in sich. Er ist eine kontinuierliche Mannigfaltigkeit, und gleichzeitig eine Gruppe; denn zu jedem Punkt gehört ein Winkel (von einem festgewählten Anfangspunkt aus gerechnet), und die Winkel kann man addieren, wobei die Summe stetig von den Summanden abhängt.

Man spricht in diesem Falle von einer t o p o l o g i s c h e n G r u p p e (und zwar einer geschlossenen, im Gegensatz etwa zur Zahlgeraden, die offen ist; ein anderes Beispiel einer geschlossenen topologischen Gruppe ist die Ringfläche).

Wir nehmen nun einen zweiten Kreis zu Hilfe und ordnen jeden Winkel im ersten Kreis den doppelt so grossen Winkel im zweiten zu; wenn also ein Punkt den ersten Kreis einmal umläuft, so umläuft der zugeordnete Punkt den zweiten Kreis zweimal — so wie der Stundenzeiger sich zweimal herumdreht, während die Sonne ihren Kreis einmal durchläuft. Eine solche Zuordnung, die weder am stetigen Zusammenhang noch an der Gruppenoperation etwas zerstört, nennt man einen C h a r a k t e r der topologischen Gruppe.

Einen andern Charakter des Kreises erhält man, wenn man statt des doppelten den dreifachen Winkel nimmt (oder den vierfachen usw.), oder indem man jedem Winkel den negativen (gleich gross im andern Drehsinn) zuordnet, oder den doppelten, dreifachen negativen Winkel usw.; man kann auch den nullfachen Winkel nehmen, d. h. jedem Winkel den Winkel 0 zuordnen. Auf diese Weise erhält man a l l e Charaktere des Kreises; dass es keine anderen gibt, das hängt eng mit der Eigenschaft des Kreises zusammen, die ich am Anfang dieser Ausführungen als zweites Beispiel erwähnt habe.

Jeder Charakter des Kreises ist also durch eine ganze Zahl 0, ±1, ±2, ... gegeben; die Gesamtheit der Charaktere bildet also etwas Diskretes, nicht Kontinuierliches, das Musterbeispiel der reinen Algebra: die Gruppe der ganzen Zahlen.

Dieses Übertragungsprinzip, das ich hier im Falle des Kreises angedeutet habe, kann man auch bei jeder andern (kommutativen) topologischen Gruppe verwenden. Man ordnet ihre Punkte in ähnlicher Weise denjenigen eines Kreises zu und nennt diese Zuordnung einen Charakter der Gruppe; die Gesamtheit der Charaktere der topologischen Gruppe bildet dann etwas rein Algebraisches, eine Gruppe, die aus abzählbar vielen isolierten Elementen besteht, also diskret ist. So gehört zu jeder geschlossenen topologischen Gruppe eine ganz bestimmte d i s k r e t e Gruppe, ihre C h a r a k t e r e n g r u p p e.

Das Hauptergebnis der Theorie besagt nun, dass die kontinuierliche und die diskrete Gruppe, die so zusammengehören, einander vollständig bestimmen. In den algebraischen Eigenschaften der diskreten Gruppe sind alle topologischen Eigenschaften der kontinuierlichen enthalten, und umgekehrt. Z. B. ist der sogenannte Rang der diskreten Gruppe gleich der Dimension der kontinuierlichen (gibt also an, ob sie eine Linie oder eine Fläche usw. ist); das Auftreten von Elementen endlicher Ordnung (die mehrmals addiert 0 ergeben) in der diskreten Gruppe bedeutet, dass das topologische Gebilde aus mehreren getrennten Stücken besteht; bis in spezillste Umgebungseigenschaften lassen sich solche Zusammenhänge verfolgen.

So werden in diesem Falle die beiden Grundbegriffe — Rechnen und Raum — in eleganter Weise ineinander übergeführt und erweisen sich gewissermassen als identisch, nur in verschiedenen Sprachen formuliert.

Lässt sich etwas Ähnliches aussagen, wenn man statt geschlossenen offene topologische Gruppen, wie z. B. die Zahlengerade, betrachtet? Hier liegen die Dinge anders. Wohl bilden auch hier die Charaktere eine Gruppe, aber sie ist selbst etwas Kontinuierliches, nämlich wieder eine offene topologische Gruppe. Zwar ist es auch hier so, dass diese die ursprüngliche vollständig bestimmt, und dass topologische in algebraische Eigenschaften übergehen; aber man kann dies jetzt nicht in derselben reinen Form erfassen wie vorher, denn gleichzeitig gehen algebraische in topologische über! Es verhält sich etwa so wie in einem allzu symmetrisch angelegten Park: man hat Mühe zu sagen, in welchem Teil man sich eigentlich gerade befindet.

10. Dies alles findet den dem Mathematiker und Naturwissenschafter wohlbekannten Ausdruck in der **harmonischen Analyse**. Dort betrachtet man — wenn wir wieder beim einfachsten Beispiel einer topologischen Gruppe, dem Kreis, bleiben — Funktionen auf dem Kreis, d. h. des Winkels, also periodische Funktionen oder Schwingungen, und löst sie auf in harmonische Schwingungen (Grundton und Obertöne, das sind besonders einfache Funktionen des Winkels, des doppelten, dreifachen Winkels usw.; es sind nichts anderes als

die Charaktere). Die Zahlen, welche angeben, wie stark die einzelnen harmonischen Schwingungen vertreten sind, die «Fourierkoeffizienten», ersetzen bekanntlich die Funktion vollständig; man ersetzt etwas Kontinuierliches, die Funktion, durch eine diskrete Reihe von Zahlen. Dagegen muss man bei gewöhnlichen, unperiodischen Funktionen — d. h. Funktionen auf der Zahlgeraden — ein kontinuierliches Spektrum von Frequenzen nehmen, wenn man sie in harmonische Schwingungen auflösen will, und sie bleibt etwas Kontinuierliches (statt der Fourierreihe hat man das Fourierintegral). Ähnlich ist es bei allen (kommutativen) topologischen Gruppen; sie bilden, zusammen mit ihren Charakterengruppen, das eigentliche Substrat der harmonischen Analyse [6]. Man beachte dabei, wie wenig eigentlich in diesem Begriff der topologischen Gruppe steckt; er resultiert aus der Vereinigung der beiden strukturbildenden Operationen, die wir aus der Fülle der Möglichkeiten ausgelesen haben: der algebraischen (Addition) und der topologischen (Umgebungsbegriff; man muss dazu nur noch das Prinzip hinzunehmen, dass jede unendliche Menge von Punkten einer Umgebung eine Häufungsstelle besitzt).

Überhaupt ergibt — man stosse sich nicht am paradoxen Klang der Worte — diese Synthese vollständig das, was man Analysis nennt. Denn wenn man von den geschlossenen topologischen Gruppen absieht, so bleiben für die offenen im wesentlichen nur die Vektorräume übrig, also die gewöhnliche analytische Geometrie. Und wenn man zur Addition in der Gruppe noch die Multiplikation hinzunimmt (Körper), so gibt es als einzige Möglichkeiten solcher topologischer Körper die «gewöhnlichen» Zahlen der Analysis, d. h. die reellen Zahlen, die komplexen Zahlen und die Quaternionen; das ist ein interessanter Satz von PONTRJAGIN [5], der neues Licht auf wichtige Fragen der Axiomatik wirft.

11. Die Gegenüberstellung von Topologie und abstrakter Algebra hat uns zu bemerkenswerten Ergebnissen geführt und eigenartige Beziehungen zwischen den beiden aufgedeckt. Verschiedenen anschaulichen und formalen Bedürfnissen entspringend, erfüllen die beiden Denkweisen verschiedene Funktionen und sind der Ausdruck zweier verschiedener Wege des Verstehens und der

Anschauung; sie ergänzen und unterstützen einander dabei aufs schönste. Aber die Verschiedenheit der beiden ist nur scheinbar; ihre Synthese ergibt nicht nur ein geschlossenes Ganzes, sondern lässt sie gewissermassen ineinander aufgehen. In der Vereinigung ist es eigentlich nur eine Frage des Standpunktes, ob man eine Eigenschaft der einen oder der andern Seite zuschreibt; ein vom Ganzen aus gesehen künstlicher Eingriff lässt zufällig die eine besser hervortreten als die andere.

Man muss hier unwillkürlich an eine ähnliche Erscheinung in der Physik denken: Licht, und auch Materie, zeigt manchmal die Natur von Wellen, manchmal von Korpuskeln (Teilchen); der modernen Quantentheorie ist es gelungen, eine Formulierung zu finden, welche beide Erscheinungsformen vereinigt, besser gesagt, welche über beiden steht (dafür allerdings auf Anschaulichkeit verzichtet); dass man einmal das eine, einmal das andere Verhalten beobachtet, das wird nur durch das Experiment veranlasst, also einen künstlichen Eingriff, welcher einmal die Wellen-, einmal die Korpuskelnatur bevorzugt und in Erscheinung treten lässt. Die Analogie mit unserem Dualismus Raum — Zahl ist bestechend; vielleicht ist es mehr als nur eine Analogie.

Es wäre leicht, an Beispielen des täglichen Lebens auf ähnliche Fälle hinzuweisen, wo man je nach dem Standpunkt einer Sache zwei ganz verschiedene Seiten abgewinnen kann — aber dann geht es meistens nicht so friedlich aus wie in der Mathematik.

Dass Raum und Zahl in ihren elementarsten Eigenschaften so eng miteinander verknüpft sind, wie ich es zu schildern versucht habe, ist gewiss überraschend. Es gibt in der Mathematik und den Naturwissenschaften noch viele andere Beispiele dieser Art, wo zwei scheinbar verschiedene Dinge sich ganz unerwartet als gleichwertig, miteinander vertauschbar, erweisen, und das scheint kein Zufall zu sein. Vielleicht darf ich es so formulieren: Um an unsere Umwelt heranzutreten, konstruieren wir verschiedene Bilder, Schemata — verschiedenen Ursprungs, mit verschiedenen Mitteln und zu verschiedenen Zwecken —, und wenn sie trotzdem gleich oder gleichwertig herauskommen, so ist offenbar der Konstrukteur daran schuld, der gleiche mathematische Geist, der bei allen tätig war; auf ihn kommt es mehr an als auf das spezielle Beispiel, an dem er sich übt. Die Beziehungen haben allgemeinere Gültigkeit als die Gegenstände, an denen sie sich äussern.

Gewiss sind und bleiben immer die einzelnen, speziellen, konkreten Probleme das eigentliche Arbeitsgebiet des Mathematikers, und er wird ihrer Bezwingung seine ganze Kraft widmen. Aber er strebt danach, sie mit möglichst wenig blinder Rechnung und möglichst viel sehenden Gedanken zu lösen. Und wenn er dabei das Allgemeine sucht und die grösste Abstraktion nicht scheut, so entfernt er sich deswegen nicht von der Wirklichkeit, sondern findet immer wieder Beziehungen und Gesetzmässigkeiten, die allen unsern Arten zu denken und die Welt zu sehen gemeinsam sind, und denen wir offenbar nicht entrinnen können.

Literaturverzeichnis

Unsere Ausführungen sind, dem Charakter des Vortrages entsprechend, allgemein gehalten, auch was die Formulierung von Sätzen und Resultaten betrifft; wegen der genauen Formulierung und Begründung, sowie der Ausführung der Einzelheiten, verweisen wir auf die folgenden Werke:

[1] D. HILBERT, Grundlagen der Geometrie (Leipzig und Berlin 1913).

[2] H. SEIFERT und W. THRELFALL: Lehrbuch der Topologie (Leipzig und Berlin 1935).

[3] H. WEYL, Die Idee der Riemannschen Fläche (Leipzig und Berlin 1913).

[4] P. ALEXANDROFF und H. HOPF, Topologie I (Berlin 1935), bes. 2. Teil.

[5] L. PONTRJAGIN, Topological groups (Princeton 1939), Kap. V.

[6] A. WEIL, L'intégration dans les groupes topologiques et ses applications (Paris 1940), Kap. VI.

Complex-analytic manifolds

Proceedings of the International Congress of Mathematicians, Vol. II (1950), 420–427

1. Generalities. 1.1. The concept of a *complex-analytic manifold* (in short *complex manifold*) is the natural generalization of the concept of a Riemann surface in the abstract sense. A $2m$-dimensional complex manifold M is a manifold of dimension $2m$ in the usual sense which is covered by a family of systems of local complex coordinates $z_1, \cdots, z_m$ (instead of the usual $2m$ real coordinates) in such a way that the relation between two such complex coordinate systems in the intersection of their existence domains is given by complex-analytic functions. Since the coordinate transformation is one-to-one, the Jacobian of these functions is not equal to 0, and the square of its absolute value is easily seen to be equal to the Jacobian of the corresponding real coordinate transformation. A complex manifold M is therefore orientable and has a natural orientation. We shall restrict ourselves throughout to closed manifolds. The set of all local complex coordinate systems which are admissible, i.e., which may be added to the given family in accordance with the analyticity condition, is called the *complex structure* of M. Concepts like analytic function, analytic map, etc. have an invariant meaning with respect to the given complex structure (i.e., are independent of special coordinate systems used to describe them).

1.2. Algebraic varieties in a complex projective space P_n (of n complex, i.e., $2n$ real, dimensions) have a natural complex structure, as well as P_n itself, and are therefore complex manifolds—provided that there are no "singularities". There exist, on the other hand, examples of complex manifolds which cannot be imbedded in a P_n (cf. 3.4). The concept of a complex manifold is therefore more general than that of an algebraic variety. It is probably also more general than that of a higher-dimensional Riemann surface in a concrete sense, defined as the space of all function elements of an algebroid function on some standard space S; but such a statement depends of course on the space S and on uniformization possibilities.

1.3. It is well-known that all orientable surfaces admit complex structures. For higher-dimensional manifolds (orientable, of even dimension) this is not the case. On the 4-dimensional sphere S^4, for example, it is not possible to define a complex structure (cf. Hopf [1], Ehresmann [2]). The general *existence problem* of a complex structure on a given manifold M, or the problem of what special properties of M are implied by such a structure, has several quite different aspects. Actually there is not much known about the complex structure itself; all consequences are deduced from assumptions which are weaker—the *almost complex* structure, or stronger—the existence of a *Kaehler metric* (or Hermitean metric without torsion). The reason for these two approaches is simple: Almost complex structure is an assumption concerning the tangent bundle of M, and therefore suitable for fiber space methods; and Kaehler metric is an assumption on the Riemannian or Hermitean geometry of M, which can be investigated by the methods of harmonic differential forms and of differential geometry. In both cases powerful existing theories can be applied.

Concerning the first approach I shall make only a few remarks, which overlap to some extent with Ehresmann's address; after that I shall discuss in more detail the second approach.

2. Almost complex manifolds. 2.1. A manifold M is called *almost complex* if for each point x of M a linear map I_x of the tangent space T_x at x into itself is defined such that $I_x^2 = -1$ (1 = identity map) for all x, and such that I_x depends continuously upon x. This is possible only if M is of even dimension $2m$. A complex manifold is almost complex; for if we use in the tangent space T_x complex vector components corresponding to local complex coordinates, multiplication of the vector components by $i = (-1)^{1/2}$ is a linear map of the required type, independent of the special coordinate system. I do not know of an example of an almost complex manifold which does not admit a complex structure, but probably there are such.

2.2. Let M be almost complex and $v(x)$ a continuous nonzero vector field on M with one possible singularity; then $I_x v(x) = w(x)$ is a second field with one possible singularity at the same point and such that $w(x)$ and $v(x)$ are independent everywhere else since $I_x^2 = -1$. Therefore there exists on M a 2-field with one isolated singularity, which is obviously of a quite special nature. This fact was used by Hopf [1] to prove that certain manifolds (the spheres S^4 and S^8 and many other examples) do not admit an almost complex structure, i.e., a field of transformations I_x; a fortiori these are not complex manifolds. Another way (cf. Ehresmann [2]) to prove existence or nonexistence of an almost complex structure on a manifold M is to apply the obstruction theory of fiber bundles; the base space of the fibering to be used is M, and the fiber at a point x is the space of all tensors at x describing transformations I_x (or related to them). For the simplest manifolds, the *spheres*, both methods give only quite special results, and in this case the existence problem seems to be related in a peculiar way to other topological questions, as I would like to indicate by some remarks.

2.3. If on a *sphere* S^n ($n = 2m$) a field of transformations I_x is given, we may assume that for all tangent vectors v at all points x, $I_x v$ is orthogonal to v. We have then for all pairs of *orthogonal* unit vectors x, y in $(n + 1)$-dimensional Euclidean space E^{n+1} a unit vector function $z = F(x, y)$, z orthogonal to both x and y, continuous in x and linear in y; this function is obtained by considering x as a point of S^n and y as a tangent vector at x, and defining $z = I_x y$. It can be extended in an obvious way to a function $\bar{F}(x, y) = z$ of two *arbitrary* vectors in E^{n+1} such that z is orthogonal to x and y and of length $(x^2 y^2 - (x \cdot y)^2)^{1/2}$ ($x \cdot y$ is the scalar product of x and y, $x^2 = x \cdot x$); $\bar{F}$ may be called a *"vector product in*

E^{n+1}," (continuous in the first factor, linear in the second). Applying this to the usual complex structure on S^2, we obtain a vector product in E^3 which is the usual one, bilinear in x, y. By fiber space methods using special homotopy groups, it can be proved that *for $n = 4k$ there exists no such vector product in E^{n+1} of the required type*, not even in the more general sense of continuity in both factors. The proof is a simple extension of the method I used previously [3] for the same purpose, but only for $n = 4$ and 8; to extend it to all $n = 4k$, one has to compute more of the homotopy groups involved, which can be done by straightforward deformations. From the argument above it follows that *there is no almost complex structure on the spheres S^{4k}* (not even if one would admit fields of transformations I_x which are only continuous, not linear, in the tangent spaces T_x, but such that $v(x)$ and $I_x v(x)$ are always independent—this is, by the way, the concept of a "manifold of type I" considered by Hopf [1]).

2.4. It is well-known [3] that a vector product in E^{n+1} leads in a natural way to a *multiplication* in E^{n+2}, i.e., a rule associating with any two vectors X, $Y \in E^{n+2}$ a product $U = X \circ Y$, with a unit and with the "norm product rule" $U^2 = X^2 \cdot Y^2$. Namely, let $X \in E^{n+2}$ be given by a pair (x_0, x) of a real number x_0 and a vector $x \in E^{n+1}$; then $X \circ Y = U = (u_0, u)$ is defined by $u_0 = x_0 y_0 - x \cdot y$, $u = x_0 y + y_0 x + \bar{F}(x, y)$. The rule $U^2 = X^2 \cdot Y^2$ is easily checked, and $E = (1, 0)$ is the unit. For the usual vector product in E^3 (i.e., the complex structure on S^2), this gives exactly the *quaternion* multiplication in E^4. The Cayley numbers in E^8 are related in the same way to a (bilinear) vector product in E^7, hence to an almost complex structure on S^6. It is not known whether it corresponds to a complex structure on S^6, nor indeed whether there is at all a complex to complex structure on S^6. [*Added in proof*: It has been proved (cf. Eckmann and Frölicher, C. R. Acad. Sci. Paris vol. 232 (1951) p. 2284) that the almost complex structure on S^6, derived from the Cayley numbers does not belong to a complex structure on S^6 by studying the "integrability conditions" relating almost complex structures.]

Furthermore, if $U = X \circ Y$ is a multiplication in E^{n+2} as described before, continuous in X and linear in Y, we have for each X with $X^2 = 1$ an orthogonal transformation $U(Y) = X \circ Y$ of E^{n+2} such that for $Y = (1, 0)$, $U(Y) = X$; i.e., we have a map Φ of S^{n+1} into the orthogonal group Ω_{n+2} of $n + 2$ variables, such that under the natural projection of Ω_{n+2} onto S^{n+1} the map Φ becomes the identity of S^{n+1}. It is easy to see that this map Φ is nothing else than a global parallelism on S^{n+1} (a system of $n + 1$ pairwise orthogonal tangent unit vector fields). Combining this with 2.3, we obtain a theorem discovered and proved first by Kirchhoff [4] in a different way: *An almost complex structure on S^n induces a global parallelism on S^{n+1}.* From this it follows again that on S^{4k} there is no almost complex structure (same result as above, here only in the sense of linear I_x).

2.5. In all preceding remarks the existence problem of an almost complex or complex structure is considered in terms of a definite differentiable structure on the manifold, and it is not clear from the method whether for another differentiable structure the answer would be the same.

It might be worthwhile to note here that the complex structure on a (differentiable) manifold, if it exists, is in general not unique. There are manifolds which admit several different complex structures—different in the sense of analytic equivalence. The 2-dimensional torus, for example, has infinitely many different conformal structures (and the analogue holds for higher-dimensional tori); for all of them, however, the corresponding complex structure of the universal covering is the same. Hirzebruch [5] discovered the interesting result that there exist also *simply connected* manifolds with infinitely many different complex structures; the simplest example is the Cartesian product of S^2 by itself.

3. Complex structure and homology. 3.1. To obtain relations between homology properties and complex structure, it seems natural to study the influence of the transformations I_z belonging to that structure on *exterior differential forms*. A differential form α of degree r in the manifold M (in short, an r-form) may be considered, at each point x of M, as a skew-symmetric multilinear function of r vectors in the tangent space T_x ; if these vectors are transformed by I_x , α becomes a new form denoted by $C\alpha$ (if not only real, but also complex forms, i.e., complex functions in T_x , are admitted, it is convenient to include in the definition of C the passage to the conjugate-complex form; an explicit expression of C is given in 3.2). Obviously the operator C verifies $CC\alpha = (-1)^r\alpha$ for all r-forms α, hence C is an isomorphi m of the linear group Φ^r of all r-forms in M onto itself. In general $dC\alpha \neq C\,d\alpha$, and $\tilde{d} = C^{-1}\,dC$ is a differential operator from Φ^r to Φ^{r+1} which is not equal to d, but has some of the same properties (e.g., $\tilde{d}\tilde{d} = 0$). $\tilde{d}$ leads to de Rham cohomology groups $\tilde{H}^r$ isomorphic to the usual ones H^r based on d. By Hodge's theorem ([6], [7]) these groups may be replaced by those of *harmonic* r-forms (with respect to a Riemann metric given on M), which are easier to handle than cohomology classes; i.e., by forms α with $\Delta\alpha = 0$ or $\tilde{\Delta}\alpha = 0$, where Δ is the generalized Laplace-Beltrami operator used by de Rham [6], and $\tilde{\Delta} = C^{-1}\Delta C$. It seems interesting to study the relation between Δ and $\tilde{\Delta}$; here we shall do this only in the case when we assume a complex structure in the strict sense, or even more than that. Before discussing this, we have to make some preliminary remarks.

3.2. On a $2m$-dimensional complex manifold M we can use, in the calculus of differential forms, instead of the $2m$ real differentials dx_ν (or vector components in T_x) the *complex differentials* dz_j, $j = 1, \cdots, m$, corresponding to admissible coordinate systems $z_1, \cdots, z_m$ *and their conjugates* $d\bar{z}_j$. The coordinate transformations, say from $z_1, \cdots, z_m$ to $\zeta_1, \cdots, \zeta_m$ being analytic, the dz_j are carried over to the $d\zeta_j$ and the $d\bar{z}_j$ to the $d\bar{\zeta}_j$ only, this splitting of the $2m$ differentials into two groups has an invariant meaning. A differential form which is homogeneous with respect to the $d\bar{z}_j$ will be called *pure*; and the corresponding degree in the $d\bar{z}_j$ only, its *type*. Any r-form α has a unique decomposition as a sum of pure forms $\alpha = \alpha_{(0)} + \cdots + \alpha_{(r)}$ of type $h = 0, 1, \cdots, r$. For a form $\alpha_{(h)}$ of type h, $C\alpha_{(h)}$ is given by $(-1)^h i^r \bar{\alpha}_{(h)}$, which is of type $r - h$; for pure forms, $\tilde{d}$ differs therefore from d only by a constant factor.

3.3. The Riemann metric, used in the definition of Δ, is called *Hermitean* if in the given complex structure the line element ds^2 may be written as $\sum_{j,k} h_{jk} \cdot$ $(dz_j \, d\bar{z}_k)$, $h_{jk} = \bar{h}_{kj}$, where $(\cdots)$ indicates the *ordinary* product of differentials. Such a metric is however not sufficient for our purposes. For it appears that, roughly speaking, one must, in the relations between harmonic forms and complex structure, consider permutations of the second covariant derivatives involved in Δ and therefore use the Riemann curvature tensor of the metric; and moreover, one should compute this tensor by the complex formalism, i.e., in the so-called Hermitean geometry. This is in general not possible: it is easy to see that the connexions (the Christoffel symbols) computed from the same Hermitean metric in Riemannian and in Hermitean geometry are *different*. There is a special case where they coincide, and this case seems of great importance: namely when the metric is a *"Kaehler metric"* [8]; i.e., has the special property that the exterior differential form $\omega = \sum_{j,k} h_{jk} \, dz_j \, d\bar{z}_k$ is *closed* ($d\omega = 0$). Actually many of the homology properties we are going to describe can be deduced from assumptions which are stated in terms of real differential geometry, without using a complex structure; these are however less natural and intuitive, and for the purpose of this address we shall stay within the frame of the Kaehler metric.

It is well-known that on the complex projective space P_n there exists a Kaehler metric, and therefore on all algebraic varieties imbedded without singularities in P_n. More generally, if a complex manifold M is imbedded analytically in a Kaehler manifold, the induced metric on M fulfills also the Kaehler condition (cf. 3.6).

3.4. In the case of a Kaehler metric, the explicit expression of Δ yields the following results (cf.[10]) on pure forms and on C. (a) If a form α is harmonic, then all the "pure components" $\alpha_{(h)}$ of α (cf. 3.2) are also harmonic. In other words: H^r is the direct sum $H^r_{(0)} + H^r_{(1)} + \cdots + H^r_{(r)}$, where $H^r_{(h)}$ is the group of pure harmonic forms of type h. If $p^r_{(h)}$ denotes the rank of $H^r_{(h)}$, we have for the Betti number $p^r = \sum_{h=0}^r p^r_{(h)}$. Of course $p^r_{(h)}$ is also defined for a general Hermitean metric; but all one can prove without the Kaehler condition is the inequality $p^r \geq \sum p^r_{(h)}$. (b) $\bar{\Delta} = \Delta$. Let $\bar{H}^r_{(h)}$ be the analogue of $H^r_{(h)}$, computed from $\bar{\Delta}$ instead of Δ, $\bar{p}^r_{(h)}$ its rank; it follows that $\bar{H}^r_{(h)} = H^r_{(h)}$ and $\bar{p}^r_{(h)} = p^r_{(h)}$. But since by definition $C\bar{\Delta} = \Delta C$, C maps $\bar{H}^r_{(h)}$ isomorphically onto $H^r_{(r-h)}$; hence $\bar{p}^r_{(h)} = p^r_{(r-h)} = p^r_{(h)}$. Therefore, *if r is odd*, $p^r = 2\sum_{h=0}^{(r-1)/2} p^r_{(h)}$, i.e., $p^r \equiv 0$ (mod 2); *if r is even*, $p^r \equiv p^r_{(r/2)}$ (mod 2). For algebraic varieties the fact that p^r is even in odd dimensions r is a well-known result already proved by Lefschetz and, using differential forms, by Hodge.

There exist complex manifolds which do not admit a Kaehler metric since their Betti numbers do not agree with the above conditions, namely the manifolds of topological structure $S^1 \times S^{2k+1}$ discovered by Hopf [1] (a complex structure on $S^1 \times S^3$, e.g., is obtained from the covering of $S^1 \times S^3$ by E^4 without the origin). From this it follows that these complex manifolds cannot be imbedded analytically in a complex projective space (nor in any Kaehler manifold).[1]

---

[1] *Added in proof*: E. Calabi and the author showed that the product of any two odd-dimensional spheres $S^{2p+1} \times S^{2q+1}$ can be given a complex structure; for $p > 0$, $q > 0$ this is an example of a *simply connected* complex manifold which does not admit a Kaehler metric.

3.5. All this is only one, somewhat special, aspect of the Kaehler metric, and there are other surprisingly strong consequences.

There is one immediate fact, which does not even use harmonic forms, but just some simple properties of the closed 2-form ω (cf. 3.3), depending only weakly upon the relation of ω to the metric. Let $n = 2m$ be the dimension of the Kaehler manifold M. The mth power ω^m is equal, up to a constant factor not equal to 0, to $| h_{jk} | \, dz_1 \, d\bar{z}_1 \cdots dz_m \, d\bar{z}_m$; since the determinant $| h_{jk} |$ is everywhere positive, $\int_M \omega^m \neq 0$ (ω^m is actually, except for a constant nonzero factor, the volume element of the metric considered as a Riemann metric). Hence ω^m is not cohomologous to 0, and the same is true for all powers ω^k, $k = 0, 1, \cdots, m$; the corresponding Betti numbers p^{2k} are therefore greater than or equal to 1.

These nonvanishing cohomology classes in even dimensions may be carried over by the duality operator D of M to homology classes. Using the same symbol α for a closed form and for the corresponding cohomology class, the dual homology class $D\alpha$ is given by $\alpha \cap M$ (here M is the fundamental $2m$-cycle of the oriented manifold, and $\cap$ the Čech-Whitney cap-product). Let us call the classes $D\omega^k = \omega^k \cap M = Z_{2(m-k)}$ of dimension $2(m - k)$ the *principal homology classes* of M. Clearly, writing Z for $Z_{2(m-1)} = D\omega$, $Z_{2(m-2)}$ is the *intersection* $Z \otimes Z$, $Z_{2(m-3)} = Z \otimes Z \otimes Z$, etc. If M is the complex projective space P_m, the principal classes Z_{2q}, $q = 0, 1, \cdots, m$, are all represented, up to certain numerical factors, by the projective planes P_q in P_m.

3.6. Let L' be a complex manifold of dimension $2l < 2m$, and f an *analytic* and locally topological *imbedding of L' in the Kaehler manifold M*, $f(L') = L \subset M$. Considering the induced dual homomorphism f^*, the image $f^*\omega$ is the form ω' in L' corresponding to the induced metric in L', which obviously is also a Kaehler metric. Thus, by the same argument as above, $0 \neq \int_{L'} \omega'^l = \int_{L'} f^*\omega^l = \int_L \omega^l$; by Stokes theorem it follows that $L \nsim 0$ (*not homologuous to 0*) *in* M.

Furthermore consider the principal homology classes $Z'_{2(l-k)}$ of L' ($k = 0, \cdots, l$), and their images $f(Z'_{2(l-k)})$ in M; they are equal to $f(\omega'^k \cap L') = f(f^*\omega^k \cap L') = \omega^k \cap f(L') = \omega^k \cap L = Z_{2(m-k)} \otimes L$, i.e., they are the intersections of L with the principal classes of M (for example, the principal classes of a manifold L imbedded in a complex projective space are given by the *plane sections of L*). *These classes are all $\nsim 0$ in M.* For $\int_{\omega^k \cap L} \omega^{l-k} = \int_L \omega^l \neq 0$, hence by Stokes theorem $\omega^k \cap L \nsim 0$.

Analogous remarks apply to "analytic cycles" in M, i.e., to the set of zeros of a distribution of local analytic functions. The problem of finding not only necessary, but also sufficient conditions for a homology class to contain analytic cycles seems to be difficult.

3.7. Very precise results on the role of ω and of principal classes in the homology structure of a Kaehler manifold M, including all of the known general homology structure of algebraic varieties, are obtained if, following the method of Hodge, harmonic forms in M are investigated in connection with special properties of

ω and of the differential geometry of the Kaehler metric. All these results are due to some simple formulas relating ω to the usual operations on differential forms: d, $*$ ($*\alpha$ is the adjoint of the r-form α, with respect to the metric, of degree $n - r = 2m - r$; cf. [6]), $\delta = \pm * d *$, $\Delta = d\delta + \delta d$, and the operator C, with $\bar{d}$, $\bar{\delta}$, etc. These formulas, most of which appear in Hodge's book [7] and in a note by A. Weil [9], have been completely established and discussed by Guggenheimer and myself [10]. They are of purely local character.

The main formula is a relation between $*$ and ω, replacing the operator $*$ for r-forms with $r \leq m$ by C and by multiplication with ω^{m-r} (and with a certain numerical factor u; I omit here all details about such factors): $*\alpha = u\omega^{m-r}C\alpha$. This is however not true for all forms α. It holds (1) if $\omega * \alpha = 0$—such forms are called *effective*; and more generally (2) if $\alpha = \omega^k\beta$, where β is effective—we shall call such a form *simple, of class k*. The nonzero factor u depends upon m, r, and k. Under condition (2) we have for $r \leq m - 2$, $\omega * (\omega\alpha) = u'\omega \cdot \omega^{m-r-2} C\omega\alpha = u'' * \alpha$, with nonzero factors u', u''. From this it follows that (a) $\omega\alpha = 0$ only if $\alpha = 0$. By an easy induction argument it can be proved that (b) *every r-form α ($r \leq m$) has a unique decomposition into a sum of simple forms of class $k = 0$, $1, \cdots, q = [r/2]$, $\alpha = \beta_0 + \omega\beta_1 + \cdots + \omega^q\beta_q$.* (a) holds therefore for *all* forms of degree $r \leq m - 2$: (c) *Multiplication by ω is an isomorphism of Φ^r, the group of all r-forms, into Φ^{r+2}.*

Further relations give $\Delta(\omega^k\alpha) = \omega^k(\Delta\alpha)$ for all forms α, and commutation of Δ with all operators involved. Hence all results apply in particular to harmonic forms, i.e., to cohomology groups H^r. From (b) above it follows that H^r is for $r \leq m$ the direct sum $H_0^r + \omega H_0^{r-2} + \cdots + \omega^q H_0^{r-2q}$, where $H_0^{r-2k} \subset H^{r-2k}$ is the group of effective harmonic $(r - 2k)$-forms; from (c), that multiplication by ω is, for $r \leq m - 2$, an *isomorphism of H^r into H^{r+2}* (cf. [7], Chap. IV). For the Betti numbers of a Kaehler manifold M we obtain therefore $p^r \leq p^{r+2}$ for $r \leq m - 2$ (and, of course, the dual statements for higher dimensions). These are strong topological conditions for the existence of a Kaehler metric; they are not fulfilled by all complex manifolds (e.g., by the examples mentioned in 3.4).

3.8. To translate the results from cohomology to homology, we take the operator D, which maps H^r isomorphically onto the rth homology group H_r of M (with complex coefficients). Since $*\omega^k = u\omega^{m-k}$, $D*\omega^k$ is, except for a constant nonzero factor, the principal homology class Z_{2k} of M, (e.g., a plane section if M lies analytically in a complex projective space). Each homology class z_r of M of dimension $r \leq m$ may be written as $z_r = D*\alpha = D*(\beta_0 + \omega\beta_1 + \cdots + \omega^q\beta_q)$, $q = [r/2]$, according to (b) in 3.7, with uniquely determined effective harmonic forms $\beta_0, \cdots, \beta_q$. The term $D*\omega^k\beta_k$ is, except for the nonzero factor u, equal to $D(C\omega^k\beta_k\omega^{m-r}) = C\omega^k\beta_k\omega^{m-r} \cap M$; that is, to $C\omega^k\beta_k \cap (\omega^{m-r} \cap M) = C\omega^k\beta_k \cap D\omega^{m-r} = C\omega^k\beta_k \cap Z_{2r} = DC\omega^k\beta_k \otimes Z_{2r}$, or to $C\beta_k \cap (\omega^{m-(r-k)} \cap M) = DC\beta_k \otimes Z_{2(r-k)}$. Therefore, (a) *all homology classes of dimension $r \leq m$ lie on the principal class Z_{2r}*, in the sense that they are intersections of something with Z_{2r}, and (b) *they have a unique decomposition into a sum $z_r = z_r^{[0]} + z_r^{[1]} + \cdots + z_r^{[q]}$, where $z_r^{[k]} = D*\omega^k\beta_k$ lies on $Z_{2(r-k)}$* (but not on Z_{2l}, $l < r - k$). Using only the assumption of a Kaehler metric we obtain thus a geometric situation of the same nature as the homology structure of algebraic varieties discovered by Lefschetz.

3.9. I would like to conclude with a remark on the differential forms themselves. If we have a simple form $\alpha = \omega^k \beta$ (cf. 3.7.), and if α is closed, it follows easily from the general relations that $\bar{\delta}\alpha = 0$; if in addition α is pure, this means $\delta\alpha = 0$, hence α is harmonic. A closed form α which is pure and simple is therefore always harmonic. If in particular $\alpha = d\gamma$, it must be equal to 0 (here we use a global result, based on the fact that M is closed), and we have the theorem: *If on a Kaehler manifold a form α is such that α and $d\alpha$ are both pure and simple, then $d\alpha = 0$, and α is harmonic.*

Let us take for example an *analytic* form α, i.e., a form in the dz_j only, with analytic coefficients. α is of type 0, and also $d\alpha$, since for all coefficients a of α we have $\partial a/\partial \bar{z}_j = 0$; it is easily seen that a form of type 0 is always effective. Therefore, by the above theorem, for *an analytic differential form α in a Kaehler manifold, $d\alpha$ is always equal to 0, and α harmonic.* Relations between the set of all analytic differential forms α in the manifold and its Betti numbers (cf. [7], [9], [10]) are therefore obtained without assuming in advance $d\alpha = 0$.

BIBLIOGRAPHY

1. H. HOPF, *Studies and essays presented to R. Courant*, New York, 1948, pp. 167–185.
2. CH. EHRESMANN, *Topologie algébrique*, Paris, 1949, pp. 3–15.
3. B. ECKMANN, Comment. Math. Helv. vol. 15 (1942–1943) pp. 318–339.
4. A. KIRCHHOFF, C. R. Acad. Sci. Paris vol. 225 (1947) pp. 1258–1260.
5. F. HIRZEBRUCH, Doctoral thesis, University of Münster, 1950.
6. G. DE RHAM and P. BIDAL, Comment. Math. Helv. vol. 19 (1946) pp. 1–49; G. DE RHAM, Annales de Grenoble vol. 22 (1946) pp. 135–152.
7. W. V. D. HODGE, *Harmonic integrals*, Cambridge, 1941.
8. E. KAEHLER, Abh. Math. Sem. Hamburgishen Univ. vol. 9 (1933) pp. 173–186.
9. A. WEIL, Comment. Math. Helv. vol. 20 (1947) pp. 110–116.
10. B. ECKMANN and H. GUGGENHEIMER, C. R. Acad. Sci. Paris vol. 229 (1949) pp. 464–466, 489–491, 503–505; B. ECKMANN, C. R. Acad. Sci. Paris vol. 229 (1949) pp. 577–579.

Homotopie et dualité

Colloque de Topologie Algébrique, CBRM Louvain (1956), 41–53

Cette conférence donne un résumé préliminaire d'un travail commun de P. J. Hilton et de l'auteur. On y définit une notion d'homotopie pour les *homomorphismes de modules*, suggérée par analogie avec des propriétés fondamentales de l'homotopie topologique. Cette notion purement algébrique conduit à des concepts tels que type d'homotopie, groupes d'homotopie absolus et relatifs, suite exacte d'homotopie, etc., tout à fait analogues aux notions topologiques correspondantes et capables de jouer un rôle semblable en algèbre. La dualité standard dans le domaine des homomorphismes de modules fait correspondre à cette homotopie algébrique une notion duale avec tous les énoncés et résultats duaux. La recherche de la dualité correspondante en *topologie* donne lieu à deux suites exactes différentes pour un seul type de groupes d'homotopie, dont l'une généralise la suite d'homotopie classique et l'autre la suite de cohomologie ; il résulte ainsi une dualité entre homotopie et cohomologie, pour des espaces convenables.

Pour les détails et démonstrations ne se trouvant pas dans cet exposé le lecteur est renvoyé au mémoire de P. J. Hilton et de l'auteur, en préparation.

I. Homomorphismes de modules

1. *Homotopie injective et projective*

Soit Λ un anneau avec 1 ; « module » signifie dans la suite Λ-module à gauche, unitaire. Pour deux modules A, B notons $(A, B) = \mathrm{Hom}_\Lambda (A, B)$ le groupe des Λ-homomorphismes de A dans B. Deux notions d'homotopie sont définies pour les éléments de (A, B) :

(1) f et $g \in (A, B)$ sont dits *i-homotopes* $(f \underset{i}{\sim} g)$, si $f\text{-}g$ peut se prolonger à tout module $A' \supset A$;

(2) f et g sont *p-homotopes* $(f \underset{p}{\sim} g)$, si f-g peut se facto-riser par tout homomorphisme sur B d'un module B' ayant B comme quotient.

Pour que $f \underset{i}{\sim} 0$, il faut et il suffit que f puisse se prolonger à un module injectif arbitraire $\overline{A} \supset A$, c'est-à-dire [1] puisse se factoriser par un monomorphisme j de A dans un module injectif $\overline{A}$; pour $f \underset{p}{\sim} 0$, que f puisse se factoriser par un épi-morphisme p d'un module projectif $\underline{B}$ sur B. Ainsi les classes d'homotopie sont pour (1) les éléments du groupe $\overline{\pi}(A, B)$ $=(A, B)/j^* (\overline{A}, B)$, pour (2) de $\pi(A, B)=(A, B)/p_* (A, \underline{B})$.

Si $f \in (A, B)$, $h \in (B, C)$, et si f ou g est $\underset{i}{\sim} 0$, il en est de même pour le composé $h \circ f$. Par suite tout $\varphi \in (A, A')$ induit par composition un homomorphisme φ^* du groupe $\overline{\pi}(A', B)$ dans $\overline{\pi}(A, B)$, et tout $\psi \in (B, B')$ un ψ_* de $\overline{\pi}(A, B)$ dans $\overline{\pi}(A, B')$. Ainsi $\overline{\pi}(A, B)$ est un *foncteur* contravariant en A et covariant en B ; il est additif. Si A est injectif, $\overline{\pi}(A, B)=0$ pour tout B, et réciproquement ; et B injectif équivaut à $\overline{\pi}(A, B)=0$ pour tout A. — Pour la *p*-homotopie on a les énoncés analogues.

2. *Type d'homotopie*

On dira que $\varphi \in (A, A')$ est une *équivalence d'homotopie injective* (*i*-équivalence), s'il existe $\psi \in (A', A)$ tel que $\psi \circ \varphi$ et $\varphi \circ \psi$ soient *i*-homotopes à l'identité de A et A' respectivement. Dans ce cas A et A' sont dits i-*équivalents* ou *du même i-type* $(A \underset{i}{\sim} A')$. Pour que A soit $\underset{i}{\sim} 0$, il faut et il suffit qu'il soit injectif. Si $A \underset{i}{\sim} A'$, $B \underset{i}{\sim} B'$, alors $A + B \underset{i}{\sim} A' + B'$. On dé-montre facilement la

PROPOSITION 2-1. — *Si un foncteur additif (co- ou contra-variant)* $\mathfrak{F}(A)$ *a la propriété d'être 0 pour tout A injectif, alors toute i-équivalence* φ *de A dans A' induit un isomor-phisme entre* $\mathfrak{F}(A)$ *et* $\mathfrak{F}(A')$ *; en d'autres termes,* $\mathfrak{F}$ *ne dépend que du i-type de A.* — *Exemples* [2]: $\overline{\pi}(A, B)$ *comme fonc-teur de A ou de B.* — $\text{Ext}^n(A, B)$ *comme foncteur de B.*

---

[1] $f \in (A, B)$ est dit monomorphe (ou un monomorphisme), si son noyau est 0; épimorphe (épimorphisme), si son conoyau $B/f(A)$ est 0, c'est-à-dire si $f(A)=B$.

[2] Pour les définitions et propriétés de Ext et Tor voir H. CARTAN and S. EILENBERG, *Homological Algebra*, Princeton, 1956.

La notion de p-équivalence et de p-type étant définie de la même façon, on a le résultat dual : Un foncteur additif qui est $=0$ pour tout A projectif *ne dépend que du p-type de la variable.* — Exemples (^{2}) : π(A, B). — Ext^n(A, B) comme foncteur de A. — Tor$_n$(A, B).

PROPOSITION 2-2 (*théorèmes de factorisation*). — *Tout homomorphisme* f $\in$ (A, B) *peut se factoriser* (a) *en un monomorphisme suivi d'une i-équivalence;* (b) *en une p-équivalence suivie d'un épimorphisme.*

En effet, soit j un monomorphisme de A dans un module injectif $\overline{A}$, f' l'homomorphisme de A dans la somme directe $\overline{A} + B$ donné par $f'(a) = (j(a), f(a))$, et f'' celui de $\overline{A} + B$ dans B donné par $f''(a, b) = b$; f'' est une i-équivalence, et $f'' \circ f' = f$. — Construction duale pour (b).

PROPOSITION 2.3. — *Si* A $\underset{i}{\sim}$ B, *il existe deux modules injectifs* A$_0$ *et* B$_0$ *tel que* A $+$ A$_0 \cong$ B $+$ B$_0$ (*et réciproquement*).

3. Suspension

Nous appelons « suspension » SA du module A le quotient $\overline{A}/A$ d'un module injectif $\overline{A} \supset A$. Bien que SA dépende du choix de $\overline{A}$, son i-type n'en dépend pas ; de façon plus précise : Soient j l'inclusion de A dans un module injectif $\overline{A}$, k celle de B dans $\overline{B}$. Pout tout $f \in$ (A, B), le composé $k \circ f \in$ (A, $\overline{B}$) peut se prolonger à $\overline{A}$ et définit ainsi un homomorphisme Sf de SA dans SB. Quel que soit le choix de ce prolongement, la classe d'i-homotopie de Sf reste la même, comme on voit aisément. De là on déduit :

PROPOSITION 3-1. — *Soit* φ *une i-équivalence entre* A *et* A', *et* SA, SA' *des suspensions arbitraires de* A *et* A'. *L'homomorphisme* Sφ (*déterminé à une i-homotopie près*) *est une i-équivalence entre* SA *et* SA'.

A l'aide de la suspension S^{n}A itérée n fois, on forme le groupe $\overline{\pi}$(S^{n}A, B). Il est déterminé par A et B (même par leurs i-types) et sera noté $\overline{\pi}_n$(A, B), $n = 0, 1, 2, \ldots$; c'est le n^{me} *groupe d'homotopie injective de* A *dans* B.

De façon duale, notons ΩB le noyau de l'épimorphisme p d'un module projectif $\underline{B}$ sur B ; le p-type de ΩB ne dépend que de celui de B, et le n^{me} *groupe d'homotopie projective de* A *dans* B *est défini par*

$$\underline{\pi}_n(A, B) = \underline{\pi}(A, \Omega^n B), \quad n = 0, 1, 2, \ldots$$

On peut aussi introduire les groupes $\overline{\pi}_n$ à l'aide d'une *résolution injective de* A, c'est-à-dire d'une suite exacte

$$0 \to A \to A_1 \to A_2 \to \ \ldots$$

où les A_i sont injectifs (le noyau-image dans A_{n+1} est $S^n A$). La suite induite

$$0 \leftarrow (A, B) \leftarrow (A_1, B) \leftarrow (A_2, B) \leftarrow \ \ldots$$

n'est en général pas exacte, et ses groupes d'homologie sont justement les $\overline{\pi}_n(A, B)$. De façon duale, on peut définir les $\overline{\pi}_n$ à l'aide d'une résolution projective de B. — Il est, du reste, clair que toutes les définitions et les raisonnements s'appliquent sans modifications à toute « catégorie exacte », et que la dualité entre *i*- et *p*-théorie est une conséquence du principe de dualité général.

4. *Groupes relatifs*

La notion de « groupe relatif » se rapporte en général à un *couple*, et par là on entend une inclusion $A_1 \subset A_2$; les considérations suivantes s'appliquent aussi bien au cas d'un homomorphisme arbitraire α de A_1 dans A_2, non nécessairement monomorphe. α donne lieu à la suite exacte

$$0 \to A_0 \to A_1 \overset{\alpha}{\to} A_2 \to A_3 \to 0 \ ,$$

déterminée à un isomorphisme près ($A_0 = $ noyau, $A_3 = $ conoyau de α). Convenons d'entendre ici par un couple α soit une telle suite exacte de quatre termes $\not\equiv 0$ au plus, soit un homomorphisme $\alpha \in (A_1, A_2)$ définissant une telle suite.

On voit sans peine que de telles suites exactes ne conviennent pas pour former une « catégorie exacte » par rapport à la notion habituelle d'homomorphisme d'une suite dans une autre ; la suite des noyaux, par exemple, avec les homomorphismes induits ne sera en général pas exacte. Passons donc tout de suite à des objets plus généraux, les *suites différentielles*.

Une suite différentielle σ est une suite

$$S_0 \to S_1 \to S_2 \to \ \ldots$$

de modules S_i et d'homomorphismes d tels que la composition de deux homomorphismes successifs soit toujours $= 0 \, (dd = 0)$. On appelle homomorphisme f d'une suite différentielle σ dans une autre τ une suite d'homomorphismes f_i de S_i dans T_i telle que le diagramme

$$\begin{array}{ccccc}
S_0 & \to & S_1 & \to & S_2 \to \ \ldots \\
\downarrow & & \downarrow & & \downarrow \\
T_0 & \to & T_1 & \to & T_2 \to \ \ldots
\end{array}$$

soit commutatif $(d \circ f = f \circ d)$. Ces homomorphismes f de σ dans τ forment un groupe (σ, τ), et on vérifie les propriétés d'une catégorie exacte ; en particulier, pour que $f \in (\sigma, \tau)$ soit monomorphe, il faut et il suffit que tous les $f_i \in (S_i, T_i)$ le soient ; si les f_i sont des inclusions, f sera dit une inclusion de σ dans τ. La notion de suite *injective* est définie comme dans le cas des modules : τ est injectif si quels que soient σ et $f \in (\sigma, \tau)$, on peut prolonger f à tout σ' contenant σ.

PROPOSITION 4.1. — *Pour qu'une suite exacte σ soit injective, il faut et il suffit que tous les S_i soient injectifs.*

PROPOSITION 4-2. — *Toute suite σ est contenue dans une suite injective $\overline{\sigma}$; c'est-à-dire il existe une suite injective $\overline{\sigma}$ et un monomorphisme de σ dans $\overline{\sigma}$.*

Comme pour les modules, on définit des groupes d'homotopie injective $\overline{\pi}(\sigma, \tau)$ pour les suites différentielles et leur homomorphisme, la notion de type d'homotopie, etc. $\overline{\pi}(\sigma, \tau)$ est le groupe $(\sigma, \tau)/j^*(\overline{\sigma}, \tau)$, où $j : \sigma \longrightarrow \overline{\sigma}$ est l'inclusion de σ dans une suite injective $\overline{\sigma}$; les homomorphismes f de σ dans τ *i-homotopes* à 0 sont ceux qu'on peut prolonger à tout σ' contenant σ. Le groupe $\overline{\pi}(\sigma, \tau)$ ne dépend que des i-types de σ et τ.

Le cas particulier qui nous intéresse spécialement est celui où la suite différentielle est un couple α (suite exacte déterminée par $\alpha \in (A_1, A_2)$). Insistons sur quelques détails relatifs à ce cas : 1° pour que α soit injectif, il faut et il suffit que A_1, A_2 *et le noyau de α soient des modules injectifs* ; 2° dire qu'un couple α est contenu dans le couple α' ne signifie pas seulement que $A_1 \subset A_1'$, $A_2 \subset A_2'$ et que α est la restriction de α' à $A_1 \subset A_1'$ (en tant qu'application dans A_2) — ce qui entraîne automatiquement $A_0 \subset A_0'$ — mais encore que $A_3 \subset A_3'$; en d'autres termes que l'*homomorphisme induit du conoyau de α dans celui de α' est monomorphe.* Cela peut encore s'exprimer sous la forme $\sigma'(A_1') \cap A_2 = \alpha(A_1)$; 3° tout couple α est contenu dans un couple injectif $\overline{\alpha}$, c'est-à-dire si dans la proposition 4.2 α est un couple, on peut choisir pour $\overline{\alpha}$ injectif un couple ; 4° les homomorphismes du couple α dans une suite σ i-homotopes à 0 sont ceux qu'on peut prolonger à tout couple α' contenant α.

Dans la suite il sera seulement question de groupes d'homotopie de couples α, β ; le groupe $\overline{\pi}(\alpha, \beta)$ peut être défini par $\overline{\pi}(\alpha, \beta) = (\alpha, \beta)/j^*(\overline{\alpha}, \beta)$, où j est l'inclusion de α dans un couple injectif $\overline{\alpha}$. Soient en particulier A un module arbitraire, ι_n l'inclusion de $S^{n-1}A$ dans $S^{\overline{n-1}}A$, et $\beta \in (B_1, B_2)$ un couple arbitraire. Le groupe $\overline{\pi}(\iota_n, \beta)$ ne dépend pas de la résolution injective de A utilisée (en effet, les couples ι_n et $\iota_n{}^*$ provenant de deux résolutions de A sont i-équivalents) ; ce groupe est noté $\bar{\pi}_n(A, \beta)$ et appelé le n^{me} *groupe relatif d'homotopie injective de* A *dans le couple* β. On voit aisément que si $B_1 = 0$ ($\beta = 0$), $\bar{\pi}_n(A, \beta)$ est isomorphe canoniquement au groupe « absolu » $\bar{\pi}_n(A, B_2)$.

Les définitions et notions duales conduisent au *groupe relatif* $\pi_n(\alpha, B)$ *d'homotopie projective d'un couple* α *dans* B ; nous n'insistons pas sur les détails.

5. *Suites exactes d'un couple. Fibrations*

Un couple $\beta \in (B_1, B_2)$ et un module A étant donnés, nous leur associons la suite d'homomorphismes (*suite d'homotopie injective de* β)

$$\Sigma_i : \ldots \to \bar{\pi}_n(A, B) \xrightarrow{\beta_*} \bar{\pi}_n(A, B_2) \xrightarrow{J} \bar{\pi}_n(A, \beta) \xrightarrow{\partial} \bar{\pi}_{n-1}(A, B_1) \to$$
$$\ldots \to \pi_0(A, B_2).$$

Les homomorphismes de Σ_i sont les suivants : 1° β_* est induit par l'homomorphisme $\beta : B_1 \to B_2$; 2° soit $\beta_0 : 0 \to B_2$, et identifions $\bar{\pi}_n(A, B_2)$ à $\bar{\pi}_n(A, \beta_0)$. Par l'application triviale $g_1 : 0 \to B_1$ et l'identité $g_2 : B_2 \to B_2$, nous définissons un homomorphisme g de β_0 dans β, et nous posons

$$J = g_* : \bar{\pi}_n(A, B_2) \cong \bar{\pi}_n(A, \beta_0) \to \bar{\pi}_n(A, \beta).$$

3° par définition $\bar{\pi}_n(A, \beta) = \bar{\pi}(\iota_n, \beta)$, où $\iota_n = $ inclusion de $S^{n-1}A \subset \overline{S^{n-1}A}$. L'homomorphisme ∂ fait correspondre à tout $f \in (i_n, \beta)$ la « composante » qui applique $S^{n-1}A$ dans B_1.

La suite d'*homotopie projective d'un couple* $\alpha \in (A_1, A_2)$, pour un module B, est définie de façon analogue :

$$\Sigma_p : \ldots \to \pi_n(A_2, B) \to \pi_n(A_1, B) \to \pi_n(\alpha, B) \to \pi_{n-1}(A_2, B) \to \ldots$$
$$\ldots \to \pi_0(A_1, B).$$

THÉORÈME 5.1. — *Les suites* Σ_i *et* Σ_p *sont exactes.*

La démonstration peut se faire en vérifiant l'exactitude à chaque étape de la suite. Un procédé plus élégant consiste à considérer, pour Σ_i par exemple, une résolution injective

$$C(A) = \{ 0 \to A \to \overline{A} \to \overline{SA} \ldots \{$$

ainsi que les suites différentielles induites

$$(C(A), B_i) = \{ 0 \leftarrow (A, B_i) \leftarrow (\overline{A}, B_i) \leftarrow (\overline{SA}, B_i) \leftarrow \ldots \{ .$$
$$i = 0, 1, 2, 3,$$

et les homomorphismes de ces suites. Leurs groupes d'homologie étant les groupes d'homotopie en question, le théorème est une conséquence d'un résultat algébrique général.

Sous des hypothèses particulières sur les couples, ces suites deviennent des « suites absolues » ne contenant pas de groupes relatifs. Par analogie à des situations topologiques bien connues, nous donnons à ces conditions le nom de « fibrations ». Un homomorphisme $f \in (A, B)$ est dit une i-*fibration*, s'il jouit de la propriété de *relèvement des i-homotopies*; c'est-à-dire si tout homomorphisme d'un module injectif dans B peut se factoriser par f.

Théorème 5.2. — *Soit* $\beta \in (B_1, B_2)$ *une i-fibration,* $\varkappa$ *le monomorphisme du noyau* B_0 *de* β *dans* B_1. Alors

a) $$\overline{\pi}_n(A, \beta) \cong \overline{\pi}_{n-1}(A, B_0)$$

par un isomorphisme sous lequel

$$\partial : \overline{\pi}_n(A, \beta) \to \overline{\pi}_{n-1}(A, B_1)$$

s'identifie à

$$\varkappa_* : \overline{\pi}_{n-1}(A, B_0) \to \overline{\pi}_{n-1}(A, B_1).$$

b) $$\overline{\pi}_n(A, \varkappa) \cong \overline{\pi}_n(A, B_2)$$

par un isomorphisme sous lequel, dans la suite Σ_i *de* $\varkappa$, *l'homomorphisme*

$$J : \overline{\pi}_n(A, B_1) \to \overline{\pi}_n(A, \varkappa)$$

s'identifie à

$$\beta_* : \overline{\pi}_n(A, B_1) \to \overline{\pi}_n(A, B_2).$$

Examinons les suites relatives Σ_i de β et de $\varkappa$.

a) La suite Σ_i de β devient

$$\Sigma_i' : \ldots \to \overline{\pi}_n(A, B_1) \xrightarrow{\beta_*} \overline{\pi}_n(A, B_2) \to \overline{\pi}_{n-1}(A, B_0) \xrightarrow{\varkappa_*} \overline{\pi}_{n-1}(A, B_1) \to \ldots$$

b) La suite Σ_i de $\varkappa$

$$\ldots \to \overline{\pi}_n(A, B_0) \xrightarrow{\varkappa_*} \overline{\pi}_n(A, B_1) \xrightarrow{J} \overline{\pi}_n(A, \varkappa) \xrightarrow{\partial} \overline{\pi}_{n-1}(A, B^0) \to \ldots$$

devient

$$\Sigma_i' : \ldots \to \overline{\pi}_n(A, B_0) \xrightarrow{\varkappa_*} \overline{\pi}_n(A, B_1) \xrightarrow{\beta_*} \overline{\pi}_n(A, B_2) \to \pi_{n-1}(A, B_0) \to \ldots$$

On obtient donc, de deux façons différentes, la même suite absolue (au signe de certains homomorphismes près).

Un homomorphisme $f \in (A, B)$ est dit une p-*fibration*, s'il jouit de la propriété d'*abaissement des p-homotopies*, c'est-à-dire si tout homomorphisme de A dans un module projectif peut se factoriser par f (si f est une inclusion, il s'agit d'une « extension des p-homotopies » !).

Théorème 5.3. — *Soit* $\alpha \in (A_1, A_2)$ *une* p-*fibration*, $\in (A_2, A_3)$ *l'épimorphisme de* A_2 *sur le conoyau* A_3 *de* α. *Alors*

a) $$\underline{\pi}_n(\alpha, B) \cong \underline{\pi}_{n-1}(A_3, B)$$

par un isomorphisme sous lequel

$$\partial : \underline{\pi}_n(\alpha, B) \to \underline{\pi}_{n-1}(A_2, B)$$

s'identifie à

$$\gamma^* : \underline{\pi}_{n-1}(A_3, B) \to \underline{\pi}_{n-1}(A_2, B) .$$

b) $$\underline{\pi}_n(\gamma, B) \cong \underline{\pi}_n(A_1, B)$$

par un isomorphisme sous lequel, dans la suite Σ_p *de* γ, *l'homomorphisme*

$$J : \underline{\pi}_n(A_2, B) \to \underline{\pi}_n(\gamma, B)$$

s'identifie à

$$\alpha^* : \underline{\pi}_n(A_2, B) \to \underline{\pi}_n(A_1, B) .$$

La suite exacte des p-fibrations qui en résulte est

$$\Sigma_p' : \ldots \to \underline{\pi}_n(A_3, B) \xrightarrow{\gamma^*} \underline{\pi}_n(A_2, B) \xrightarrow{\alpha^*} \underline{\pi}_n(A_1, B) \to \underline{\pi}_{n-1}(A_3, B) \to \ldots$$

On pourrait ainsi poursuivre la liste des procédés et résultats topologiques qui se retrouvent de façon tout à fait analogue dans l'homotopie purement algébrique introduite ici. Nous y reviendrons ailleurs, ainsi que sur des applications diverses en algèbre, dont certaines font intervenir des hypothèses particulières sur l'anneau Λ opérant sur les modules et des propriétés des groupes d'homotopie qui en résultent.

II. Applications continues

6. *Homotopie*

Interprétons tous les concepts et énoncés de I en topologie, en remplaçant partout les modules par des espaces topologiques (avec points de base notés *), et les homomorphismes par des applications continues (respectant les points de base ; de même, les homotopies — au sens habituel — seront sous-entendues relatives aux points de base). Il s'agit d'un procédé heuristique, et les propositions topologiques obtenues se démontrent de façon directe ([3]).

([3]) Et séparément pour la partie « injective » et la partie « projective »; on ne dispose — tout au moins sous la forme présente — pas d'un principe de dualité disant que si un résultat est établi, l'énoncé dual est automatiquement valable.

Les notions d'homotopie injective et projective suggérées par cette analogie paraissent toutefois trop restrictives pour la suite, et nous les modifions comme ceci : L'application f d'un espace A dans B sera dite $\underset{i}{\sim} 0$, si elle peut s'étendre à tout espace A′ ⊃ A *ayant la propriété d'extension des homotopies* (au sens habituel); et f sera dit $\underset{p}{\sim} 0$, si on peut la factoriser par toute projection p de B′ sur B ayant la propriété de *relèvement des homotopies* (au sens habituel). Or sous cette forme il est immédiat que $f \underset{i}{\sim} 0$ équivaut à l'homotopie habituelle $f \sim 0$ (extension de f au *cône sur* A, noté $\overline{A}$). Il en est de même pour $f \underset{p}{\sim} 0$: en effet, $f \sim 0$ entraîne que f peut se relever dans B′ ; réciproquement, si $f \underset{p}{\sim} 0$, relevons f dans l'*espace* $\underline{B}$ *des chemins* dans B d'extrémité fixe *, qui est contractile, et il résulte $f \sim 0$. Ainsi on est amené à identifier les notions d'homotopie injective et projective avec l'homotopie ordinaire. Désignons par $\pi(A, B)$ l'ensemble des classes d'applications homotopes de A dans B ; il peut être doué d'une structure de *groupe*, si B est un H-*espace* (espace muni d'une multiplication, c'est-à-dire d'une application $B \times B \to B$ avec identité, existence d'un inverse et loi associative — à des homotopies près), ou si A est un H′-*espace* (par là nous entendons un espace muni d'une « comultiplication », c'est-à-dire d'une application $A \to A \vee A$ vérifiant les propriétés duales). En particulier, l'*espace* ΩX *des lacets dans* X (en *) est un H-espace, et la *suspension* SX un H′-espace, relatifs à des structures H, H′ évidentes. On établit facilement les deux propositions :

a) *Si* A *est un* H′-*espace et* B *un* H-*espace, les deux structures de groupe définies dans* $\pi(A, B)$ *sont identiques, et abéliennes.*

b) $$\pi(SA, B) \cong \pi(A, \Omega B)$$

par un isomorphisme canonique.

7. *Suites exactes*

Dans la correspondance modules-espaces le cône $\overline{X}$ sur X joue, pour l'homotopie, le rôle d'un module injectif contenant un module donné, et l'espace $\underline{X}$ des chemins (d'extrémité *) dans X celui d'un module projectif de quotient donné. Appelons, pour poursuivre cette analogie, *noyau de l'application* $f : A \to B$ le sous-espace $f^{-1}(*)$ de A, et *conoyau* l'espace $B/f(A)$ obtenu en identifiant $f(A)$ en un seul point *. La *suspension* SX = conoyau de l'inclusion X ⊂ $\overline{X}$ et l'*espace des* *lacets* ΩX = noyau de la projection $\overline{X \longrightarrow} X$ correspondent alors aux modules notés par les mêmes symboles. Pour les groupes d'homotopie

$$\pi_n(A, B) = \pi(S^n A, B) \quad et \quad \pi_n(A, B) = \pi(A, \Omega^n B)$$

il s'ensuit de 6. qu'ils sont *isomorphes*, et abéliens pour $n \geqslant 2$ (pour $n = 0$ une structure de groupe n'est définie que si A est un H'-espace ou B un H-espace); on les notera $\pi_n(A, B)$. Les groupes relatifs sont définis comme avant : Un « couple » α est une application continue $A_1 \longrightarrow A_2$, une application f du couple $\alpha : A_1 \longrightarrow A_2$ dans le couple $\beta : B_1 \longrightarrow B_2$ une paire d'applications $f_1 : A_1 \longrightarrow B_1$ et $f_2 : A_2 \longrightarrow B_2$ telles que

$$\begin{array}{ccc} A_1 & \to & A_2 \\ \downarrow & & \downarrow \\ B_1 & \to & B_2 \end{array}$$

soit commutatif; rappelons qu'il s'agit toujours d'applications respectant les points de base $*$. On a alors une application induite $f_0 : A_0 \longrightarrow B_0$, où $A_0 = \alpha^{-1}(*)$ est le noyau de α, B_0 celui de β, et une application induite $f_3 : A_3 \longrightarrow B_3$ des conoyaux. Le « groupe d'homotopie » $\pi(\alpha, \beta)$ est l'ensemble des classes d'homotopie de ces applications $f : \alpha \longrightarrow \beta$, muni d'une structure de groupe (abélien) sous des conditions convenables.

En particulier, soit ι_n l'homéomorphisme de $\overline{S^{n-1}A}$ dans $\overline{S^{n-1}A}$; on définit le n^{me} *groupe relatif d'homotopie de* A *dans le couple* β par

$$\pi_n(A, \beta) = \pi(\iota_n, \beta).$$

De même, si φ_n est la projection de $\overline{\Omega^{n-1}B}$ sur $\Omega^{n-1}B$, le n^{me} *groupe relatif du couple* α *dans* B est $\overline{\text{défini}}$ par

$$\pi_n(\alpha, B) = \pi(\alpha, \varphi_n).$$

Ce sont des groupes si $n \geqslant 2$, abéliens pour $n \geqslant 3$. Les suites relatives d'homotopie, Σ_i du couple $\beta : B_1 \longrightarrow B_2$ et Σ_p du couple $\alpha : A_1 \longrightarrow A_2$, sont construites comme dans le cas des modules (cf. 4) :

$$\Sigma_i : \ldots \to \pi_n(A, B_1) \xrightarrow{\beta_*} \pi_n(A, B_2) \xrightarrow{J} \pi_n(A, \beta) \xrightarrow{\partial} \pi_{n-1}(A, B_1) \to \ldots$$

$$\Sigma_p : \ldots \to \pi_n(A_2, B) \xrightarrow{\alpha^*} \pi_n(A_1, B) \xrightarrow{J} \pi_n(\alpha, B) \xrightarrow{\partial} \pi_{n-1}(A_2, B) \to \ldots$$

THÉORÈME 7.1. — *Les suites* Σ_i *et* Σ_p *sont exactes.*

Cas particuliers :

1. Soit S_n la sphère à n dimensions. Pour $A = S_0$,

$$\pi_n(S_0, B) = \pi(S_n, B)$$

est le n^{me} groupe d'homotopie ordinaire $\pi_n(B)$; notons $\pi_n(\beta)$ le groupe $\pi_n(S_0, \beta)$ (dans le cas d'un homéomorphisme de B_1 dans B_2, c'est le n^{me} groupe d'homotopie relatif ordinaire de B_2 mod. B_1). La suite Σ_i pour $A = S_0$

$$\Sigma_i^0 : \ldots \to \pi_n(B_1) \xrightarrow{\beta_*} \pi_n(B_2) \xrightarrow{J} \pi_n(\beta) \xrightarrow{\partial} \pi_{n-1}(B_1) \to \ldots$$

généralise ainsi la suite d'homotopie relative classique.

Il faut encore noter que pour un espace A localement compact, mais autrement arbitraire, la suite Σ_i d'un couple $\beta : B_1 \longrightarrow B_2$ peut aussi s'interpréter comme suite Σ_i^0 du couple induit $\beta^A : B_1{}^A \longrightarrow B_2{}^A$ (espaces des applications continues de A dans B_1 et B_2)

$$\ldots \to \pi_n(B_1{}^A) \to \pi_n(B_2{}^A) \to \pi_n(\beta^A) \to \pi_{n-1}(B_1{}^A) \to \ldots$$

2. Σ_p du couple $\alpha : A_1 \longrightarrow A_2$ généralise la *track-group sequence* de M. Barratt [4]; si A_1 et A_2 sont localement compacts, Σ_p est identique à la suite Σ_i^0 du couple B^α:

$$B^{A_2} \to B^{A_1}.$$

3.
$$\pi_n(A, S_{n+k}) = \pi(S^n A, S^n S_k) \cong \pi(A, S_k)$$

pour A compact, dim $A \leqslant 2k - 2$. Ce dernier groupe étant le groupe de cohomologie de Borsuk-Spanier, Σ_p contient comme cas particulier et avec les restrictions habituelles sur les dimensions la suite de cohomotopie.

8. *Fibrations, cohomologie*

A la lumière des observations des paragraphes 5 et 6, il convient d'appeler i-*fibration* une application $f : A \longrightarrow B$ avec *relèvement des homotopies*; p-*fibration* une application $f : A \longrightarrow B$ avec « *abaissement des homotopies* », c'est-à-dire avec la propriété que si une application $g : A \longrightarrow X$ peut se factoriser par f, alors il en est de même pour toute homotopie de g (dans le cas d'un homéomorphisme de A dans B c'est la propriété d'*extension des homotopies* de A à B).

Pour ces fibrations, on établit les analogues des théorèmes 5.2 et 5.3. Les énoncés correspondants s'obtiennent en remplaçant dans 5.2 et 5.3, modules par espaces, homomorphismes $\beta : B_1 \longrightarrow B_2$ et $\alpha : A_1 \longrightarrow A_2$ par applications continues, $\varkappa$ par l'homéomorphisme du noyau $f^{-1}(*) = B_0$ dans B_1, γ par l'identification $A_2 \longrightarrow A_3 = A_2/\alpha(A_1)$, etc. Bornons-nous à donner ici les suites exactes « absolues » qui en résultent (et qui peuvent s'obtenir de deux manières différentes, en partant

---

[4] M. G. BARRATT, *Track groups*, I, II (*Proc. London Math. Soc.*, V, 1955).

de la suite relative soit du couple β soit du couple x dans le cas de Σ_i, et du couple α ou γ dans le cas de Σ_p) :

THÉORÈME 8.1. — *Soit* β : $B_1 \longrightarrow B_2$ *une i-fibration*, x *l'homéomorphisme de* $B_0 = f^{-1}(*)$ *dans* B_1. *On a, pour un espace* A *arbitraire, la suite d'homotopie injective exacte*

$$\Sigma_i' : \ldots \to \pi_n(A, B_0) \overset{x_*}{\to} \pi_n(A, B_1) \overset{\beta_*}{\to} \pi_n(A, B_2) \to \pi_{n-1}(A, B_0) \to \ldots$$

THÉORÈME 8.2. — *Soit* α : $A_1 \longrightarrow A_2$ *une p-fibration*, γ *l'identification* $A_2 \longrightarrow A_3 = A_2/\alpha(A_1)$ *de l'image de* A_1 *en un point. On a, pour un espace* B *arbitraire, la suite d'homotopie projective exacte*

$$\Sigma_p' : \ldots \to \pi_n(A_3, B) \overset{\gamma^*}{\to} \pi_n(A_2, B) \overset{\alpha^*}{\to} \pi_n(A_1, B) \to \pi_{n-1}(A_3, B) \to \ldots$$

Donnons une application de 8.2 qui présente un intérêt particulier. Soit B un espace d'Eilenberg-Mac Lane $K(G, q)$, où G est un groupe abélien et q un entier positif

$$[\pi_n(K(G, q)) = 0 \text{ pour } n \neq q, \text{ et } = G \text{ pour } n = q].$$

Rappelons que $K(G, q)$ possède une H-structure, unique à l'homotopie près, et posons

$$\pi(X, K(G, q)) = H^q(X, G)$$

pour tout espace X. On sait que $\Omega K(G, q)$ est un $K(G, q-1)$, donc

$$\pi_n(X, K(G, n+q)) = \pi_n(X, \Omega^n K(G, n+q))$$
$$= \pi(X, K(G, q)) = H^q(X, G).$$

En substituant $K(G, n+q)$ pour B dans Σ_p, pour une p-fibration α : $A_1 \longrightarrow A_2$, on obtient

$$\ldots \to H^q(A_3, G) \overset{\gamma^*}{\to} H^q(A_2, G) \overset{\alpha^*}{\to} H^q(A_1, G) \to H^{q+1}(A_3, G) \to \ldots ;$$

cette suite correspond entièrement à la suite exacte de cohomologie, $H^q(A_3, B)$ jouant le rôle du groupe de cohomologie relatif [sans l'hypothèse que α est une p-fibration, on définirait ce groupe relatif simplement par $\pi_n(\alpha, K(G, q+n))$]. Si α est un homéomorphisme de A_1 dans A_2, on constate sans peine que les autres axiomes d'Eilenberg-Steenrod ([5]) pour la cohomologie sont satisfaits par les $H^q(X, G)$. Il résulte que pour la classe des polyèdres X, $H^q(X, G)$ *est le* q^{me} *groupe de cohomologie de* X *à coefficients dans* G, et Σ_p contient donc comme cas particulier la suite de cohomologie pour les polyèdres.

Dans la dualité entre *i*- et *p*-théories topologiques, les groupes d'homotopie classiques correspondent donc aux groupes de cohomologie. Les premiers se basent sur la suite d'espaces standard $S^n S_0 = S_n$, les seconds sur les $K(G, q)$ définis par

$$\pi(S_n, K(G, q)) = 0 \text{ pour } n \neq q, \text{ et } = G \text{ pour } n = q.$$

Pour compléter la symétrie on peut définir pour un groupe abélien G des espaces ([6]) $K'(G, n)$ en posant

$$H^q(K'(G, n), Z) = \pi(K'(G, n), K(Z, q)) = 0$$

pour $n \neq q$, $= G$ pour $n = q$ (Z = groupe des entiers); S_n est donc un $K'(Z, n)$. Ensuite on définit des groupes d'homotopie à coefficients dans G par

$$\pi_n(X; G) = \pi(K'(G, n), X).$$

Ils vérifient les axiomes duaux de ceux des $H^q(X, G)$; en particulier, on voit que la suite exacte des i-fibrations correspond à la suite exacte de cohomologie, et elle peut jouer un rôle analogue dans l'axiomatique des groupes d'homotopie.

On traitera ailleurs d'autres aspects de cette dualité qui s'applique à beaucoup de concepts et résultats topologiques ou en suggère de nouveaux.

([5]) S. Eilenberg and N. E. Steenrod, *Foundations of Algebraic Topology*, Princeton, 1952 (« Single space » theory, chap. X).
([6]) Nous n'entrons pas ici dans la discussion de l'existence et de la construction de ces espaces.

Groupes d'homotopie et dualité

Bull. Soc. Math. France 86 (1958), 271–281

I. — INTRODUCTION.

1. — Dans cet exposé je me propose de montrer comment il est possible, en partant du concept classique d'*homotopie*, d'obtenir une grande partie des notions et lois fondamentales de la topologie algébrique en réduisant l'appareil algébrique à un minimum ; au lieu de l'introduire *a priori* sous forme de complexes, cochaines, etc., on fait intervenir les structures algébriques de façon directe, par des définitions topologiques et naturelles au sens de l'homotopie. Je n'affirme nullement que cette façon de faire soit nécessaire ou supérieure aux autres, ou même qu'elle permette de se passer des procédés classiques dans les exemples les plus élémentaires. Il s'agit simplement de montrer qu'elle est possible et donne une vue d'ensemble unifiée sur la topologie algébrique, et avant tout qu'elle conduit tout naturellement à une *dualité* qui, tout au moins, a une valeur heuristique. La forme générale de cette dualité s'exprime dans *deux suites exactes* d'homotopie très générales qui contiennent comme cas particuliers presque toutes les suites exactes connues.

Il existe, dans l'algèbre des modules — ou plus généralement dans toute catégorie abélienne vérifiant certains axiomes — deux « suites d'homotopie » purement algébriques tout à fait analogues ; j'en ai parlé ailleurs [1] et je ne reviens pas ici sur cette analogie. Je tiens à remarquer que le présent exposé est une introduction sommaire à des recherches faites en collaboration avec Peter J. HILTON.

2. — On va donc considérer les *applications* (continues) d'un espace A dans un espace B et les *classes d'homotopie* de ces applications, au sens habituel; par exemple A étant un cercle S_1, les classes des chemins fermés dans B. Le rôle essentiel d'un point de base, dont on a besoin pour définir une opération de groupe parmi ces classes, est bien connu. Je précise que dans ce qui suit, « espace » signifie toujours *espace avec point-base* déterminé, noté o, « application » une application $A \to B$ envoyant o de A dans o de B, et « homotopie » une déformation respectant cette condition (pour toute valeur du paramètre de déformation). L'ensemble des applications $A \to B$ sera noté (A, B), celui des classes d'homotopie de ces applications $\Pi(A, B)$; dans ce dernier il y a un élément distingué o, la classe des applications « homotopes à o », celle qui contient l'application triviale f_0, $f_0(A) = o$. L'ensemble $\Pi(A, B)$ est l'objet essentiel de ce qui suit; en essayant d'y introduire des *structures de groupe naturelles*, dans un sens précis, on sera amené aux groupes d'homotopie généraux et aux deux suites exactes duales qui constituent un accès « homotopique » à la topologie algébrique.

Pour $g : A \to A'$, notons comme d'habitude g^* l'application de (A', B) dans (A, B) induite, donnée par $g^*(f) = f \circ g : A \to A' \to B$ pour tout $f \in (A, B)$, et aussi celle de $\Pi(A', B)$ dans $\Pi(A, B)$ correspondante. De même pour $h : B \to B'$, par h_* l'application induite de (A, B) dans (A, B') ou de $\Pi(A, B)$ dans $\Pi(A, B')$, donnée par $h_*(f) = h \circ f : A \to B \to B'$, pour tout $f \in (A, B)$.

II. — GROUPES D'HOMOTOPIE.

3. — Le procédé le plus simple pour définir une structure de groupe dans $\Pi(A, B)$ se présente lorsque B est un *groupe topologique* (par exemple S_1) : On multiplie deux fonctions f et g en multipliant les valeurs, c'est-à-dire les images $f(a).g(a)$, et les axiomes de groupe sont satisfaits pour cette opération $f.g$. Il en est de même lorsque B est un groupe « à l'homotopie près » seulement, ce qu'on appelle un *H-espace*. C'est un espace B muni d'une *multiplication M* associant à b, $b' \in B$ un produit $M(b, b')$ continu, c'est-à-dire d'une application continue $M : B \times B \to B$, de telle façon que les axiomes de groupe soient satisfaits « à l'homotopie près » (par exemple l'application $M_0 : b \to M(b, o)$ est homotope à l'identité I de B, etc.); M est aussi appelée une *H-structure dans B*. Le produit $f.g$ de deux applications $A \to B$ est alors défini par $(f.g)(a) = M(f(a), g(a))$; il établit dans $\Pi(A, B)$ une structure de groupe.

Notons que dans ce cas, pour tout $h : A \to A'$, l'application induite $h^* : \Pi(A', B) \to \Pi(A, B)$ respecte le produit $f.g$. Ce procédé pour définir une structure de groupe dans $\Pi(A, B)$ pour un H-espace B et tout A, est donc « naturel »; par là nous entendons que les applications induites h^* sont toujours des *homomorphismes*.

Nous allons examiner de façon générale toutes les structures de groupe naturelles qu'on peut définir dans $\Pi(A, B)$. Soit donc donnée une règle qui, pour B fixe, fait de $\Pi(A, B)$ un groupe pour tout A, de telle façon que tous les h^* induits par $h : A \to A'$ soient des homomorphismes. L'opération dans ces groupes est notée multiplicative. Alors la classe triviale $o \in \Pi(A, B)$ est l'élément neutre de ce groupe. En effet, il en est ainsi pour $A = o$ se réduisant à un point ; pour A arbitraire et $h : A \to o$, l'image de cette classe par h^* qui est égale à $o \in \Pi(A, B)$ doit être l'élément neutre, h^* étant un homomorphisme. Prenons ensuite $A = B \times B$; soient $p_1, p_2 : B \times B \to B$ les projections sur les deux facteurs, $p_1(b, b') = b$, $p_2(b, b') = b'$, et soit $p = p_1.p_2$. On remarque que pour $l : B \to B \times B$ défini par $l(b) = (b, o)$, on a $l^*p = l^*(p_1.p_2) = l^*p_1.l^*p_2 = I.o = I \in \Pi(B, B)$; en effet, $l^*p_1 : b \to b \times o \to b$ est l'identité de B, $l^*p_2 : b \to b \times o \to o$ est l'application triviale. Choisissons une application M dans la classe $p : B \times B \to B$. L'application $M_0 = l^*M : b \to b \times o \to M(b, o)$ est donc homotope à l'identité I de B. De façon analogue on vérifie les autres axiomes : M est une H-structure dans B.

Soient f, g deux applications $A \to B$, $f \times g$ l'application $A \to B \times B$ donnée par $a \to f(a) \times g(a)$. On a pour les classes correspondantes $\in \Pi(A, B)$, $f \cdot g = (f \times g)^*p$, donc $(f.g)(a) = M(f(a), g(a))$. En effet, $(f \times g)^*p = (f \times g)^*p_1 . (f \times g)^*p_2$; or $(f \times g)^*p_1$ est l'application $a \to f(a) \times g(a) \to f(a)$, c'est f, et $(f \times g)^*p_2 = g$. Il résulte ainsi :

Si, pour un espace B, une structure de groupe naturelle est donnée dans $\Pi(A, B)$ pour tout A, alors B est un H-espace et l'opération dans $\Pi(A, B)$ est celle définie par la multiplication (H-structure) dans B.

4. — La question qui s'impose alors est de déterminer les structures de groupe naturelles dans $\Pi(A, B)$ pour A fixe et B variable, dualité consistant à renverser le sens des flèches (des applications) ou, ce qui revient au même, à permuter le rôle de A et de B.

Supposons une telle règle donnée, définissant dans $\Pi(A, B)$ pour tout B une opération naturelle, que nous désignerons par $+$. Alors, de façon tout à fait analogue à ce qui précède, il s'ensuit que A est muni d'une *comultiplication (H'-structure)*, et que $f + g$ est définie par cette comultiplication.

Que faut-il entendre par là ? Choisissons $B = A \vee A$, la réunion de deux exemplaires de A ayant o en commun (par exemple $A \times o \cup o \times A \subset A \times A$) ; appelons p_1 la classe de l'application $a \to a \times o$, p_2 celle de $a \to o \times a$, et soit $M' : A \to A \vee A$ dans la classe $p_1 + p_2$. Cette application M' vérifie des axiomes correspondant par dualité à ceux d'une H-structure ; par exemple, pour $l : A \vee A \to A$ [donnée par $l(a \times o) = a$, $l(o \times a) = o$], $l.M' : A \xrightarrow{M'} A \vee A \xrightarrow{l} A$ est homotope à l'identité I de A, etc. Nous l'appellerons une H'-structure dans A. On montre comme avant, que pour $f, g : A \to B$ on a $f + g = (f \vee g).M'$; c'est-à-dire $f + g$ est donnée par l'application $A \xrightarrow{M'} A \vee A \xrightarrow{f \vee g} B$ [$f \vee g$ désignant l'application $a \times o \to f(a)$, $o \times a \to g(a)$]. Comme avant, la classe triviale o est l'élément neutre de $\Pi(A, B)$.

Si, pour un espace A, une structure de groupe naturelle est donnée dans $\Pi(A, B)$ pour tout B, alors A est un H'-espace et l'opération dans $\Pi(A, B)$ est celle définie par la comultiplication (H'-structure) dans A.

Si A est un H'-espace et B un H-espace, on a donc deux opérations naturelles $+$ et $.$ dans $\Pi(A, B)$. Pour quatre éléments $f, f_1, g, g_1 \in (A, B)$ effectuons $(f+g).(f_1+g_1)$:

C'est évidemment la même chose que $f.f_1 + g.g_1$, donc

$$(f+g).(f_1+g_1) = f.f_1 + g.g_1.$$

Choisissons $g = f_1 = o = $ classe triviale $\in \Pi(A, B)$:

$$(f+o).(o+g_1) = (o+g_1).(f+o) = f+g_1,$$
$$f+g_1 = f.g_1 = g_1.f,$$

c'est-à-dire : *Si A est un H'-espace et B un H-espace, les deux structures de groupe dans $\Pi(A, B)$ coïncident et sont abéliennes.* La structure de groupe dans $\Pi(A, B)$ ne dépend donc, dans ce cas, ni de la H- ni de la H'-structure choisie dans B et A ; il y a une seule structure de groupe naturelle dans $\Pi(A, B)$, et elle est abélienne.

5. — Il existe des procédés standard pour associer à des espaces donnés des H-espaces et des H'-espaces.

a. ΩB, *l'espace des lacets dans B* au point o [c'est-à-dire des application continues ω du segment $s = (o \leq t \leq 1)$ dans B avec $\omega(o) = \omega(1) = o$] est un H-espace par rapport à une multiplication classique, la composition habituelle des chemins.

b. ΣA, *la suspension de A* (le cône double sur la base A) est définie par $A \times s$, avec $A \times (o) \cup A \times (1) \cup o \times s$ identifié en un seul point o ; elle possède une comultiplication M' évidente.

On voit sans peine qu'il y a correspondance biunivoque canonique entre les éléments de $\Pi(\Sigma A, B)$ et $\Pi(A, \Omega B)$, et qu'elle respecte les opérations $+$ et $.$ induites par ces structures : $\Pi(\Sigma A, B) \cong \Pi(A, \Omega B)$ par un isomorphisme canonique.

Pour deux espaces A et B et un entier n, définissons le $n^{\text{ième}}$ *groupe d'homotopie de A dans B* par

$$\Pi_n(A,\, B) = \Pi(\Sigma^n A,\, B) = \Pi(A,\, \Omega^n B) = \Pi(\Sigma^{n-k} A,\, \Omega^k B),$$

$o \leq k \leq n$. Si n est > 1, $\Pi_n(A,\, B)$ est abélien. Ces groupes coïncident, dans le cas particulier où A est une sphère S_k à k dimensions, donc $\Sigma^n A = S_{n+k}$, avec les groupes d'homotopie de Hurewicz $\pi_{n+k}(B)$. Nous reviendrons sur d'autres choix particuliers importants de A ou de B.

6. — La dualité qui s'est imposée ci-dessus consiste en un procédé systématique pour obtenir, à partir d'un énoncé (définition, théorème, ...) l'énoncé dual. Sans trop insister, notons qu'on « renverse les flèches » (les directions des applications); remplacer le produit cartésien $B \times B$ par la somme $A \vee A$, ΩB par ΣA, etc. peut aussi être interprété dans ce sens.

Pour une application $f : A \to B$, appelons $N = f^{-1}(o) \subset A$ le noyau, et $C = B/f(A)$, obtenu par identification de $f(A)$ en un point o, le conoyau; N et C se correspondent dans la dualité. Notons que la suite

$$N \to A \xrightarrow{f} B \to B/f(A)$$

n'est pas « exacte » : l'image $f(A)$ n'est pas équivalente à A/N. La dualité n'a qu'un caractère heuristique, dans ce sens que le dual d'un énoncé n'est pas automatiquement valable, mais doit être redémontré.

III. — GROUPES RELATIFS, SUITES EXACTES.

7. — Les groupes d'homotopie de Hurewicz $\pi_m(B)$, cas particuliers de nos $\Pi_n(A,\, B)$, sont liés entre eux par une suite exacte relative, instrument fondamental et bien connu pour les raisonnements où interviennent ces groupes. Nous allons généraliser cette suite exacte dans le cas de nos groupes $\Pi_n(A,\, B)$.

Pour cela, rappelons que le lien essentiel entre les $\pi_m(B)$ des différentes dimensions m est le *groupe relatif* $\pi_m(\beta)$ d'une paire β d'espaces $B_1 \subset B_2$: les éléments de $\pi_m(\beta)$ sont les classes d'homotopie des applications $(V_m,\, S_{m-1}) \to (B_1,\, B_2)$, où V_m désigne la boule à m dimensions, S_{m-1} sa frontière (où bien $V_m =$ cône sur S_{m-1}). Plus exactement, il s'agit d'une paire d'applications $f_1 : S_{m-1} \to B_1$, $f_2 : V_m \to B_2$ avec $f_2 \mid S_{m-1} = f_1$, c'est-à-dire telles que le diagramme

$$
\begin{array}{ccc}
S_{m-1} & \xrightarrow{f_1} & B_1 \\
\downarrow{\scriptstyle j} & & \downarrow{\scriptstyle \beta} \\
V_m & \xrightarrow{f_2} & B_2
\end{array}
$$

($j =$ plongement de S_{m-1} dans V_m, $\beta =$ plongement de B_1 dans B_2) soit commutatif.

Il convient pour la suite de nos raisonnements de ne pas se restreindre au cas d'un plongement $\beta : B_1 \to B_2$; nous entendrons par *une « paire » simplement une application arbitraire* β de B_1 dans B_2. En passant de la *catégorie des espaces à celle des paires*, on peut refaire les raisonnements de la partie II de cet exposé; nous allons les exquisser, sans entrer dans les détails.

Soient $\alpha : A_1 \to A_2$, $\beta : B_1 \to B_2$ deux paires. Une application $f : \alpha \to \beta$ est formée par deux applications $f_i : A_i \to B_i$, $i = 1$, 2, telles que

$$
\begin{array}{ccc}
A_1 & \xrightarrow{f_1} & B_1 \\
\downarrow{\scriptstyle\alpha} & & \downarrow{\scriptstyle\beta} \\
A_2 & \xrightarrow{f_2} & B_2
\end{array}
$$

soit commutatif. Une homotopie de $f : \alpha \to \beta$ et $g : \alpha \to \beta$ est formée par une homotopie de f_1 et une de f_2 telles que pour toute valeur du paramètre de déformation la commutativité soit conservée. L'ensemble $\Pi(\alpha, \beta)$ de ces classes d'homotopie contient un élément trivial $[f_i(A_i) = o \in B_1]$. On peut établir une structure de groupe naturelle dans $\Pi(\alpha, \beta)$ pour β fixe et α variable, si β est une *H-paire* et seulement dans ce cas; pour α fixe et β variable, si α est une *H'-paire*. Une *H*-paire $b : B_1 \to B_2$ est une application d'un *H*-espace B_1 dans un *H*-espace B_2 qui est un homomorphisme pour les deux *H*-structures; de façon analogue pour une *H'*-paire α. *Si α est une H'-paire et β une H-paire, les deux structures de groupe dans $\Pi(\alpha, \beta)$ coïncident et sont abéliennes.*

En particulier, pour $\beta : B_1 \to B_2$ arbitraire, l'application induite $\Omega\beta : \Omega B_1 \to \Omega B_2$ des lacets de B_1 dans ceux de B_2 est une *H*-paire. De même, pour $\alpha : A_2 \to A_2$, l'application correspondante des suspensions $\Sigma\alpha : \Sigma A_1 \to \Sigma A_2$ est une *H'*-paire, et l'on a $\Pi(\Sigma\alpha, \beta) \cong \Pi(\alpha, \Omega\beta)$; on pourrait définir des groupes $\Pi_n(\alpha, \beta)$, etc.

8. Ici nous allons nous borner à ce qui est nécessaire pour définir les groupes « relatifs » tels qu'ils interviennent dans les suites exactes reliant les $\Pi_n(A, B)$.

Soit β une application $B_1 \to B_2$ arbitraire, et, pour A arbitraire, ι_1 l'application qui plonge A comme base dans son cône CA. Comme $\Sigma^{n-1} CA$ est équivalent à $C\Sigma^{n-1} A$, l'application $\iota_n = \Sigma^{n-1} \iota_1 : \Sigma^{n-1} A \to \Sigma^{n-1} CA$ peut aussi s'interpréter comme plongement de $\Sigma^{n-1} A$ dans $C\Sigma^{n-1} A$. Le groupe $\Pi(\iota_n, \beta) = \Pi(\Sigma^{n-1}\iota_1, \beta)$ est noté $\Pi_n(A, \beta)$ et dit le $n^{\text{ième}}$ *groupe d'homotopie relatif de A dans la paire β*. Pour qu'il y ait effectivement une structure de groupe, il faut supposer $n > 1$; ce groupe est abélien pour $n > 2$. Les éléments de $\Pi_n(A, \beta)$ sont, dans le cas d'un plongement $\beta : B_1 \subset B_2$, les classes d'applications de $C\Sigma^{n-1} A$ dans B_2 envoyant $\Sigma^{n-1} A$ dans B_1. Si

$B_1 = o \in B_2$, on peut identifier ces applications à celles de $\Sigma^n A = C \Sigma^{n-1} A / \Sigma^{n-1} A$ dans B_2, et le groupe relatif $\Pi_n(A, \beta)$ s'identifie au groupe absolu $\Pi_n(A, B_2)$. De façon analogue, si B_2 se réduit à un point o ($\beta : B_1 \to o$), le groupe $\Pi_n(A, \beta)$ s'identifie à $\Pi_{n-1}(A, B_1)$.

Pour $\beta : B_1 \to B_2$ arbitraire, on a alors une suite d'homomorphismes

$$S_*(\beta) : \quad \ldots \to \Pi_n(A, B_1) \xrightarrow{\beta_*} \Pi_n(A, B_2) \xrightarrow{J} \Pi_n(A, \beta) \xrightarrow{\partial} \Pi_{n-1}(A, B_1) \to \ldots,$$

J s'obtient en interprétant $\Pi_n(A, B_2)$ comme $\Pi_n(A, \mathring{\beta})$, $\mathring{\beta} : o \to B_2$, et en prenant l'homomorphisme de $\Pi_n(A, \mathring{\beta})$ dans $\Pi_n(A, \beta)$ induit par l'application de $\mathring{\beta}$ dans β qui est l'identité sur B_2. L'homomorphisme ∂ est donné par la restriction à $\Sigma^{n-1} A \to B_1$ des applications de ι_n dans β. On vérifie directement que *cette suite $S_*(\beta)$ est exacte*.

De façon duale, on définit pour une application $\alpha : A_1 \to A_2$ arbitraire et un espace B les *groupes relatifs* $\Pi_n(\alpha, B)$, $n > 1$ (ensemble sans structure de groupe pour $n = 1$), abéliens pour $n > 2$: On entend par ρ_1 la projection naturelle de EB [espace des chemins ouverts ω d'origine o dans B, $\rho_1(\omega) = $ extrémité de ω] dans B, de noyau $\rho_1^{-1}(o) = \Omega B$, et l'on pose $\rho_n = \Omega^{n-1} \rho_1$, $n > o$. Alors ρ_n s'interprète aussi comme projection de $E \Omega^{n-1} B$ dans $\Omega^{n-1} B$, de noyau $\Omega^n B$. Le groupe $\Pi_n(\alpha, B)$ est défini par $\Pi(\alpha, \rho_n)$; si $A_2 = o$, $\Pi_n(\alpha, B)$ s'identifie à $\Pi_n(A_1, B)$, et si $A_1 = o$, à $\Pi_{n-1}(A_2, B)$. On associe à α arbitraire la *suite exacte*

$$S^*(\alpha) : \quad \ldots \to \Pi_n(A_2, B) \xrightarrow{\alpha^*} \Pi_n(A_1, B) \xrightarrow{J} \Pi_n(\alpha, B) \xrightarrow{\partial} \Pi_{n-1}(A_2, B) \to \ldots$$

les homomorphismes J et ∂ étant définis de façon tout à fait duale de ceux dans $S_*(\beta)$.

Si A est une sphère, et $\beta : B_1 \subset B_2$ un plongement, la suite $S_*(\beta)$ se réduit à la suite relative classique des groupes d'homotopie π_m de Hurewicz. On constate également que $S^*(\alpha)$ contient comme cas particulier, avec les restrictions habituelles sur les dimensions, la suite de *cohomotopie* (Borsuk-Spanier), et d'autres suites connues. Le fait que la suite de Hurewicz se réduit, sous des conditions particulières (fibration), à une suite *absolue*, se retrouve pour notre suite générale $S_*(\beta)$ et de façon duale, pour $S^*(\alpha)$; entrons dans quelques détails de cette question.

9. — Appelons *fibration* une application $\beta : B_1 \to B_2$ qui jouit de la propriété de *relèvement des homotopies* (pour les applications dans B_2 d'une certaine classe d'espaces précisée suivant le cas) : pour toute application $f_2 : X \to B_2$ qui se factorise par β, $f_2 = \beta \circ f_1$, toute homotopie de f_2 se factorise également par β. Dualement, appelons *cofibration* une application $\alpha : A_1 \to A_2$ avec la propriété d' « *abaissement des homotopies* » : Pour toute application $f_1 : A_1 \to X$ qui se factorise par α, $f_1 = f_2 \circ \alpha$, toute homotopie

de f_1 se factorise également par α. En diagrammes :

Si α est un plongement $A_1 \subset A_2$, alors cette dernière propriété n'est autre chose que celle d'*extension des homotopies*, valable en particulier pour tout polyèdre A_2 et sous-polyèdre A_1.

Pour une *fibration* $\beta : B_1 \to B_2$, soit $B_0 = \beta^{-1}(o)$ la fibre-type et $\nu : B_0 \subset B_1$ le plongement de B_0 dans B_1. On démontre alors sans peine que, pour A arbitraire,

$$\Pi_n(A, \nu) \cong \Pi_n(A, B_2) \qquad \text{et} \qquad \Pi_n(A, \beta) \cong \Pi_{n-1}(A, B_0)$$

par des isomorphismes canoniques; le premier fait de $S_*(\nu)$ une suite absolue

$$\ldots \to \Pi_n(A, B_0) \to \Pi_n(A, B_1) \to \Pi_n(A, B_2) \to \Pi_{n-1}(A, B_0) \to \ldots,$$

et le second de $S_*(\beta)$ la même suite (à des signes près).

Pour une *cofibration* $\alpha : A_1 \to A_2$, appelons *cofibre* le quotient $A_2/\alpha A_1 = A_3$ (le conoyau de α), et γ l'application $A_2 \to A_3$ donnée par l'identification de αA_1 en un point o. On démontre alors que, pour B arbitraire,

$$\Pi_n(\gamma, B) \cong \Pi_n(A_1, B) \qquad \text{et} \qquad \Pi_n(\alpha, B) \cong \Pi_{n-1}(A_3, B),$$

et l'on obtient une suite exacte absolue

$$\ldots \to \Pi_n(A_3, B) \to \Pi_n(A_2, B) \to \Pi_n(A_1, B) \to \Pi_{n-1}(A_3, B) \to \ldots.$$

Dans le cas d'une cofibration $\alpha : A_1 \subset A_2$, l'isomorphisme

$$\Pi_n(\alpha, B) \cong \Pi_{n-1}(A_3, B)$$

peut s'interpréter ainsi : Le groupe relatif $\Pi_n(\alpha, B)$ de $\alpha : A_1 \to A_2$ ne dépend que du quotient A_2/A_1, c'est-à-dire ne change pas si l'on enlève de $A_1 \subset A_2$ les points intérieurs. C'est donc simplement la *propriété d'excision* comme on la rencontre en homologie et cohomologie. La propriété duale $\Pi_n(A, \beta) \cong \Pi_{n-1}(A, B_0)$ dans le cas d'une fibration $\beta : B_1 \to B_2$ exprime que le groupe relatif $\Pi_n(A, \beta)$ ne dépend que de $B_0 = \beta^{-1}(o)$.

IV. — COHOMOLOGIE ET HOMOTOPIE A COEFFICIENTS.

10. La suite exacte $S_*(\beta)$ se réduit, si l'on prend pour A la sphère S_{m-n}, à la suite exacte des groupes de Hurewicz π_m.

Soit maintenant B un espace d'Eilenberg-MacLane $K(G, m)$, où G est un groupe abélien et m un entier ≥ 0; c'est-à-dire $\pi_n(B) = 0$ pour $n \neq m$, et $\pi_m(B) = G$. On sait que pour tout G et m il existe un polyèdre de ce genre, et que son type d'homotopie est déterminé par G et m. De plus, il est immédiat que $\Omega K(G, m+1)$ est un espace $K(G, m)$; si A est un polyèdre, les éléments de $\Pi(A, K(G, m))$ et de $\Pi(A, \Omega K(G, m+1))$ se correspondent de façon biunivoque. $\Pi(A, K(G, m))$ est donc un groupe, $\cong \Pi_n(A, K(G, m+n))$ pour tout $n \geq 0$, noté $H^m(A; G)$. Si A n'est pas un polyèdre, sous-entendons qu'on remplace A par son polyèdre singulier.

De façon analogue, définissons pour $\alpha : A_1 \to A_2$, un groupe abélien G et un entier m, le groupe *relatif* $H^m(\alpha; G)$ par

$$H^m(\alpha; G) = \Pi_1(\alpha, K(G, m)) = \Pi_n(\alpha, K(G, m+n-1));$$

pour $\alpha : 0 \to A_2$ il se confond avec $\Pi_{n-1}(A_2, K(G, m+n-1)) = H^m(A_2; G)$.

Avec ces définitions, la suite exacte $S^*(\alpha)$ avec $B = K(G, m+n)$ prend la forme

$$\ldots \to H^m(A_2; G) \overset{\alpha^*}{\to} H^m(A_1; G) \overset{J}{\to} H^{m+1}(\alpha; G) \overset{\partial}{\to} H^{m+1}(A; G) \to \ldots.$$

On y reconnaît la *suite de cohomologie*, lorsque α est un plongement $A_1 \subset A_2$, J étant le « cobord » et ∂ induit par l'application

$$
\begin{array}{ccc}
0 & \overset{f_1}{\longrightarrow} & A_1 \\
\mathring{\alpha} \downarrow & & \downarrow \alpha \\
A_2 & \overset{f_2}{\longrightarrow} & A_2
\end{array}
$$

de $\mathring{\alpha}$ dans α donnée par $f_2 =$ Identité de A_2.

Quant aux autres axiomes de cohomologie (EILENBERG-STEENROD), on vérifie sans peine qu'ils sont satisfaits pour les H^m ici définis; pour cela il est entre autre, nécessaire de se débarrasser de la condition $\alpha(0) = 0$ pour les applications $\alpha : A_1 \to A_2$ et définir l'homomorphisme induit α^* dans le cas général. L'axiome d'*excision* est satisfait pour toute cofibration $A_1 \subset A_2$ (donc pour des polyèdres). On est donc en présence d'une théorie de cohomologie « légitime »; elle coïncide pour les polyèdres avec la cohomologie habituelle — cela peut, du reste, aussi être vérifié directement. La convention de remplacer, si A est un espace arbitraire, dans la définition $H^m(A, G) = \Pi(A, K(G, m))$, l'espace A par son polyèdre singulier conduit naturellement à la *théorie singulière*. Il est clair comment il faudrait procéder pour obtenir d'autres théories de cohomologie.

Si G est un groupe topologique, on peut donner une *topologie* à $\Pi(A, K(G, m))$, compacte si G est compact. Cela permet de définir les groupes d'*homologie* par rapport à un groupe de coefficients discret G par $H_m(A; G) = \mathrm{Car}\, H^m(A; \mathrm{Car}\, G)$, où Car désigne le passage au groupe des caractères (PONTRJAGIN). Ces groupes H_m coïncident avec les groupes d'homologie habituels (simpliciaux, singuliers, etc., suivant le cas).

11. — Notre dualité pour les groupes $\mathbf{\Pi}_n(A, B)$, qui s'exprime en particulier dans les deux suites exactes $S_*(\mathfrak{Z})$ et $S^*(\alpha)$, contient ainsi comme cas spécial une *dualité entre homotopie* (*groupes de Hurewicz* π_m) *et cohomologie* H^m. Pour la rendre plus complète, il convient de définir des *groupes d'homotopie à coefficients*, par rapport à un groupe de coefficients abélien G arbitraire.

Soit $K'(G, m)$ un espace à homologie entière triviale excepté en dimension m, où $H_m = G$, et avec $\pi_1 = H_1$ (espace de Moore); m est un entier > 0. Si c'est un polyèdre, et si $m > 1$, son type d'homotopie est déterminé par G et m; la suspension $\Sigma K'(G, m-1)$ est un $K'(G, m)$. En vertu de cette H'-structure, $\mathbf{\Pi}(K'(G, m), B)$ est un groupe pour $m > 1$, abélien pour $m > 2$, que nous notons $\pi_m(G; B)$. Pour $\mathfrak{Z} : B_1 \to B_2$ arbitraire, le groupe relatif $\mathbf{\Pi}_1(K'(G, m), \beta)$ est noté $\pi_{m+1}(G; \beta)$; ce groupe s'identifie à $\pi_{m+1}(G; B_2)$ pour $\beta : 0 \to B_2$. Notons que

$$\pi_m(G; B) = \mathbf{\Pi}_n(K'(G, m-n), B) \qquad \text{et} \qquad \pi_m(G; \beta) = \mathbf{\Pi}_n(K'(G, m-n), \beta)$$

où n est arbitraire dans des limites évidentes; de la suite exacte $S_*(\beta)$ on déduit la suite exacte des groupes d'homotopie à coefficients

$$\ldots \to \pi_m(G; B_1) \xrightarrow{\beta_*} \pi_m(G; B_2) \xrightarrow{j} \pi_m(G; \beta) \xrightarrow{\partial} \pi_{m-1}(G; B_1) \to \ldots,$$

De cette façon, les $\pi_m(G; B)$ vérifient des axiomes duaux de ceux des $H^m(A; G)$; en particulier, la propriété duale de l'excision exprime que pour une fibration β le groupe relatif $\pi_m(G; \beta)$ ne dépend que de $\beta^{-1}(0) = B_0 \subset B_1$. On peut baser une axiomatique des groupes d'homotopie à coefficients sur ces axiomes duaux, où le cas d'un plongement $\beta : B_1 \subset B_2$ ne joue aucun rôle, mais doit être remplacé par celui d'une fibration $\beta : B_1 \subset B_2$.

12. — Les deux suites générales $S_*(\mathfrak{Z})$ et $S^*(\alpha)$ pour les groupes $\mathbf{\Pi}_n(A, B)$ contiennent également les *suites exactes* de cohomologie et d'homotopie *relatives aux coefficients*. En effet, soit $h : G_1 \to G_2$ un homomorphisme de groupes abéliens; il peut être « réalisé » par une application

$$\beta : K(G_1, m) \to K(G_2, m)$$

telle que l'homomorphisme des π_m induit par β soit égal à h, la classe d'homotopie de β étant déterminée par h (on peut même choisir pour β une fibration). L'homomorphisme β_* de $H^m(A; G_1) = \mathbf{\Pi}(A, K(G_1, m))$ dans $H^m(A; G_2) = \mathbf{\Pi}(A, K(G_2, m))$ est déterminé par h et noté h_*. Soit alors $0 \to G_0 \xrightarrow{j} G_1 \xrightarrow{h} G_2 \to 0$ une suite exacte de groupe abéliens; si l'on réalise h par une fibration $\beta : K(G_1, m+n) \to K(G_2, m+n)$, la fibre est un $K(G_0, m+n)$ et la suite exacte $S_*(\beta)$ donne

$$\ldots \to H^m(A; G_0) \xrightarrow{j_*} H^m(A; G_1) \xrightarrow{h_*} H^m(A; G_2) \to H^{m+1}(A; G_0) \to \ldots,$$

la suite de coefficients en cohomologie.

De façon duale, on peut « réaliser » $o \to G_0 \overset{j}{\to} G_1 \overset{h}{\to} G_2 \to o$ par une cofibration $\alpha : K'(G_0, m - n) \subset K'(G_1, m - n)$ de cofibre $K'(G_2, m - n)$, et la suite $S^*(\alpha)$ donne la suite de coefficients pour l'homotopie

$$\ldots \to \pi_m(G_2; B) \overset{h^*}{\to} \pi_m(G_1; B) \overset{j^*}{\to} \pi_m(G_0; B) \to \pi_{m-1}(G_2; B) \to \ldots$$

Il faut toutefois remarquer que pour les groupes d'homotopie à coefficients les homomorphismes notés h^* et j^* ci-dessus ne sont en général pas déterminés entièrement par h et j. Comme en cohomologie, on déduit de cette dernière suite exacte un « *théorème de coefficients universels* » pour les groupes $\pi_m(G; B)$: Pour $G = Z$ (groupe cyclique infini), $\pi_m(Z; B) = \pi_m(B)$ est le $m^{\text{ième}}$ groupe de Hurewicz, et l'on a une suite exacte

$$o \to \text{Ext}(G, \pi_{m+n}(B)) \to \pi_m(G; B) \to \text{Hom}(G, \pi_m(B)) \to o.$$

$\pi_m(G; B)$ *est donc déterminé par les groupes* $\pi_m(B)$, $\pi_{m+1}(B)$ *et* G, à une extension près (qui n'est en général pas triviale, même si B est un polyèdre).

13. — La dualité entre cohomologie et homotopie telle que je viens de l'exposer suggère beaucoup de problèmes et méthodes faisant intervenir de façon systématique les groupes d'homotopie à coefficients. Mentionnons pour terminer quelques-uns de ces aspects, qui seront traités ailleurs :

a. Opérations homotopiques pour les groupes $\pi_m(G; B)$ dans les cas d'un groupe de coefficients $G \neq Z$ (par exemple $G = Z_2$ ou Z_p).

b. Notion générale de transgression, contenant comme cas particuliers la transgression cohomologique dans les espaces fibrés et son dual homotopique pour les cofibrations.

c. Classification des cofibrations $A_1 \subset A_2$ de cofibre $A_2/A_1 = K'(G, m)$, et classes caractéristiques de ces cofibrations.

d. Décomposition homologique d'un espace simplement connexe (dual de la décomposition de Postnikov) : on peut caractériser le type d'homotopie (singulier) d'un tel espace par ses groupes d'homologie entière et certains éléments de groupes d'homotopie à coefficients.

BIBLIOGRAPHIE.

[1] ECKMANN (B.). — *Homotopie et dualité* (*Colloque de Topologie algébrique*, Louvain, Centre belge de Recherches mathématiques, 1956, p. 41-53).

[2] ECKMANN (B.) et HILTON (Peter J.). — *Groupes d'homotopie et dualité* ... (*C. R. Acad. Sc.*, t. 246, 1958, p. 2444-2447, 2555-2558 et 2991-2993).
ainsi que les Mémoires cités dans [1] et dans [2].

[3] ECKMANN (B.) et HILTON (Peter J.) — *Transgression homotopique et cohomologique* (*C. R. Acad. Sc.*, t. 247, 1958, p. 620-623).

Homotopy and cohomology theory

Proc. Int. Congress of Math. 1962, Institut Mittag-Leffler (1963), 59–73

1. Introduction

Algebraic topology has become a quite complex discipline. This is not only due to many new ramifications and to highly developed, refined and powerful algebraic techniques, but also to the topic itself. Its aim is to solve geometrical problems by algebraic computation; but these problems come from very different sources ranging from point-set theory in Euclidean space to differential geometry and Lie group theory and to algebraic geometry and complex analysis. The algebraic-topological methods have in many cases been suggested by the respective fields, and they have, conversely, strongly influenced them and contributed to their recent development.

Nevertheless there has been in recent years a trend to unification, which clarifies the common ideas and is helpful for applications. At its origin is, on one hand, the functorial formulation due to Eilenberg and MacLane and the axiomatic approach to homology due to Eilenberg and Steenrod; and, on the other hand, the successful efforts made by many authors in homotopy and fiber bundle theory, including operations, manifold theory, extraordinary cohomology etc., where it would be difficult to list all the names. It appeared more and more clear that there is a common framework for the various concepts and methods, simple at least in its basic ideas. It is this simplicity I wish to emphasize in the following, without attempting any degree of completeness.

The unified approach I would like to present in this expository lecture is based on the simple and intuitive geometrical concept of *homotopy*. Thus our starting point for the algebraization is the set $\Pi(X, Y)$ of homotopy classes of maps of the space X into the space Y. The following conventions are valid throughout: All spaces are assumed to be provided with a base-point written o, all maps $f\colon X \to Y$ carry base-points into base-points, and all homotopies (i.e., continuous one-parameter families of maps $X \to Y$) respect base-points for all values of the deformation parameter; in other words, the image of the base-point remains fixed during the deformation. The set $\Pi(X, Y)$ contains a distinguished element 0, namely the homotopy class of the constant map $X \to o \in Y$, the maps in that class being called "contractible" or null-homotopic. A map $h\colon Y \to Y'$ induces, by composition, a map h_* of the set $\Pi(X, Y)$ into $\Pi(X, Y')$, with $h_*(0) = 0$; and a map $g\colon X' \to X$ induces a map g^* of $\Pi(X, Y)$ into $\Pi(X', Y)$, with $g^*(0) = 0$. The set $\Pi(X, Y)$ is, of course, a topological invariant of both X and Y; more precisely, it is a functor from based spaces to based sets, covariant in Y and contravariant in X, the map h_* induced by a homeomorphism $h\colon Y \to Y'$ (or g^* induced by a homeomorphism $g\colon X' \to X$) being an equivalence of sets. In fact $\Pi(X, Y)$ depends only on the homotopy type of X and of Y: if h: $Y \to Y'$ is a homotopy equivalence (i.e., if there is a map h': $Y' \to Y$ such that hh' and $h'h$ are homotopic to the respective identities), then h_* is an equivalence of the sets $\Pi(X, Y)$ and $\Pi(X, Y')$; and similarly for a homotopy equivalence g: $X' \to X$.

Our starting point thus seems to make use of only a weak part of the structure present in the spaces involved, and to eliminate all those aspects which are not homotopy invariant. The situation, however, is as follows: In general neither the problems nor the results are homotopy invariant; but whenever it comes to actual algebraization and computation, the construction depends only on the homotopy type—not of the original space, manifold variety, etc., but of other spaces associated with them by procedures (which need not be homotopy invariant, nor even of topological nature) which use as much of the given structures as possible.

2. Two-space homotopy groups

For a fixed space Y the set $\Pi(X, Y)$ is a contravariant functor of the space X which will be used to investigate X; and vice versa, for a fixed X, a co-variant functor of Y used to investigate Y. The structure "set with distinguished element 0" is, of course, not good enough for algebraic purposes, and one wants to give this set a *natural group structure*, with 0 being the neutral element of the group. "Natural" means that the structure in $\Pi(X, Y)$, for fixed Y, turns $\Pi(X, Y)$ into a functor from spaces to groups, the induced maps g^* being homomorphisms; and similarly for fixed X. We do not go here into the general question of finding all test-spaces Y or X respectively for which this is possible (cf. [4, 5]), but simply list the two cases which will be used in the following.

(*a*) For an arbitrary space B, let ΩB be the space of loops in B beginning and ending at o, with the constant loop as base-point; ΩB is provided with a multiplication (a map $\Omega B \times \Omega B \to \Omega B$), namely the usual composition of loops, fulfilling the group axioms up to homotopy. Two maps f, f': $X \to \Omega B$ can be multiplied by multiplying the images $f(x)$ and $f'(x)$ in ΩB, and this establishes in $\Pi(X, \Omega B)$ a natural group structure.

(*b*) For a space A, let ΣA be the suspension of A, i.e., the double cone with basis A; more precisely, ΣA is the product space of A with the unit interval $I = \{0 \leqslant t \leqslant 1\}$, where the union $A \times (0) \cup A \times (1) \cup o \times I$ is identified to single point, the base-point of ΣA. There is a "comultiplication" μ in ΣA, i.e. a map of ΣA into $\Sigma A \vee \Sigma A$ (union with identified base-points); it essentially consists in pinching $A \times (\frac{1}{2})$ to a point. For an arbitrary space Y, μ can be used to define a multiplication for maps f, f': $\Sigma A \to Y$, as follows: Let F: $\Sigma A \vee \Sigma A \to Y$ be the map given by f on the first copy ΣA and by f' on the second, and define $F\mu$: $\Sigma A \to Y$ to be the product of f and f'. It is easily checked that this establishes a group structure in $\Pi(\Sigma A, Y)$, natural with respect to Y.

Moreover, the usual identification of maps from $A \times I$ to B with maps of A into the mapping space B^I immediately yields a natural equivalence between $\Pi(\Sigma A, B)$ and $\Pi(A, \Omega B)$ which is a group isomorphism. We thus identify the groups $\Pi(\Sigma A, B) = \Pi(A, \Omega B)$ and define *two-space homotopy groups* $\Pi_n(A, B), n \geqslant 0$, by

$$\Pi_n(A, B) = \Pi(\Sigma^n A, B) = \Pi(\Sigma^{n-k} A, \Omega^k B), \quad 0 \leqslant k \leqslant n,$$

where Σ^n is the n-fold suspension and Ω^n the n-fold loop-space. For $n \geqslant 2$, we can use in $\Pi_n(A,B) = \Pi(\Sigma^{n-1}A, \Omega B)$ either the multiplication in ΩB or the comultiplication in $\Sigma^{n-1}A$ for defining the group structure; by a simple and general argument the two group structures coincide and are Abelian (cf. [4, 5]). We thus obtain, for $n \geqslant 2$, functors $\Pi_n(A,B)$ from (based) spaces to Abelian groups, covariant in B and contravariant in A, with homotopic maps inducing the same homomorphisms. We sometimes write $\Pi_*(A,B)$ for the direct sum of all $\Pi_n(A,B)$, $n \geqslant 2$. If A is a sphere S_{m-n}, then $\Pi_n(A,B) = \Pi(\Sigma^n S_{m-n}, B) = \Pi(S_m, B) = \pi_m(B)$ is the usual mth Hurewicz homotopy group of B; the well-known and important algebraic feature of these groups is expressed by an exact sequence. In the next section we show more generally that there are two dual exact sequences for the $\Pi_n(A,B)$, one for a fixed A and one for a fixed B. The "duality" appearing here and in other instances consists in exchanging the role of A and B, i.e., in reversing the directions of all maps; it can be given an abstract foundation, cf. [5] and [6].

3. Exact sequences

For a fixed space A, we consider the group $\Pi_n(A,B)$ and, for a given map $\beta : B \to B'$, the induced homomorphism $\beta_* : \Pi_n(A,B) \to \Pi_n(A,B')$, $n \geqslant 2$. In the usual manner, the deviation of these β_* from being isomorphisms is measured by "relative" groups; this means that we can define groups $\Pi_n(A,\beta)$ attached to the map β (and to A) and two homomorphisms J: $\Pi_n(A,B') \to \Pi_n(A,\beta)$ and $\partial : \Pi_n(A,\beta) \to \Pi_{n-1}(A,B)$ such that the sequence

$$\ldots \to \Pi_n(A,B) \xrightarrow{\beta} \Pi_n(A,B') \xrightarrow{J} \Pi_n(A,\beta) \xrightarrow{\partial} \Pi_{n-1}(A,B) \to \ldots$$

is exact. We do not give here the full details of these definitions, but only some indications to show that everything is obtained in a quite elementary geometrical fashion.

In order to define $\Pi_n(A,\beta)$, we have to consider maps $f : \alpha \to \beta$ "of the map α into the map β". We write CX for the cone over X (i.e., the space $X \times I$ with $X \times (1) \cup o \times I$ identified to a single point o) and take for α the inclusion map $\Sigma^{n-1}A \to C(\Sigma^{n-1}A)$ of the basis into the cone. A map $f : \alpha \to \beta$ is then given by two "components" $f_0 : \Sigma^{n-1}A \to B$ and $f_1 : C(\Sigma^{n-1}A) \to B'$ subject to the condition that the diagram

$$\begin{array}{ccc} \Sigma^{n-1}A & \xrightarrow{f_0} & B \\ \alpha \downarrow & & \downarrow \beta \\ C(\Sigma^{n-1}A) & \xrightarrow[f_1]{} & B' \end{array}$$

be commutative. By a homotopy of f we understand a pair of homotopies of f_0 and f_1 respectively such that the above diagram is commutative for all values of the deformation parameter. The homotopy classes of these maps $f : \alpha \to \beta$ are the elements of the group $\Pi_n(A,\beta)$; the group structure is again defined with the help of the suspension Σ (for $n \geqslant 2$; it is Abelian for $n \geqslant 3$). If B' consists of a single point o, one has $\Pi_n(A,\beta) \cong \Pi_{n-1}(A,B)$ by an obvious isomorphism (omitting the component f_1). If B consists of a single point, one has $\Pi_n(A,\beta) \cong \Pi_n(A,B')$, since the space $C(\Sigma^{n-1}A)/\Sigma^{n-1}A$, obtained from the cone by identifying its basis to a single point o, is homeomorphic to $\Sigma^n A$.

In order to define J and ∂, we consider the commutative diagram

$$
\begin{array}{ccccccc}
 & & & & & o & \\
 & & & & & \downarrow \gamma & \\
o & \to & o & \to & B & \overset{1}{\to} & B \\
\downarrow \gamma & & \downarrow \gamma & & \downarrow \beta & & \downarrow \bar{\gamma} \\
B & \overset{\beta}{\to} & B' & \overset{1}{\to} & B' & \to & o
\end{array}
$$

where β is the given map $B \to B'$, γ and γ' are the embeddings of the base-point into B and B' respectively, and $\bar{\gamma}$ is the trivial map $B \to o$. The horizontal maps define three maps: $\gamma \to \gamma'$, $\gamma' \to \beta$ and $\beta \to \bar{\gamma}$, with induced homomorphisms $\Pi_n(A, \gamma) \to \Pi_n(A, \gamma') \to \Pi_n(A, \beta) \to \Pi_n(A, \bar{\gamma}) \cong \Pi_{n-1}(A, \gamma)$; replacing the first group by $\Pi_n(A, B)$, the second by $\Pi_n(A, B')$ and the fourth by $\Pi_{n-1}(A, B)$ we get the three homomorphisms β_*, J and ∂ of the sequence above. From these definitions, exactness is easily obtained.

[If we take for A a sphere S_{m-n}, $\Pi_n(A, B) = \pi_m(B)$ is the mth Hurewicz homotopy group of B; if $\beta : B \to B'$ is an inclusion map of a subspace $B \subset B'$, $\Pi_n(A, \beta)$ is the relative Hurewicz homotopy group $\pi_m(B' \bmod B)$, and the sequence is the classical relative Hurewicz homotopy sequence, which thus appears as the first natural case of algebraization. Obviously the assumption that β be an inclusion map is irrelevant.]

From the above considerations it is clear that the groups appearing in the exact sequence are in fact special cases of relative groups, and the whole argument generalizes easily to arbitrary relative groups, as follows. In the diagram above we replace the two first vertical maps by non-trivial maps $\gamma : B_0 \to B$, $\gamma' : B_0 \to B'$ of a space B_0 into B and B':

$$
\begin{array}{ccccccc}
 & & & & & B_0 & \\
 & & & & & \downarrow \gamma & \\
B_0 & \overset{1}{\to} & B_0 & \overset{\gamma}{\to} & B & \overset{1}{\to} & B \\
\downarrow \gamma & & \downarrow \gamma & & \downarrow \beta & & \downarrow \bar{\gamma} \\
B & \underset{\beta}{\to} & B' & \underset{1}{\to} & B' & \to & o
\end{array}
$$

Commutativity of the diagram then requires that $\gamma' = \beta \gamma$. The group $\Pi_n(A, \bar{\gamma}) = \Pi_{n-1}(A, B)$ can no longer be identified with the group $\Pi_{n-1}(A, \gamma)$, but there is an obvious homomorphism η into it. Thus the three homomorphisms induced by the vertical maps (the third one composed with η) yield a sequence

$$\ldots \to \Pi_n(A, \gamma) \to \Pi_n(A, \beta\gamma) \to \Pi_n(A, \beta) \to \Pi_{n-1}(A, \gamma) \to \ldots.$$

of relative groups for the maps γ, β and $\beta\gamma : B_0 \to B \to B'$. As before the sequence is easily seen to be exact. It is called the *triple sequence* for the "triple" of spaces $B_0 \overset{\gamma}{\to} B \overset{\beta}{\to} B'$, *relating the three relative groups of γ, of β, and of the composition $\beta\gamma$*, for a fixed A. It establishes a precise relation between $\Pi_*(A, \gamma)$ and $\Pi_*(A, \beta)$ on one hand, and $\Pi_*(A, \beta\gamma)$ on the other; this relation does not determine $\Pi_*(A, \beta\gamma)$ but expresses it in a well-known fashion "up to a group extension". The ordinary exact sequence is a special case of the triple sequence; the latter has the advantage of making use of the additional structure present in the category of maps between spaces, namely the composition of maps.

The next step then is obviously the take a composition of more than two maps $\beta = \beta_1\,\beta_2...\beta_r$ (for some integer $r \geqslant 2$; the infinite case could be handled similarly, but is omitted here for simplicity):

$$B_r \xrightarrow{\beta_r} B_{r-1} \to ... \xrightarrow{\beta_2} B_1 \xrightarrow{\beta_1} B$$

and to ask for the relationship between the $\Pi_*(A,\beta_i), i = 1,...,r$ on one hand and $\Pi_*(A,\beta)$ on the other. The answer is clearly given by a *spectral sequence*: one starts from the direct sum $\sum_{i-1}^r \Pi_*(A,\beta_i)$ and amalgamates the triple sequences corresponding to the successive compositions into exact couples; the spectral sequence thus obtained (cf. Eckmann–Huber [6]) begins with $E_1 = \sum_{i-1}^r \Pi_*(A,\beta_i)$ and ends with $\Pi_*(A,\beta)$ in the usual sense that E_∞ is the graded group of $\Pi_*(A,\beta)$ for an obvious filtration (given by the factorization $\beta = \beta_1\,\beta_2...\beta_r$). The differentials of the spectral sequence are obtained from the homomorphisms appearing in the various triple sequences.

Just a few words about the dual situation: In the groups $\Pi_n(A,B)$, a fixed test-space B is considered; for a map $\alpha : A \to A'$ with induced homomorphisms $\alpha^* : \Pi_n(A',B) \to \Pi_n(A,B)$ a relative group $\Pi_n(\alpha,B)$ is defined and homomorphisms $J : \Pi_n(A,B) \to \Pi_n(\alpha,B), \partial : \Pi_n(\alpha,B) \to \Pi_{n-1}(A',B)$ such that the sequence

$$... \to \Pi_n(A',B) \xrightarrow{\alpha^*} \Pi_n(A,B) \xrightarrow{J} \Pi_n(\alpha,B) \xrightarrow{\partial} \Pi_{n-1}(A',B) \to ...$$

is exact. The elements of $\Pi_n(\alpha,B)$ are the homotopy classes of maps of α into β, where β designates the projection map $E\Omega^{n-1}B \to \Omega^{n-1}B$ (EX denoting the space of all open paths in X ending at o, with projection $\beta : EX \to X$, which is a fiber space over X with fiber $\beta^{-1}(o) - \Omega X$):

$$A \to E\Omega^{n-1}B$$
$$\alpha \downarrow \qquad \downarrow \beta$$
$$A' \to \Omega^{n-1}B$$

More generally, for the composition $\gamma\alpha$ of two maps $A \xrightarrow{\alpha} A' \xrightarrow{\gamma} A_0$ one has an exact *triple sequence*

$$... \to \Pi_n(\gamma,B) \to \Pi_n(\gamma\alpha,B) \to \Pi_n(\alpha,B) \to \Pi_{n-1}(\gamma,B) \to ...,$$

and for the composition of r maps $\alpha = \alpha_r\,\alpha_{r-1}...\alpha_1$ a *spectral sequence* beginning with the direct sum $E_1 = \sum_{i-1}^r \Pi_*(\alpha_i,B)$ and ending with $\Pi_*(\alpha,B)$ for a filtration corresponding to the factorization.

The two dual results, incidentally, need not be established separately, but can be obtained from a general categorical argument using semi-simplicial constructions (cf. P. J. Huber [7]).

Before going into the applications of this general algebraic framework, all based upon a special choice of the test-spaces A or B respectively, we describe a further general feature known under the name of "excision". It again manifests itself in two perfectly dual forms.

4. Excision

A map $\beta : E \to B$ is called a *fiber map*, with fiber $F = \beta^{-1}(o)$, if the *covering homotopy property* holds for arbitary maps $f : X \to B$; i.e., if f is factored through

β as $f=\beta f'$, then any homotopy of f can be factored through β. For the Hurewicz homotopy groups it is then a well-known fact that $\pi_n(\nu)$ is naturally isomorphic to $\pi_n(B)$, where ν denotes the inclusion $F \subset E$; this is easily seen to be equivalent to the isomorphism $\pi_n(\beta) \cong \pi_{n-1}(F)$, and in this form a general statement can be formulated, expressing relative groups $\Pi_n(A,\beta)$ of a fiber map β by "absolute" groups of the fiber F: The commutative diagram

$$F \xrightarrow{\gamma} E$$
$$\beta_0 \downarrow \qquad \downarrow \beta$$
$$o \rightarrow B$$

considered as a map of β_0 into β induces a homomorphism $\Pi_n(A,\beta_0) \rightarrow \Pi_n(A,\beta)$, called excision; from the fiber map property it follows easily that this excision homomorphism is an isomorphism. The left-hand side being $\cong \Pi_{n-1}(A,F)$, one has a natural isomorphism $\Pi_{n-1}(A,F) \cong \Pi_n(A,\beta)$. [Equivalently one might say that for a fiber map β the group $\Pi_n(A,\beta)$ remains unchanged if everything but the fiber is removed from E; i.e., if β is replaced by β_0. The exact sequence then gives the above isomorphism.]

Dually a map $\alpha:A \rightarrow E$ is called a *cofiber map*, with cofiber $F = E/\alpha(A)$ (the image in E identified to a single point o), if the *homotopy extension property* holds for arbitrary maps $f:A \rightarrow X$; i.e., if f is factored through α as $f=f'\alpha$, then any homotopy of f can be factored through α. For an inclusion $A \subset E$ this is the ordinary homotopy extension property. The diagram

$$A \rightarrow o$$
$$\alpha \downarrow \qquad \downarrow \alpha_0$$
$$E \rightarrow F$$

then induces an excision isomorphism $\Pi_n(\alpha_0, B) \rightarrow \Pi_n(\alpha, B)$, where the right-hand side is $\cong \Pi_{n-1}(F, B)$. Thus the relative group $\Pi_n(\alpha, B)$ of a cofibration is naturally isomorphic to the "absolute" group $\Pi_{n-1}(F, B)$ of the cofiber. In other words, the relative group of a cofibration only depends on the cofiber $F = E/\alpha(A)$ and remains unchanged if the interior of $\alpha(A)$ is removed.

These two excision isomorphisms for fiberings and cofiberings, converting relative groups into absolute ones, prove in certain cases useful for computation.

5. Spectra. Cohomology

From the general set-up the groups normally used in algebraic topology are obtained by special choices of the test-spaces A and B respectively. As said before, taking for A a sphere (e.g., S_0) yields the Hurewicz homotopy groups with their exact sequences, triple sequences and spectral sequences, and with the excision property for fiberings. In the dual situation, cohomology theory is obtained by using test-spaces B belonging to a "spectrum", as explained below (there is a certain lack of duality appearing at this stage, which will be discussed later).

An Ω-*spectrum* $\mathfrak{B}$, or in short a spectrum, is a sequence of spaces B_k given for all integers k, together with maps $w_k:B_k \rightarrow \Omega B_{k+1}$ which are *homotopy equivalences* for all k. Given such a spectrum $\mathfrak{B} = (B_k, w_k)$, there is an isomorphism (given by the respective w_k) between the groups $\Pi_n(A, B_{m+n}) = \Pi(A, \Omega^n B_{m+n})$ and $\Pi(A, B_m)$, for any $n > 0$. Thus the two groups can be identified; thus $\Pi_n(A, B_{m+n}) = \Pi(A, B_m)$ is independent of $n > 0$ and we denote this Abelian group by $h^m(A)$, m being an arbitrary integer. Similarly, relative groups $h^m(\alpha)$ for a map $\alpha : A \to A'$ are defined by $h^m(\alpha) = \Pi_{n+1}(\alpha, B_{m+n}) = \Pi_1(\alpha, B_m)$ for all m; for $\alpha^0 : o \to A$ one has $h^m(\alpha^0) = h^m(A)$, for $\alpha_0 : A \to o$, one has $h^m(\alpha_0) = h^{m-1}(A)$. The exact sequence of the groups Π_n

$$\ldots \to \Pi_n(A', B_{m+n}) \xrightarrow{\alpha^*} \Pi_n(A, B_{m+n}) \xrightarrow{J} \Pi_n(\alpha, B_{m+n}) \xrightarrow{\partial} \Pi_{n-1}(A', B_{m+n}) \to \ldots$$

then translates into an exact sequence

(S) $$\ldots \to h^m(A') \xrightarrow{\alpha^*} h^m(A) \xrightarrow{J} h^{m+1}(\alpha) \xrightarrow{\partial} h^{m+1}(A') \to \ldots$$

which looks like a cohomology sequence, where the upper indices correspond to dimensions and where J is usually denoted by δ. The excision property for cofibrations $\Pi_{n+1}(\alpha, B_{n+m}) \cong \Pi_n(F, B_{m+n})$, F being the cofiber of α, translates into the isomorphism $h^m(F) \cong h^m(\alpha)$, again in agreement with cohomology excision; the isomorphism is induced by

$$\begin{array}{ccc} A & \to & o \\ \alpha \downarrow & & \downarrow \varphi \\ A' & \to & F \end{array}$$

considered as a map of α into φ. A third cohomology property to be mentioned here, obvious from our homotopical definitions, is the fact that homotopic maps α induce the same homomorphisms $\alpha^* : h^m(A') \to h^m(A)$.

To sum up: We have obtained a sequence of contravariant functors $h^m(\alpha)$ from the category of based maps α of spaces, to the category of Abelian groups, defined for all integers m and such that for any space A the two maps $\alpha^0 : o \to A$ and $\alpha_0 : A \to o$ are related by a natural "connecting" isomorphism $h^m(\alpha^0) \cong h^{m+1}(\alpha_0)$; this group is written $h^m(A)$, defining a contravariant functor from spaces to Abelian groups. These functors have the three properties (I) Exactness, (II) Excision, (III) Homotopy; (I) means that the sequence (S) induced by the diagram

$$\begin{array}{ccccccc} o & \to & A & \xrightarrow{1} & A & \xrightarrow{\alpha} & A' \\ \downarrow & & \downarrow \alpha & & \downarrow & & \downarrow \\ A' & \xrightarrow{1} & A' & \to & o & \to & o \end{array}$$

and the "connecting isomorphism" is exact; (II) means that for a cofibration α the group $h^m(\alpha)$ depends only on the cofiber, and (III) that homotopic maps induce the same homomorphism. Any system of functors with these properties is called a *general cohomology theory*. We thus summarize the above considerations by saying that, *given a spectrum* $\mathfrak{B} = (B_m, w_m)$, *our homotopical definitions yield a general cohomology theory* $h^m(\alpha)$. We will limit ourselves here entirely to such homotopically defined cohomology theories; by a recent representation theorem of R. Brown [2] this is, to a large extent not a real restriction: the theorem states that (under certain countability assumptions) any abstractly given cohomology theory can be realized homotopically by means of a spectrum $\mathfrak{B}$.

Here the following remarks should be added for completeness or in view of later application.

(1) We have restricted ourselves to "reduced" cohomology theories h^*, with $h^m = 0$ for a one-point space and with groups and homomorphisms defined for spaces with base-points and based maps only. "Unreduced" theories h_+^* for free spaces[1] can be obtained from these reduced ones (and vice versa) by standard procedures (cf., e.g., [8]), and there are no essentially different features. In particular, if we write A^+ for the topological sum of A with an extra point which is taken as base-point of A^+, the groups $h_+^m(A)$ of h_+^* are given by $h^m(A^+)$; the map $A^+ \to A$ which is the identity on A induces a monomorphism $h^m(A) \to h_+^m(A)$.

(2) The exactness property (I) includes, in fact, the triple sequence for the composition of two maps; and consequently also the *spectral sequence* for the composition of $r \geqslant 2$ maps $\alpha = \alpha_r \alpha_{r-1} \dots \alpha_1$, with $E_1 = \sum_{i=1}^r h^*(\alpha_i)$ and E_∞ equal to the graded group associated with the filtration of $h^*(\alpha)$ determined by the factorization.

(3) From the definitions it follows immediately that $\Pi(\Sigma A, B_{m+1}) \cong \Pi(A, \Omega B_{m+1}) = \Pi(A, B_m)$, i.e., $h^{m+1}(\Sigma A) \cong h^m(A)$. Thus, for example, the cohomology h^* of the k-sphere S_k can be expressed by $h^*(S_0)$, with $h^m(S_k) \cong h^{m-k}(S_0)$ for all m.

(4) The "dimension axiom" of cohomology requires that $h^m(S_0) = 0$ for all $m \neq 0$, or equivalently, that for any $k \geqslant 0$ $h^m(S_k) = 0$ for all $m \neq k$. We note that $h^m(S_k) = \Pi(S_k, B_m) = \pi_k(B_m)$, and there may very well be spectra for which *this dimension axiom does not hold*. In order that this axiom holds, it is necessary and sufficient that, for each integer m, $\pi_k(B_m) = 0$ for $k \geqslant 0$, $k \neq m$; this means for $m \geqslant 0$ that B_m is an Eilenberg–MacLane space $K(G_m, m)$ for the dimension m and the Abelian group $G_m = \pi_m(B_m)$, and for $m < 0$ that B_m is (homotopy equivalent to) a one-point space o. Moreover the homotopy equivalence $w_m : B_m \to \Omega B_{m+1}$ shows that $G_m \cong G_{m+1}, m \geqslant 0$. Thus the *cohomology with respect to a spectrum $\mathfrak{B}$ fulfills the dimension axiom if and only if $\mathfrak{B}$ is the Eilenberg-MacLane spectrum*: $B_m = K(G, m)$ for $m \geqslant 0$ and a fixed Abelian group G, $B_m = o$ for $m < 0$ (hence $h^m(A) = 0$ for all $m < 0$).

It is to be emphasized that the above homotopical definition of cohomology (without or with dimension axiom) and the whole algebraic machinery including the spectral sequence is entirely independent of any type of cellular structure of the spaces involved or of "cohomology" in any of the classical meanings.

6. Polyhedra, simplicial cohomology

The special role of polyhedral spaces in the context of cohomology lies, roughly speaking, in the particular "possibilities for effective computation" given by the cellular structure. The precise meaning of this feature, in addition to the general algebraization outlined above, can be described as follows.

By a "polyhedron" we mean here, for simplicity, a finite simplicial complex (the finiteness is only used for part of the following considerations,

---

[1] The notations normally used are h^m for our h_+^m and $\bar{h}^m$ for our h^m.

and moreover everything can be carried over to spaces having the homotopy type of CW-complexes). For such a polyhedron A, let $A^{(m)}$ denote its m-skeleton, $A^{(-1)} = o \subset A^{(0)}$, and α_m the embedding $A^{(m-1)} \to A^{(m)}$ for $m = 0, 1, ..., n (= \dim. A)$. Then the map $\alpha: o \to A$ is factored as $\alpha = \alpha_n \alpha_{n-1} ... \alpha_1 \alpha_0$. The spectral sequence for $h^*(\alpha) = h^*(A)$ corresponding to that factorization begins with $E_1 = \sum_{m=0}^n h^*(\alpha_m)$. By the excision property for cofibrations we have $h^q(\alpha_m) = h^q(A^{(m)}/A^{(m-1)})$; now $A^{(m)}/A^{(m-1)}$, the m-skeleton where the $(m-1)$-skeleton is identified to a single point o, is the union of a certain number of spheres S_m with one common point o, each sphere corresponding to an m-simplex σ_m of A. The elements of the group $h^q(\alpha_m) \cong \Pi(A^{(m)}/A^{(m-1)}, B_q)$ can thus be considered as functions associating to each simplex σ_m of A an element of $\Pi(S_m, B_q) = \pi_m(B_q) = h^q(S_m) \cong h^{q-m}(S_0)$; in other words, $h^q(\alpha_m)$ is the group $C^m(A; \pi_m(B_q)) = C^m(A; h^{q-m}(S_0))$ of all simplicial m-cochains of A with values in $\pi_m(B_q) \cong h^{q-m}(S_0)$. Thus the spectral sequence begins with

$$E_1 = C^*(A; h^*(S_0));$$

more precisely, writing p for $q - m$, $E_1 = \sum_{m=0}^n \sum_{p=-\infty}^{+\infty} E_1^{m,p}$ with $E_1^{m,p} = h^{m+p}(\alpha_m) = C^m(A; h^p(S_0))$ [contributing through the spectral sequence to $h^{m+p}(\alpha) = h^{m+p}(A)$]. The differential d_1 in E_1 is easily computed: it is the ordinary simplicial coboundary in $C^m(A; h^p(S_0))$. Therefore we have

$$E_2 = H^*(A; h^*(S_0)),$$

more precisely $E_2 = \sum_{m,p} E_2^{m,p}$ with $E_2^{m,p} = H^m(A; h^p(S_0))$. Here H^m(and H^*) denote the (reduced) *elementary simplicial cohomology* of the polyhedron A, with the respective coefficient groups, computed as usual from incidence numbers, cocycles and coboundaries.

We summarize the result: *For a cohomology theory h^** (with respect to a given spectrum $\mathfrak{B}$) *and a polyhedron A there is a spectral sequence for $h^*(A)$ with $E_2 = H^*(A; h^*(S_0))$, where all $H^m(A; h^p(S_0))$ with $m + p = q$ contribute to $h^q(A)$. Thus the first step of the computation of $h^*(A)$ is given by the simplicial cohomology of A with coefficients in $h^*(S_0)$.*

The remainder of this lecture (sections 7 and 8) is devoted to a rapid description of some of the numerous applications of this fundamental fact of polyhedral algebraic topology; I have to choose from a large variety of results which increases every day. Some of them use further "partly computable" spectral sequences obtained from other factorizations of maps (the factorization being defined with the help of skeletons), e.g. for fiber maps. A different approach to such spectral sequences, for abstractly given cohomology is due to Dold (an outline can be found in [3]); it includes multiplicative structures and interesting general properties of cohomology for polyhedra.

7. Special cases and remarks

(1) We first consider, for a polyhedron A, the case of a cohomology theory h^* *with dimension axiom*: $h^q(S_0) = 0$ for $q \neq 0$, and we write G for the group $h^0(S_0) \cong h^m(S_m)$ for arbitrary $m \geqslant 0$ (the isomorphism being given by suspension). Then in the spectral sequence for $h^*(A)$ we have $E_2 = \sum_{m=0}^n E_2^{m,p}$ $(n = \dim. A)$ with $E_2^{m,p} = H^m(A; h^p(S_0)) = 0$ for $p \neq 0$ and $= H^m(A; G)$ for $p = 0$. It follows that $d_r = 0$ for all $r \geqslant 2$, hence

$$E_2^{m,p} = E_r^{m,p} = E_\infty^{m,p} = 0 \text{ for } p \neq 0, \; = H^m(A;G) \quad \text{for} \quad p=0, \; m=0,\ldots,n,$$

and $h^q(A) \cong E_\infty^{q,0}$; we thus obtain

$$h^q(A) \cong H^q(A;G), \quad \text{with} \quad G = h^0(S_0),$$

for $0 \leqslant q \leqslant n$ and 0 otherwise. The isomorphism (of degree 0) $h^*(A) \cong H^*(A; h^0(S_0))$ is the *uniqueness theorem for cohomology with dimension axiom* in the category of polyhedra: any such theory is isomorphic to the simplicial cohomology relative to the coefficient group $G = h^0(S_0)$. It may also be called the simplicial computation theorem for cohomology with dimension axiom, thus making the whole classical machinery available for applications where the explicit structure of cohomology groups is needed.

In terms of the homotopical definitions the dimension axiom tells that the spectrum $\mathfrak{B}$ of the cohomology theory h^* consists of Eilenberg–MacLane spaces $K(G,m)$, i.e., $h^m(A) = \Pi(A, K(G,m))$. Then the above isomorphism $\Pi(A, K(G,m)) \cong H^m(A; G)$ yields the "simplicial computation of homotopical Eilenberg-MacLane cohomology". It should be added that in the whole argument above the spectral sequence is not really needed, since the term E_2 only occurs; the triple sequence for two maps $A^{(m-1)} \to A^{(m)} \to A^{(m+1)}$ would be sufficient. A close inspection of course provides an explicit homotopical description not only of simplicial cohomology classes, but also of simplicial cocycles and coboundaries (useful, for example, in obstruction theory).

In the following H^* always denotes cohomology with dimension axiom, which on polyhedra is identified with simplicial cohomology.

(2) The spectral sequence for $h^*(A)$, where h^* is a general cohomology theory and A a polyhedron, depends functorially upon A and upon h^*. As an application of the functorial dependence upon A, we consider a map $\alpha: A \to A'$ and assume that it induces an isomorphism $H^*(A'; G) \cong H^*(A; G)$ for any G (or, equivalently, that α induces an isomorphism of simplicial integral homology). Then α induces isomorphisms of all the terms E_r, $r \geqslant 2$, of the spectral sequence for $h^*(A')$ onto those for $h^*(A)$, and thus an isomorphism $h^*(A') \cong h^*(A)$. [An alternative proof for this result, not using the spectral sequence, is given by the fact that the map α yields a homotopy equivalence of ΣA with $\Sigma A'$.] The functorial behavior with respect to h^* appears in the next item.

(3) Let $\mathfrak{B} = (B_m, w_m)$ and $\mathfrak{B}' = (B'_m, w'_m)$ be two spectra, and $t: \mathfrak{B} \to \mathfrak{B}'$ a map of $\mathfrak{B}$ into $\mathfrak{B}'$, i.e., a sequence of maps $t_m: B_m \to B'_m$ compatible with the homotopy equivalences w_m and w'_m. Let h^*, k^* denote the cohomology theories relative to $\mathfrak{B}$ and $\mathfrak{B}'$ respectively. For an arbitrary space A, the maps T_A of $h^m(A) = \Pi(A, B_m)$ into $k^m(A) = \Pi(A, B'_m)$ induced by the t_m are then homomorphisms. They constitute a system of natural transformations of the functors h^m into k^m for all m, compatible with the "connecting isomorphism" (cf. section 4). Such a system is called in short a *transformation of the cohomology theory h^* into k^**; any such transformation can be given by maps of the corresponding spectra. As particular cases, these transformations include (additive) cohomology operations, coefficient relations in ordinary cohomology, etc. The spectral sequences for h^* and k^* can be applied to such transformations (there is a possibility of applying also the dual exact and spectral sequences of the $\Pi_n(A, B)$ to $t_m: B_m \to B'_m$, but we do not describe it here).

We first consider a transformation $T: h^* \to k^*$ of cohomology theories which is an *isomorphism on S_0*. Then, *for any polyhedron A, T induces isomorphisms* of the terms $E_r, r \geqslant 2$, of the spectral sequence for $h^*(A)$ into those for $k^*(A)$, and thus T is *an isomorphism $h^*(A) \cong k^*(A)$*. Again it should be observed here that the result can also be obtained by homotopical arguments not using the spectral sequence; namely, it follows easily from the assumption that the map $t: \mathfrak{B} \to \mathfrak{B}'$ corresponding to the transformation is for each m a weak homotopy equivalence $B_m \to B_m'$.

(4) The following procedure due to Dold [3] yields an interesting transformation T relating an arbitrary cohomology theory h^* to ordinary cohomology H^* relative to the coefficient group $h^*(S_0) \otimes \mathbb{Q}$ (tensor product with the group of rationals).

Let k^* be a cohomology theory, with spectrum $\mathfrak{B} = (B_m, w_m)$, and consider the composition of maps

$$\Sigma^N A \xrightarrow{f} S_{N+p} \xrightarrow{g} B_{N+p+q}.$$

For N sufficiently large the homotopy class of f is (under the suspension isomorphism) independent of N and defines an element of the stable cohomotopy group $\pi_{st}^p(A)$, while g defines, for an arbitrary N, an element of $k^q(S_0)$ and gf an element of $k^{p+q}(A)$. We thus get a transformation T' given by

$$T_A': \pi_{st}^p(A) \otimes k^q(S_0) \to k^{p+q}(A).$$

Now $\pi_{st}^p(S_0)$ is finite for $p \neq 0$ and infinite cyclic for $p = 0$; therefore, if we assume $k^*(S_0)$ to be a module over the ring $\mathbb{Q}$ of rationals, T_{S_0}' will be an isomorphism and thus T_A' an isomorphism for polyhedra A. This result can be applied (a) to $k^q = h^q \otimes \mathbb{Q}$, where h^* is an arbitrary cohomology theory, and (b) to $k^q = \sum_{m+p=q} H^m(A; h^p(S_0) \otimes \mathbb{Q})$, and thus yields an isomorphism

$$T_A'': h^q(A) \otimes \mathbb{Q} \cong \sum_{m+p=q} H^m(A; h^p(S_0) \otimes \mathbb{Q})$$

for polyhedra A. Combining this with the natural embedding $\tau_A: h^q(A) \rightarrow h^q(A) \otimes \mathbb{Q}$ *we get a transformation T given by*

$$T_A: h^q(A) \to \sum_{m+p=q} H^m(A; h^p(S_0) \otimes \mathbb{Q})$$

which reduces to τ_{S_0} on the 0-sphere and which is, for any polyhedron A, an isomorphism if tensored by $\mathbb{Q}$. Thus for polyhedra the difference between H^* and an arbitrary cohomology theory h^* lies essentially only in torsion elements. The transformation T is, by the way, uniquely determined by the fact that T_{S_0} is the embedding $\tau_{S_0}: h^*(S_0) \to h^*(S_0) \otimes \mathbb{Q}$.

(5) There is a famous example (Grothendieck-Atiyah-Hirzebruch) of a cohomology theory without dimension axiom and the corresponding transformation into ordinary rational cohomology. It is given by the *unitary spectrum* $\mathfrak{B} = (B_m, w_m)$ with $B_m = \Omega U$ if m is even, $B_m = U$ if m is odd, where U denotes the infinite unitary group and ΩU its loop-space; by the Bott periodicity theorem ΩU can be used as classifying space for U (there is a (weak) homotopy equivalence $w: U \to \Omega \Omega U$; hence we can put $w_m = 1$ for m

even, $w_m = w$ for m odd). The corresponding cohomology groups $\Pi(A, \Omega U)$ and $\Pi(A, U)$ are denoted by $K^m(A)$ (reduced, usually written $\tilde{K}^m$); one has $K^m(S_0) = \pi_0(\Omega U)$ or $\pi_0(U)$ respectively, i.e. $K^m(S_0) = \mathbf{Z}$ if m is even, $= 0$ if m is odd. The transformation T of (4) yields isomorphisms for polyhedra A

$$K^q(A) \otimes \mathbf{Q} \cong \sum_{m+p=q} H^m(A; K^p(S_0) \otimes \mathbf{Q})$$

$$= \sum_{m \text{ even}} H^m(A; \mathbf{Q}) \quad \text{if} \quad q \text{ is even,}$$

$$= \sum_{m \text{ odd}} H^m(A; \mathbf{Q}) \quad \text{if} \quad q \text{ is odd,}$$

thus expressing $K^q(A) \otimes \mathbf{Q}$ by the Betti numbers of A. Moreover one has, of course, the whole spectral sequence of section 6 for $K^*(A)$, with $E_2 = H^*(A; K^*(S_0))$, as given originally by Atiyah-Hirzebruch. All this and related further facts have been used successfully to compute K^* and K^*_+ (unreduced theory, cf. section 5) for complex and real projective spaces, and "truncated" ones; and also to compute, in these cases, cohomology operations $K^* \rightarrow K^*$ defined by linear representations of the unitary groups. On the basis of these computations, Adams has given the complete solution of the real and of the unitary vector field problem on spheres (determination of the maximum number of independent fields; for the real case, see [1]).

We have not so far mentioned the relationship between $K^*(A)$ and unitary *vector bundles* over A: Since B_m is the classifying space of the unitary group (m even) or its loop-space (m odd), the group $K^m(A) = \Pi(A, B_m) = \Pi(\Sigma A, B_{m+1})$ is the group of stable unitary vector bundles over A (m even) or over ΣA (m odd); and $K^m_+(A)$ the group of vector bundles with specified fiber dimension. The transformation $T: K^*(A) \rightarrow H^*(A; K^*(S_0) \otimes \mathbf{Q})$ above, regarded as a transformation into $H^*(A; \mathbf{Q})$ is nothing else than the "Chern character": for $\xi \in K^*(A)$, $T(\xi) = \text{ch } \xi$ is the Chern character belonging to the total Chern class of ξ (ch ξ and the total Chern class determine each other).

This relation to bundle theory is irrelevant for the computations above, but of course it plays an important part in many other connections; in particular when multiplicative structures are involved, see section 8.

(8) We close this section with some remarks on *duality*. The duality aspects of the two-space homotopy groups $\Pi_n(A, B)$, including their algebraic properties as described in sections 2–4, are obtained by interchanging A and B, i.e, by reversing the direction of all maps. The Hurewicz homotopy groups $\pi_m(B) = \Pi(S_m, B)$ and the cohomology groups $h^m(A) = \Pi(A, B_m)$ for some spectrum $\mathfrak{B} = (B_m, w_m)$ behave, in some respects, dually to each other: from $\Pi(A, B)$ one obtains the π_m by choosing $A = S_m$, the h^m by choosing $B = B_m$ (with equivalences $w'_m : \Sigma S_m \rightarrow S_{m+1}$ and $w_m : B_m \rightarrow \Omega B_{m+1}$, hence $\pi_m(B) = \Pi_n(S_{m-n}, B)$ and $h^m = \Pi_n(A, B_{m+n})$, independent of n), and thus many dual features obviously are carried over from the general $\Pi_n(A, B)$ to a dual behavior of the π_m and the h^m (exactness, spectral sequence for composition of maps, excision, etc.). It is, however, clear that cohomology goes beyond the general set-up and that there is no obvious dualization of the special properties of h^* described in sections 5 and 6. In this context we add two simple comments.

(a) The simplicial computation theorem for h^* relies on the special structure of polyhedral spaces and on the excision property for cofibrations. In order to obtain a similar computation scheme for π_* one has to take, instead of polyhedra, complicated spaces constructed artificially (by certain fibrations with Eilenberg–MacLane fibers). There is no evidence so far that these spaces are of geometrical significance.

(b) If one considers, in a spectrum $\mathfrak{B} = (B_m, w_m)$ a fixed B_m, one has in $\mathfrak{B}$ a "trivial" part below B_m, in the sense that all B_n with $n < m$ can be replaced by $\Omega^{m-n} B_m$, and a non-trivial part above B_n. The system of spheres S_m together with $w_m' : \Sigma S_m = S_{m+1}$ contains for $m = 0$ only the trivial part above S_0, $S_n = \Sigma^n S_0$ for $n > 0$, and it cannot be extended to a non-trivial part below S_0. Thus the π_m do not go beyond the general $\Pi_n(A, B)$, except that a special space A is being used; other choices of A such as Moore spaces (π_m with coefficients, cf. [4]) do not essentially change the situation. However a more systematic use of spectra may lead to further duality.

It is well-known that the π_m and the $\Pi_n(A, B)$ in general, even if there is no simple geometrical class of spaces for which they are easily computable, have been and will continue to be of great interest. This is due both to their intuitive geometrical meaning and to their algebraic structure as explained above, which through the spectra directly leads to cohomology. There are various ways to bring also *homology* and related duality theorems into the picture but we do not describe them here (cf. G. W. Whitehead [8]).

8. Multiplicative cohomology

In this section we sketch some aspects of general cohomology which play a central role in current developments of algebraic topology and which, at least in their algebraic background, belong to the methods outlined in this lecture. No details will be given, and very few references and definitions will be included.

A cohomology theory is *multiplicative* if h^* is provided with a natural product turning this graded group into a graded anti-commutative ring, with $h^*(A^+) = h_+^*(A)$ having a unit $e \in h^0(A^+)$, and if, moreover, the relative cohomology $h^*(\alpha)$, $\alpha : A' \to A$, is a module over $h^*(A)$. (In particular, if P denotes a one-point space, $h^*(P^+) - h^*(S_0)$ has a unit $e \in h^0(S_0)$; under suspension it corresponds to an element γ^n of $h^n(S_n)$.)

The spectral sequence arguments for polyhedra can be carried over to multiplicative cohomology theories. In particular, a transformation T: $h^* \to k^*$ where h^* and k^* are multiplicative cohomology theories and k^* a Q-module, preserves products if it does so on S_0. For example, $H^*(\ ; \mathbf{Q})$ is multiplicative with respect to the cup-product; and K^* with respect to the product given in K^0 by the tensor product of bundles and given between any K^m and K^n by means of standard suspension properties. Then the Chern character ch: $K^* \to H^*(\ ; \mathbf{Q})$ is a ring homomorphism on S_0, namely the embedding of $K^*(S_0)$ into $K^*(S_0) \otimes \mathbf{Q}$, and ch is a product preserving transformation on polyhedra.

As one very well knows, further very effective structure properties of ordinary cohomology H^* (with a coefficient ring, for example) are available if one confines attention to *manifolds*; most important are of course those connected with Poincaré duality and the Hopf "inverse homomorphism" $f_* : H_+^*(A) \to H_+^*(A')$ associated with a map $f : A \to A'$ of manifolds. In particular, if f is the trivial map $A \to P$ of the n-dimensional orientable manifold A onto the one-point space P, then $f_* a = [a] \cdot e \in H_+^0(P)$, where $[a]$ is the "top-component" $\in H^n(A)$ of a, evaluated on the manifold-cycle of A. Such special properties valid for manifolds can be expressed with the help of the tangent bundle or normal bundle of A (for differentiable manifolds; for non-differentiable ones the new concept of "micro-bundle" has to be used, cf. the lecture of John Milnor). In order to obtain similar further structure properties for general cohomology h^*, it is therefore natural to start from *real vector bundles over A*.

Let ξ be such a bundle, with fiber dimension n, and h^* a multiplicative cohomology theory. Let further A^ξ denote the Thom space of ξ (the associated solid sphere bundle E over A modulo its boundary S_{n-1}-bundle $\dot{E}$), p the projection $E \to A$. Using the multiplicative properties of the spectral sequence (cf. Dold [3]) one proves: *If there is a $v \in h^n(A^\xi)$ which on a fiber of A^ξ reduces to an element corresponding to $\gamma^n \in h^n(S_n)$, then the map $\varphi_h : h_+^m(A) \to h^{m+n}(A^\xi)$ given by $\varphi_h(a) = p^*(a) \cdot v$, $a \in h_+^m(A)$ is an isomorphism* (Thom–Gysin isomorphism). The element v, and therefore also φ_h, are not unique in general. [For ordinary cohomology $H^*(\ ;\mathbf{Z})$ the element v exists (and is unique) if and only if ξ is orientable (and oriented); Poincaré duality can be obtained as a special case. For K^* special conditions for existence and uniqueness of v have been given by Atiyah–Hirzebruch.] If one now takes for A a (differentiable) manifold and for ξ the stable normal bundle ν of A, the Thom–Gysin isomorphism $\varphi_h : h_+^*(A) \to h^*(A')$ can be used to define an inverse homomorphism: A' is a model of the S-dual $\hat{A}$ of A, and if $f : A \to A'$ is a map of differentiable manifolds, there is an S-dual map $\hat{f} : \hat{A}' \to \hat{A}$ and an induced homomorphism $\hat{f}^* : h^*(\hat{A}) \to h^*(\hat{A}')$; the inverse homomorphism f_* is then defined by $f_* = \varphi_h'^{-1} \hat{f}^* \varphi_h :$ $h_+^*(A) \to h^*(\hat{A}) \to h^*(\hat{A}') \to h_+^*(A')$. Here φ_h' of course denotes the Thom–Gysin isomorphism $h^*(A') \to h^*(A'')$; note that $f_* : h_+^*(A) \to h_+^*(A')$ depends on the choice of $v \in h^*(A')$ and $v' \in h^*(A'')$. This inverse homomorphism in h^*-theory has properties analogous to those of the Hopf inverse homomorphism for H^*. The element $f_* a \in h_+^*(A')$ can be considered as the generalized analogue of the "top-component" of a, in particular for $A' = P$.

Using the Thom–Gysin isomorphism $\varphi_h : h_+^*(A) \to h^*(A')$, if it exists, one further defines, for a product-preserving transformation $T : h^* \to k^*$ of multiplicative cohomology theories, a sort of "genus" $\mathfrak{T}(A)$ of the manifold A: one puts

$$T \varphi_h(e) = \varphi_k(\mathfrak{T}(A)),$$

where $e \in h_+^*(A), \mathfrak{T}(A) \in k_+^*(A)$. It then easily follows that

$$T \varphi_h(a) = \varphi_k(T(a) \cdot \mathfrak{T}(A))$$

for all $a \in h_+^*(A)$. Combining this with the inverse homomorphisms $f_*^h : h_+^*(A) \to h_+^*(A')$ and $f_*^k : k_+^*(A) \to k_+^*(A')$ for a map of manifolds $f : A \to A'$, a simple computation yields the formula

$$f_*^k(T(a) \cdot \mathfrak{T}(A)) = T(f_*^h a) \cdot \mathfrak{T}(A'),$$

i.e., a relation for the "top-components" $f_*^k(Ta)$ and $T(f_*^h a)$, $\mathfrak{T}(A)$ and $\mathfrak{T}(A')$ giving the necessary corrections between the two terms. [Of course $\mathfrak{T}(A)$ and $\mathfrak{T}(A')$ depend on the same choices of the Thom–Gysin isomorphisms φ_h, φ_k on A and A' as the f_*^h and f_*^k]. This general relation is a formal expression for the various Riemann–Roch theorems, and also for the analogous formulae for cohomology operations, according to the choices of h^*, k^* and T, and to the special properties of f. It was communicated to the author by E. Dyer, but seems already to belong to the folklore of algebraic topology. In the different interesting cases to which this general formula can be applied, deeper results are based on explicit relations of $\mathfrak{T}(A)$ to characteristic classes.

REFERENCES

[1]. ADAMS, J. F., Vector fields on spheres. *Ann. Math.*, 75 (1962), 603–632.

[2]. BROWN, E. H., Cohomology theories. *Ann. Math.*, 75 (1962).

[3]. DOLD, A., Relations between ordinary and extraordinary homology. *Algebraic Topology Colloquium Aarhus*, 1962, 2–9.

[4]. ECKMANN, B. & HILTON, P. J., Groupes d'homotopie et dualité. *C. R. Acad. Sci. Paris*, 246 (1958), 2444–2447; 2555–2558; and 2991–2993. See also: Eckmann, B., Groupes d'homotopie et dualité. *Bull. Soc. Math. France*, 86 (1958), 271–281.

[5]. ECKMANN, B. & HILTON, P. J., Group-like structures in general categories. I. Multiplications and comultiplications. *Math. Ann.* 145 (1962), 227–255.

[6]. ECKMANN, B. & HUBER, P. J., Spectral sequences for homology and homotopy groups of maps. *Report Battelle Memorial Institute Geneva*, (1960).

[7]. HUBER, P. J., Homotopy theory in general categories. *Math. Ann.*, 144 (1961), 361–385.

[8]. WHITEHEAD, G. W., Generalized homology theories. *Trans. Amer. Math. Soc.*, 102 (1962), 227–283.

Simple homotopy type and categories of fractions

Istituto Nazionale di Alta Matematica Symposia Mathematica 5 (1970), 286–299

The purpose of this lecture is to give a categorical description for the group of simple homotopy types which can be assigned to a polyhedron. The description is valid both for the case of finite (compact) polyhedra and of infinite, locally finite polyhedra (in the latter case all maps and homotopies being proper). The definition of the group is entirely independent of algebraic tools such as K-theory and torsion: we show that the set of simple homotopy types of a polyhedron can be given an Abelian group structure, and moreover this defines a homotopy functor $E: \mathfrak{CW} \to \mathfrak{Ab}$ from the appropriate category of CW-complexes to Abelian groups. In the case of a finite CW-complex X it can be shown *a posteriori* that the group $E(X)$ is isomorphic to the Whitehead group of $\pi_1 X$. In the infinite case the fundamental group π_1 is not sufficient for an algebraic description of $E(X)$.

The general categorical construction reported here has been given independently by Siebenmann (mimeographed notes), and by the author and Bolthausen (diploma thesis ETH 1968). A purely geometrical approach is due to the author and Maumary [1].

In section 1 below we outline the topological motivation for the categorical construction. The simple isomorphisms and the simple morphism classes, in a category of fractions $\mathfrak{C}(\Sigma^{-1})$ relative to a family of morphisms Σ in $\mathfrak{C}$, are introduced and discussed in section 2. Section 3 deals with the functorial nature of the set $A(X)$ of simple morphism classes starting from an object X, and of the set $E(X)$ of simple isomorphism classes; the main tool is a « push-out axiom » in $\mathfrak{C}$ relative to Σ. In section 4 natural Abelian monoid and group structures are defined in $A(X)$ and $E(X)$ respectively, using the same axiom. In some cases the category of fractions $\mathfrak{C}(\Sigma^{-1})$ can be identified with

I risultati contenuti in questo lavoro sono stati esposti nella conferenza tenuta il 13 aprile 1970.

more familiar categories; conditions for this are given in section 5, in view of the application to simple homotopy type of polyhedra. In section 6 we briefly describe the topological situation in the case of locally finite and of finite CW-complexes.

1. Introduction.

1.1. We first recall the notion of simple homotopy type for finite polyhedra. Given a finite CW-complex X, one considers all homotopy equivalences $f\colon X \to Y$; these are divided into classes $\langle f \rangle$ by defining $\langle f \rangle = \langle g \rangle$, $g\colon X \to Z$ if there is a « simple » homotopy equivalence $h\colon Y \to Z$ such that g is homotopic to hf. The simple equivalences are defined as follows: Among the cellular inclusion maps of CW-complexes there are special ones called *elementary expansions* (cf. [1], [5], for example), and their finite compositions are called *expansions*; they are homotopy equivalences. A *simple* homotopy equivalence is a map $h\colon X \to Y$ of finite CW-complexes which is homotopic to a finite composition of expansions and of homotopy inverses of expansions. For example, the inclusion of Y into the mapping cylinder M_f of $f\colon X \to Y$ is simple, and so is the identity map of a finite W-complex X onto a subdivision of X.

The classes $\langle f \rangle$ described above, for a finite CW-complex X, are the « simple homotopy types of complexes homotopy equivalent to X »; the set of these classes will be written $E(X)$. It is well known that the Whitehead torsion τ establishes a 1-1-correspondence between the elements of $E(X)$ and those of $\mathrm{Wh}\,(\pi_1 X)$, the Whitehead group of $\pi_1 X$; thus $E(X)$ can be given an Abelian group structure. We will give a description of this group structure which is elementary and categorical in the sense that it does not use Whitehead methods and K-theory etc., but only categorical properties. It applies to other similar categorical situations, and it exhibits the functorial nature of the group $E(X)$. More generally we will consider the set $A(X)$ of all classes $\langle f \rangle$ defined as above, where $f\colon X \to Y$, $g\colon X \to Z$ are arbitrary maps (not homotopy equivalences), and show that $A(X)$ has a natural functorial *Abelian monoid* structure, $E(X)$ being a *subgroup* of the monoid $A(X)$. Both $A(X)$ and $E(X)$ are homotopy type invariants of X.

1.2. Due to the role of special inclusions in these concepts, the categorical description cannot be given in $\mathfrak{CW}^0$ (the category of finite CW-complexes and continuous maps), nor in its homotopy category $\mathfrak{CW}^0_{\text{homot.}}$, but in an intermediate category obtained as follows. Let $\mathfrak{CW}^0_{\text{incl.}}$ be the *inclusion category* of $\mathfrak{CW}^0$, the morphisms being the cellular maps $X \to Y$ which are isomorphisms of X onto a subcomplex of Y. In this category, let Σ be the class of all expansions, and $\mathfrak{CW}^0_{\text{incl.}}(\Sigma^{-1})$ the corresponding *category of fractions*. It is a special feature of polyhedral topology that $\mathfrak{CW}^0_{\text{incl.}}(\Sigma^{-1})$ can be identified with $\mathfrak{CW}^0_{\text{homot.}}$. Thus the isomorphisms in the category of fractions correspond to homotopy equivalences; the finite compositions of morphisms of Σ and of their inverses, to simple homotopy equivalences.

From this the general categorical construction of $E(X)$ and $A(X)$ becomes clear. Under suitable axioms on Σ, all fulfilled in the case of finite CW-complexes and of $\Sigma = $ expansions, $A(X)$ and $E(X)$ define functors to Abelian monoids and groups respectively, which factor through the category of fractions. The whole set-up can be applied to the category $\mathfrak{CW}$ of locally finite CW-complexes, with the class Σ of expansions modified as follows: it is the smallest class of inclusions containing all elementary expansions and closed with respect to finite compositions, arbitrary disjoint unions and extensions (cf. 6.1). The category of fractions $\mathfrak{CW}_{\text{incl.}}(\Sigma^{-1})$ can then be identified with the homotopy category of $\mathfrak{CW}$, all maps and homotopies being *proper*. $E(X)$ and $A(X)$ define functors as before; $E(X)$ is the *group of proper simple homotopy types of CW-complexes properly homotopy equivalent to X*.

2. Categories of fractions and simple morphism type.

2.1. Let $\mathfrak{C}$ be a category, Σ a class of morphisms of $\mathfrak{C}$ containing all isomorphisms and closed with respect to composition. We recall (cf. [2]) that the *category of fractions* of $\mathfrak{C}$ with respect to Σ is defined as a category $\mathfrak{C}(\Sigma^{-1})$ together with a functor $Q: \mathfrak{C} \to \mathfrak{C}(\Sigma^{-1})$ which turns all morphisms in Σ into isomorphisms and which is universal with respect to that property; i.e., if $F: \mathfrak{C} \to \mathfrak{D}$ is any functor which maps the morphisms of Σ to isomorphisms in $\mathfrak{D}$, then F factors uniquely through Q as $F = F'Q$. It is easily seen that $\mathfrak{C}(\Sigma^{-1})$, if it exists, is unique up to category isomorphism. The existence can be exhibited by an explicit construction: The objects of $\mathfrak{C}(\Sigma^{-1})$ are those of $\mathfrak{C}$, and a morphism $\alpha: X \to Y$ in $\mathfrak{C}(\Sigma^{-1})$ is an equivalence class of diagrams in $\mathfrak{C}$ of the form

$$(*) \qquad X \to X_1 \to X_2 \leftarrow \ldots \to X^r \leftarrow \ldots \to X_1 \leftarrow Y$$

of arbitrary length, with arrows pointing from X to Y or from Y to X, the latter being taken from Σ. The equivalence relation is generated by replacing $\to \cdot \to$ or $\leftarrow \cdot \leftarrow$ by their composition, and $Z \xrightarrow{s} W \xleftarrow{s} Z$ by 1_Z, $W \xleftarrow{s} Z \xrightarrow{s} W$ by 1_W. Composition of these morphisms is simply given by composing the diagrams (compatible with the equivalence relation). The functor Q is given by $Q(f) = \{f\}$ (single arrow diagram), and all the properties above are easily checked.

2.2. The isomorphisms in $\mathfrak{C}(\Sigma^{-1})$ which are finite compositions of $Q(s)$ and $Q(t)^{-1}$, $s, t, \ldots \in \Sigma$ are called *simple*. Two morphisms α, β in $\mathfrak{C}(\Sigma^{-1})$, $\alpha\colon X \to Y$, $\beta\colon X \to Z$ belong to the same *simple morphism class* of X, if $\beta = \gamma\alpha$, where $\gamma\colon Y \to Z$ is simple; the class of α is denoted by $\langle\alpha\rangle$. We assume in the following that the simple morphism classes of X form a *set* $A(X)$. This set has a distinguished element $\langle 1_x \rangle$. If α is an isomorphism in $\mathfrak{C}(\Sigma^{-1})$, all other morphisms in $\langle\alpha\rangle$ are also isomophisms, and these classes form a subset $E(X)$ of $A(X)$ containing the distinguished element.

2.3. We will now formulate additional axioms on Σ (and $\mathfrak{C}$) which imply special properties of $A(X)$ and $E(X)$, in particular the functoriality of A and E.

(1) For $f\colon X \to Y$, $s\colon X \to Z$ in $\mathfrak{C}$, $s \in \Sigma$, there is a commutative square

$$\begin{array}{ccc} X & \xrightarrow{f} & Y \\ {\scriptstyle s}\downarrow & & \downarrow{\scriptstyle s'} \\ Z & \xrightarrow{f'} & W \end{array}$$

with $s' \in \Sigma$.

(2) For f and g in $\mathfrak{C}$, $Q(f) = Q(g)$ implies the existence of $s, t \in \Sigma$ such that $sf = tg$

$$X \underset{g}{\overset{f}{\rightrightarrows}} Y \underset{t}{\overset{s}{\rightrightarrows}} Z$$

(3) $\mathfrak{C}$ has push-outs, and the push-out of f and s can be taken as commutative square in (1).

The axioms are used individually at the different stages of our procedure. (1) is of general importance in categories of fractions, (2) provides a characterization of simple morphism classes, and (3) is used for the functoriality and for the monoid and group operation. It should be noted that the axioms do not provide a « calculus of fractions » [2] which requires an axiom stronger than (2).

2.4. Axiom (1) yields the well-known fraction representation of morphisms in $\mathfrak{C}(\Sigma^{-1})$:

PROPOSITION 2.1. If (1) holds, all morphisms α in $\mathfrak{C}(\Sigma^{-1})$ can be written in the form $\alpha = Q(s)^{-1}Q(f)$, the simple isomorphisms in the form $Q(s-)^1Q(t)$, $s, t \in \Sigma$.

For any diagram (∗) representing α can be shortened by apply-ing (1) and the equivalence relation, until it is reduced to $X \xleftarrow{t} Z \xrightarrow{s} Y$ or $X \xleftarrow{t} Z \xrightarrow{s} Y$.

Axiom (2) allows to better characterize the simple morphism classes.

PROPOSITION 2.2. If (1) and (2) hold, all simple morphism classes $\langle \alpha \rangle$ in $\mathfrak{C}(\Sigma^{-1})$ are of the form $\langle \alpha \rangle = \langle Q(f) \rangle$, f in $\mathfrak{C}$.

For $\alpha = Q(s)^{-1}Q(f)$, and $Q(s)\alpha = Q(f)$ is in the same class as α. Since $\langle \alpha \rangle = \langle Q(f) \rangle$ is determined by f, we will simply write $\langle f \rangle$ for $\langle Q(f) \rangle$; in other words, the morphisms in $\mathfrak{C}$ are divided into simple classes, $\langle f \rangle = \langle g \rangle$ being defined by $Q(f) = \gamma Q(g)$, $\gamma =$ simple isomor-phism in $\mathfrak{C}(\Sigma^{-1})$.

PROPOSITION 2.3. If (1) and (2) hold, then $\langle f \rangle = \langle g \rangle$ for two mor-phisms f, g in $\mathfrak{C}$ if and only if there exist morphisms $s, t \in \Sigma$ such that

$$sf = tg:$$
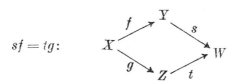

PROOF. $\langle f \rangle = \langle g \rangle$ means $Q(f) = \gamma Q(g)$ with $\gamma = Q(s_0)^{-1}Q(t_0)$, $s_0, t_0 \in \Sigma$. Hence $Q(s_0)Q(f) = Q(t_0)Q(g)$, i.e., $Q(s_0 f) = Q(t_0 g)$, and hence there are $s_1, t_1 \in \Sigma$ such that

$$s_1 s_0 f = t_1 t_0 g.$$

The converse is obvious.

3. The functor A.

3.1. We assume that the class of morphisms Σ in $\mathfrak{C}$ fulfills (2), and that the push-out axiom (3) holds (and whence (1)). Then the functoriality of the set function $A(X)$ can be established as follows.

Let $f: X \to Y$ and $h: X \to Z$ be in $\mathfrak{C}$. In the push-out square

$$\begin{array}{ccc} X & \xrightarrow{f} & Y \\ h\downarrow & & \downarrow h' \\ Z & \xrightarrow{f'} & W \end{array}$$

we write temporarily $h_* f$ for f'. Then we have, for a composition $f_1 f: X \to Y \to Y_1$,

$$h_*(f_1 f) = h'_* f_1 \circ h_* f.$$

To prove this we consider the diagram

$$X \xrightarrow{f} Y \xrightarrow{f_1} Y_1$$
$$\downarrow{\scriptstyle h} \quad \downarrow{\scriptstyle h'} \quad \downarrow$$
$$Z \xrightarrow{j'} W \xrightarrow{j''} W_1$$

consisting of two push-out squares. It yields $f_1'f'$ for the right-hand side; but since the whole square is also a push-out diagram, $f_1'f'$ is equal to $h_*(f_1 f)$.

If $f_1 \in \Sigma$, then $f_1' \in \Sigma$. Assuming now $\langle f \rangle = \langle g \rangle$, i.e., according to Proposition 2.3, $sf = tg$ for suitable $s, t \in \Sigma$, it follows that

$$h_*(sf) = h_*'s \circ h_* f = h_*(tg) = h_*''t \circ h_* g \ ,$$

(with obvious notations h', h''). Since $h_*'s$ and $h_*''t \in \Sigma$, we have $\langle h_* f \rangle = \langle h_* g \rangle$. Thus we are in a position to *define*, for $\langle f \rangle \in A(X)$ and $h \colon X \to Z$

$$A(h) \langle f \rangle = \langle h_* f \rangle \in A(Z) \ .$$

The functor properties of $A(h)$ follow easily from simple push-out arguments. Moreover $A(h) \langle 1_X \rangle = \langle 1_r \rangle$. Thus we get a *functor A*: $\mathfrak{C} \to Sets$, the category of pointed sets.

3.2. We will now show that A factors through $\mathfrak{C}(\Sigma^{-1})$. We first prove

PROPOSITION 3.1. If $s \in \Sigma$, $s \colon X \to Y$, then the map $A(s)$ is a bijection of the sets $A(X) \to A(Y)$.

PROOF. A map $\varphi \colon A(Y) \to A(X)$ is defined by

$$\varphi \langle g \rangle = \langle gs \rangle$$

$g \colon Y \to Z$, obviously independent of the choice of g in $\langle g \rangle$. Then $\varphi A(s) \langle f \rangle$, for $\langle f \rangle \in A(X)$ is given by $f's$

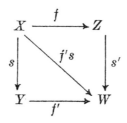

in the push-out square. Since $\langle f's \rangle = \langle s'f \rangle = \langle f \rangle$, we see that $\varphi A(s) =$ identity of $A(X)$.

The composition $A(s)\varphi\langle g \rangle$, for $\langle g \rangle \in A(Y)$ is given by t in the push-out square

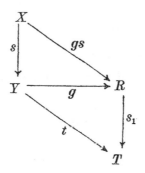

with $ts = s_1 gs$, hence $Q(t) = Q(s_1)Q(g)$, and thus $\langle t \rangle = \langle g \rangle$. Hence $A(s)\varphi$ is the identity of $A(Y)$, and $A(s)$ is a bijection.

Since the functor $A \colon \mathfrak{C} \to Sets$ maps all $s \in \Sigma$ to isomorphisms, it factors uniquely through the category of fractions $\mathfrak{C}(\Sigma^{-1})$ as $A = A'Q$, $A' \colon \mathfrak{C}(\Sigma^{-1}) \to Sets$. In particular we have

PROPOSITION 3.3. Let h be a morphism in C such that $Q(h)$ is an isomorphism in $\mathfrak{C}(\Sigma^{-1})$. Then $A(h)$ is a bijection.

4. The monoid structure in $A(X)$ and the group $E(X)$.

4.1. The assumptions on $\mathfrak{C}$ and Σ are the same as in section 3 above.

For $\langle f \rangle$, $\langle g \rangle \in A(X)$ we *define*

$(+)$ $\qquad\qquad\qquad \langle f \rangle + \langle g \rangle = \langle f'g \rangle = \langle g'f \rangle$

where $\langle f' \rangle = A(g)f$ and $\langle g' \rangle = A(f)\langle g \rangle$; i.e., $f'g = g'f$ is the diagonal in the push-out square

$$X \overset{f}{\longrightarrow} Y$$
$$g\downarrow \qquad \downarrow g'$$
$$Z \underset{f'}{\longrightarrow} W$$

Obviously $\langle f \rangle + \langle g \rangle = \langle g \rangle + \langle f \rangle$.

4.2. The two following propositions are straightforward consequences of general properties of push-out squares and of their compositions. We leave the proofs to the reader and give only a short sketch of one argument.

PROPOSITION 4.1. The addition $(+)$ turns $A(X)$ into an Abelian monoid, $\langle 1_x \rangle$ being the neutral element 0.

PROPOSITION 4.2. For any $h: X \to X'$ in $\mathfrak{C}$, $A(h)$ is a monoid homomorphism. In other words, $A: \mathfrak{C} \to Sets$ factors through the category of Abelian monoids.

As an example of proof, we point out that the formula

$$A(h)(\langle f \rangle + \langle g \rangle) = A(h)\langle f \rangle + A(h)\langle g \rangle$$

is obtained from the diagram

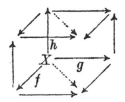

where all squares are push-outs; then the two dotted arrows together with h belong also to a push-out square.

4.3. The subset $E(X)$ of $A(X)$ defined in 2.2. consists of those classes $\langle \alpha \rangle$ containing isomorphism in $\mathfrak{C}(\Sigma^{-1})$. Under our assumption on $\mathfrak{C}$ and Σ it can be described as consisting of those $\langle f \rangle$, f in $\mathfrak{C}$, for which $Q(f)$ is an ismorphism in $\mathfrak{C}(\Sigma^{-1})$; this property is independent of the representative f of $\langle f \rangle$. The key lemma for dealing with such $\langle f \rangle$ is the following.

LEMMA 4.3. For a morphism f in $\mathfrak{C}$, $Q(f)$ has a left-inverse if and only if there is an $\bar{f}$ in $\mathfrak{C}$ such that $\bar{f}f \in \Sigma$.

PROOF. $a)$ Assume that $Q(f)$ has a left-inverse $\alpha = Q(s)^{-1}Q(g)$, with $s \in \Sigma$. Then $Q(s)^{-1}Q(g)Q(f) = 1$, or $Q(gf) = Q(s)$. Hence there are $r, t \in \Sigma$ such that $rgf = ts$, and $\bar{f} = rg$ has the required property.

$b)$ If conversely there is an $\bar{f}$ with $\bar{f}f = s \in \Sigma$, then $Q(s)^{-1}Q(\bar{f})Q(f) = 1$, and thus $Q(f)$ has a left-inverse $\alpha = Q(s)^{-1}Q(\bar{f})$.

REMARK: If $Q(f)$ has an *inverse* α, it is of course the $Q(s)^{-1}Q(\bar{f})$ above. Therefore, if $Q(f)$ is an isomorphism and if $\bar{f}$ is chosen according to the Lemma $(\bar{f}f \in \Sigma)$, then $Q(\bar{f})$ is also an isomorphism in $\mathfrak{C}(\Sigma^{-1})$.

PROPOSITION 4.4. Let f, h be in $\mathfrak{C}$, and $\langle f' \rangle = A(h)\langle f \rangle$. If $Q(f)$ is an isomorphism, then $Q(f')$ is also an isomophism.

PROOF. Since $Q(f)$ has a left-inverse, we choose $\bar{f}$ such that $\bar{f}f = s \in \Sigma$ and consider the push-out diagram

Since $\bar{f}f = s$ is in Σ, so is $\bar{f}'f' = s'$. Hence $Q(f')$ has a left-inverse. Moreover $Q(\bar{f}')Q(f') = Q(s')$, hence

$$Q(\bar{f}')Q(f')Q(s'^{-1}) = 1 .$$

Thus $Q(\bar{f}')$ has a *right-inverse*.

$Q(f)$ being an isomorphism, so is $Q(\bar{f})$, and the above argument applied to $\bar{f}$ instead of f shows that $Q(\bar{f}')$ has a *left-inverse*. Therefore $Q(\bar{f}')$ is an isomorphism. The relation $\bar{f}'\bar{f}' = s'$ implies that $Q(f')$ is also an isomorphism.

COROLLARY 4.5. If $\langle f \rangle$, $\langle g \rangle \in E(X)$, then $\langle f \rangle + \langle g \rangle \in E(X)$.

For $\langle f \rangle + \langle g \rangle = \langle f'g \rangle$, $\langle f' \rangle = A(g)\langle f \rangle$, and since $Q(f)$ and $Q(g)$ are isomorphisms, so are $Q(f')$ and $Q(f')Q(g) = Q(f'g)$.

Moreover $\langle 1_x \rangle \in E(X)$. We see thus that $E(X)$ is a submonoid of $A(X)$ for all $X \in \mathfrak{C}$. The $E(X)$ together with the maps $E(h)$ induced by $h : X \to Y$, $E(h) =$ restriction of $A(h)$ to $E(X)$, define a functor E from $\mathfrak{C}$ to the category of Abelian monoids; it factors through $\mathfrak{C}(\Sigma^{-1})$.

4.4. Let $\langle f \rangle$ be an element of $E(X)$, $f : X \to Y$. We choose $\bar{f} : Y \to Z$ such that $\bar{f}f = s \in \Sigma$. We know (Prop. 3.3) that $A(f)$ is a bijection $A(X) \to A(Y)$; hence there is an element $\langle g \rangle \in A(X)$ such that $A(f)\langle g \rangle = \langle \bar{f} \rangle \in E(Y)$. This element $\langle g \rangle$ is in $E(X)$. For we can assume that g belongs to a push-out square

$$\begin{array}{ccc} X & \xrightarrow{f} & Y \\ g\downarrow & & \downarrow \bar{f} \\ W & \xrightarrow{f'} & Z \end{array}$$

with $f'g = \bar{f}f = s \in \Sigma$; hence $Q f' Q g = Q s$. On the other hand $\langle f' \rangle = A(g)\langle f \rangle$ is in $E(W)$, according to Proposition 4.4; hence $Q(f')$ is an isomorphism, and so is $Q(g)$.

For this element $\langle g \rangle \in E(X)$ we now have

$$\langle f \rangle + \langle g \rangle = \langle \bar{f}f \rangle = \langle s \rangle = 0 \ ,$$

i.e., $\langle f \rangle$ is invertible in the monoid structure of $E(X)$: The submonoid $E(X)$ of $A(X)$ is an *Abelian group*.

THEOREM 4.6. The addition $\langle f \rangle + \langle g \rangle$ defined in $E(X)$, $X \in \mathfrak{C}$, turns the functor E into a functor from $\mathfrak{C}$ to $\mathfrak{Ab}$, the category of Abelian groups. This functor factors through $\mathfrak{C}(\varSigma^{-1})$ as $E = E'Q$: $\mathfrak{C} \to \mathfrak{C}(\varSigma^{-1}) \to \mathfrak{Ab}$.

5. Categories isomorphic to the category of fractions.

5.1. We consider a functor $H: \mathfrak{C} \to \mathfrak{D}$ from $\mathfrak{C}$ to a category $\mathfrak{D}$ and assume that H is bijective on objects and maps all $s \in \varSigma$ to isomorphisms in $\mathfrak{D}$. By the universal property of $\mathfrak{C}(\varSigma^{-1})$ we then have $H = H'Q$: $\mathfrak{C} \xrightarrow{Q} \mathfrak{C}(\varSigma^{-1}) \xrightarrow{H'} \mathfrak{D}$. Under suitable conditions on H we show that H' is a *category isomorphism*. These conditions are patterned after the case of polyhedra and passage H to homotopy classes; we do not attempt here to go into further generality.

The following assumptions on H, obviously suggested by homotopy, will be sufficient for our purpose.

(A) For any morphism φ in $\mathfrak{D}$ there exists f, s in $\mathfrak{D}$, $s \in \varSigma$, such that $H(s)\varphi = H(f)$.

(B) To any object X in $\mathfrak{C}$ there are assigned an object Z and two morphisms $r, t: X \to Z$ in $\varSigma$ with the property: If $f, g: X \to Y$ are morphisms with $H(f) = H(g)$, there exist s in $\varSigma$ and h in $\mathfrak{C}$ such that

$$Q(sf) = \beta Q(hr) \quad \text{and} \quad Q(sg) = \beta Q(ht), \quad \text{with} \quad \beta \in \varSigma \ ,$$

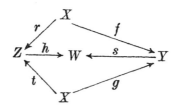

PROPOSITION 5.1. If $H: \mathfrak{C} \to \mathfrak{D}$ fulfills (A), then H' is surjective on morphisms.

PROOF. Let φ be in $\mathfrak{D}$. Then there are f, s in $\mathfrak{C}$, s in Σ such that $\varphi = H(s)^{-1} H(f) = H'(Q(s)^{-1} Q(f)) = H'(\alpha)$, α in $\mathfrak{C}(\Sigma^{-1})$.

5.2. A first consequence of assumption (B) is that H' is injective on those morphisms of $\mathfrak{C}(\Sigma^{-1})$ which are of the form $Q(f)$.

PROPOSITION 5.2: If $H: \mathfrak{C} \to \mathfrak{D}$ fulfills (B), then $H(f) = H(g)$, f, g in $\mathfrak{C}$, implies $Q(f) = Q(g)$.

PROOF. Using (B), we deduce from $H(f) = H(g)$, $f, g: X \to Y$, the relations

$$Q(sf) = \beta Q(h) Q(r) = \beta Q(h) Q(t) \alpha ,$$

where $\alpha = Q(t)^{-1} Q(r)$, hence

$$Q(sf) = \beta Q(ht) \alpha = Q(sg) \alpha .$$

Since α depends on X only, this can be applied to $f = g$ and yields

$$Q(sg) = Q(sg) \alpha ,$$

hence $Q(sf) = Q(sg)$. Since $Q(s)$ is invertible, this implies $Q(f) = Q(g)$.

From this we deduce that H' is injective on all morphisms of $\mathfrak{C}(\Sigma^{-1})$:

PROPOSITION 5.3. If $H: \mathfrak{C} \to \mathfrak{D}$ fulfills (B), then $H'(\alpha) = H'(\beta)$, α, β in $\mathfrak{C}(\Sigma^{-1})$, implies $\alpha = \beta$.

PROOF. Let $\alpha, \beta: X \to W$ be in $\mathfrak{C}(\Sigma^{-1})$ such that $H'(\alpha) = H'(\beta)$. Using axiom (1), we know that α, β are of the form $\alpha = Q(s)^{-1} Q(f)$, $\beta = Q(t)^{-1} Q(g)$, with $f: X \to X_1$, $s: W \to X_1$, $g: X \to X_2$, $t: W \to X_2$, $s, t \in \Sigma$. s, t can be completed to a commutative square

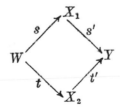

with $s', t' \in \Sigma$. We put $f' = s' f$, $g' = t' g$, both: $X \to Y$.

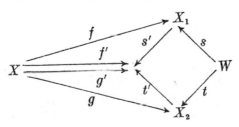

Then

$$H'(\alpha) = H(s)^{-1}H(f) = H(s)^{-1}H(s')^{-1}H(s')H(f) = H(s's)^{-1}H(f')$$

and similarly

$$H'(\beta) = H(t't)^{-1}H(g') \; .$$

Since $s's = t't \in \Sigma$, this implies $H(f') = H(g')$, and thus $Q(f') = Q(g')$, or

$$Q(s')Q(f) = Q(s's)Q(s)^{-1}Q(f) = Q(t')Q(g) = Q(t't)Q(t)^{-1}Q(g) \; ,$$

hence $\alpha = \beta$.

REMARK: Among the axioms on Σ (cf. 2.3), only (1) has been used. If one knows that $H: \mathfrak{C} \to \mathfrak{D}$ above (without even assuming (A) or (B)) has the property: $H(f) = H(g)$ implies $sf = tg$ for suitable $s, t \in \Sigma$, then it is clear that $Q(f) = Q(g)$ implies the same; i.e., (2) holds.

6. **Simple homotopy type of polyhedra.**

6.1. We now return to our starting point and show—without detailed proof—how the general procedure for obtaining functors A and E can be applied to $\mathfrak{C} = \mathfrak{CW}_{incl.}$, the category of locally finite CW-complexes and their inclusion maps (cellular maps which are isomorphisms onto a subcomplex). The distinguished class Σ of morphisms is the family of all (generalized) *expansions*: i.e., the smallest class of inclusion maps containing all elementary expansions and closed with respect to finite composition, to arbitrary disjoint union and to « extension ». The last property means that if $s: X \to Y$ is in Σ and if $f: X \to Z$ is any inclusion map, then the inclusion s' of Z into $W = Y \bigcup_X Z$ (union with images of X identified) is also in Σ; we note that

$$
\begin{array}{ccc}
X & \xrightarrow{\;s\;} & Y \\
{\scriptstyle f}\downarrow & & \downarrow{\scriptstyle f'} \\
Z & \xrightarrow[\;s'\;]{} & W
\end{array}
$$

is a push-out square in $\mathfrak{CW}_{incl.}$.

All these expansions $s \in \Sigma$ are proper homotopy equivalences. Those $s \in \Sigma$ which are in $\mathfrak{CW}^0_{incl.}$, the category of finite CW-complexes and their inclusion maps, are finite compositions of elementary expansions, i.e., expansions in the usual sense; we write Σ_0 for this subclass. If $f\colon X \to Y$ is in $\mathfrak{CW}_{\text{incl.}}$, the inclusion map of Y into the mapping cylinder M_h is in Σ. The proof of these statements is easily obtained along the usual lines of homotopy theory; in certain instances, here and in the following, classical arguments have to be carried over to proper maps and proper homotopies.

The class Σ *fulfills the push-out* ([1]) *axiom* (3) of 2.3; this is true for generalized expansions by definition, and for the class Σ_0 of ordinary expansions in $\mathfrak{CW}^0_{\text{incl.}}$ by the remark above. Since (3) implies axiom (1), the fraction representation of Proposition 2.1 for $\mathfrak{CW}_{\text{incl.}}(\Sigma^{-1})$ and for $\mathfrak{CW}^0_{\text{incl.}}(\Sigma_0^{-1})$ can be applied.

6.2. Let $\mathfrak{CW}_{\text{homot.}}$ denote the category of all locally finite CW-complexes and of the proper homotopy classes of proper continuous maps between them. Let further H be the functor $\mathfrak{CW}_{\text{incl.}} \to \mathfrak{CW}_{\text{homot.}}$ which is bijective on objects and which places each cellular inclusion map into its proper homotopy class. Since H turns all $s \in \Sigma$ into isomorphisms, it factors through $Q\colon \mathfrak{CW}_{\text{incl.}} \to \mathfrak{CW}_{\text{homot.}}(\Sigma^{-1})$ as $H = H'Q$.

By a mapping-cylinder and push-out argument one establishes the following property of H: If two inclusion maps $f, g\colon X \to Y$ are properly homotopic, then there exist $s, t \in \Sigma$ such that $sf = tg$. In other words, $H(f) = H(g)$ implies $sf = tg$. The same conclusion holds, of course, if $Q(f) = Q(g)$. Hence Σ *fulfills axiom* (2) of 2.3.

6.3. The properties (A) and (B) of 5.1. hold for the functor H. The proof of (A) is given by the usual mapping cylinder argument: For φ in $\mathfrak{CW}_{\text{homot.}}$, let $h\colon X \to Y$ be a continuous map representing the homotopy class p, and factor h as pf, $f =$ top inclusion of X into the mapping cylinder M_h, $p =$ projection of M_h into Y. The inclusion $s\colon Y \to M_h$ is in Σ. Then $sh = spf$ is properly homotopic to f, i.e., $H(s)\varphi = H(f)$.

To prove (B), one takes $r, t\colon X \to Z$ to be the top and bottom inclusion of X into the cylinder over X. For two inclusion maps $f, g\colon X \to Y$ which are properly homotopic, one takes W to be the mapping cylinder of the homotopy between f and g, and $h\colon Z \to W$ the top inclusion, $s\colon Y \to W$ the bottom inclusion ($\in \Sigma$). A slight modification of the mapping cylinder then yields the commutativity formulas of (A).

From 5.1. and 5.2. it follows that $H'\colon \mathfrak{CW}_{\text{incl.}}(\Sigma^{-1}) \to \mathfrak{CW}_{\text{homot.}}$ is a *category isomorphism*. The same holds for $H'\colon \mathfrak{CW}^0_{\text{incl.}}(\Sigma^{-1}) \to \mathfrak{CW}^0_{\text{homot.}}$.

PROPOSITION 6.1. There are natural category isomorphisms $\mathfrak{CW}_{\text{incl.}}(\Sigma^{-1}) \cong \mathfrak{CW}_{\text{homot.}}$ and $\mathfrak{CW}^0_{\text{incl.}}(\Sigma_0^{-1}) \cong \mathfrak{CW}^0_{\text{homnt.}}$.

([1]) The push-out of two inclusions is to be taken in $\mathfrak{CW}$, but it lies in $\mathfrak{CW}_{\text{incl.}}$. This slight complication does not affect the general arguments.

6.3. The general construction of the functors A to Abelian monoids and E to Abelian groups can be applied to $\mathfrak{CW}_{incl.}$. One only has to check that for any locally finite CW-complex X the simple morphism classes form a set $A(X)$, and this can easily be done.

The meaning of the functors A and E from $\mathfrak{CW}_{incl.}$ to monoids and groups can be interpreted, by Proposition 6.1, in the *homotopy category* $\mathfrak{CW}_{homot.}$, as follows.

a) The simple isomorphism in $\mathfrak{CW}_{incl.}(\Sigma^{-1})$ correspond to those maps between locally finite CW-complexes which are properly homotopic to finite compositions of expansions and of homotopy inverses of expansions; they are called *simple homotopy equivalences*.

b) For a locally finite CW-complex X, the elements of $A(X)$ can be regarded as classes $\langle f \rangle$ of proper continuous maps from X to any Y, two maps f, g being in the same class if there is a simple homotopy equivalence h such that hf is properly homotopic to g.

c) The functor A from $\mathfrak{CW}_{incl.}$ to Abelian monoids factors through $\mathfrak{CW}_{incl.}(\Sigma^{-1}) \simeq \mathfrak{CW}_{homot.}$. It therefore defines a *homotopy functor from* $\mathfrak{CW}$ to the *category of Abelian monoids*.

d) The elements of $E(X)$ can be regarded as the proper simple homotopy types of locally finite CW-complexes properly homotopy equivalent to X.

e) The functor E from $\mathfrak{CW}_{incl.}$ to $\mathfrak{Ab}$, the category of Abelian groups, factors through $\mathfrak{CW}_{incl.}(\Sigma^{-1}) \simeq \mathfrak{CW}_{homot.}$ and thus defines a homotopy functor from $\mathfrak{CW}$ to $\mathfrak{Ab}$.

Similar statements are valid for $\mathfrak{CW}^0$, the category of finite CW-complexes; they are left to the reader.

6.4. As for the functor $E \colon \mathfrak{CW} \to \mathfrak{Ab}$ or its restriction to $\mathfrak{CW}^0 \to \mathfrak{Ab}$, it has the particular property that $E(X)$ *depends on the 2-skeleton of* X *only*. More precisely, the inclusion map of the 2-skeleton X^2 into X induces an isomorphism $E(X^2) \simeq E(X)$.

The proof of this fact is given, in the finite case $X \in \mathfrak{CW}^0$, in [1]; it uses a canonical form of the elements of $E(X)$ obtained by the Whitehead technique (cf. [6] or [4]). For infinite CW-complexes the result has been announced by Siebenmann.

For finite CW-complexes, this result implies that $E(X)$ depends on the fundamental group $\pi_1 X$ only; more precisely, $E \colon \mathfrak{CW} \to \mathfrak{Ab}$ factors through the fundamental group functor π_1 as $E = \bar{E}\pi_1$ (cf. [1]), and it can be shown by means of the « Whitehead torsion » that $\bar{E}$ is the functor Whitehead group Wh from groups to Abelian groups.

Such a conclusion is not possible in the infinite case, where other algebraic and geometrical methods are needed for a description of $E(X)$ in terms of invariants of X.

REFERENCES

[1] B. Eckmann et S. Maumary, *Le groupe des types simples d'homotopie*, Essays on Topology..., dedicated to G. de Rham (Springer, Berlin, 1970).

[2] P. Gabriel and M. Zisman, *Calculus of fractions and homotopy theory*, Ergebn. Math., Bd. 35 (Springer, Berlin, 1967), Kap. I.

[3] J. Milnor, *Whitehead torsion*, Bull. Am. Math. Soc., 72 (1966).

[4] G. de Rham, S. Maumary and M. Kervaire, *Torsion et type simple d'homotopie*, Lecture Notes in Math., Vol. 48 (Springer, Berlin, 1967).

[5] C. T. C. Wall, *Formal deformations*, Proc. Lond. Math. Soc., XVI (1966).

[6] J. H. C. Whitehead, *Simple homotopy type*, Amer. Journ. Math., 72 (1950).

Some recent developments
in the homology theory of groups

J. Pure Appl. Algebra 19 (1980), 61–75

For Saunders MacLane, on the occasion of his 70th birthday

Introduction

The homology theory of groups originated from the observation that it was possible to apply to a group G algebraic methods used in topology. At first this was done with the help of an aspherical polyhedron with fundamental group G (an "Eilenberg–MacLane space" for G), and soon afterwards by translating the whole resulting approach into purely algebraic terms. One had originally in view applications to topological problems. But it soon became evident, and Saunders MacLane's influence in this development need hardly be emphasized, that there was a new way to attack the problem of finding invariants of groups, and of characterizing group-theoretic properties; some of these were partially known before, but in a more complicated and less intuitive form.

Then came the rapid development of homological algebra which made the underlying algebraic methods a general tool for almost all of mathematics, similar in power to linear algebra and in fact related to it. Homological methods in group theory, as a special branch of this new field, remained active, though not always fully recognized by "pure" group-theorists. The last decade has witnessed, however, a striking return of the original trend. This is largely due to the work initiated by Serre and by Stallings and continued by many others. For interesting classes of groups, topological concepts or their algebraic translation have proved remarkably effective. This is not too surprising, since many of these groups stem from a geometrical context (transformation groups, subgroups of Lie groups or algebraic groups, etc.) and often come equipped with a natural Eilenberg–MacLane space which is finite-dimensional or even a manifold. Such groups are "of finite dimension", which algebraically means finite cohomology dimension; they are necessarily torsion-free, and their dimension and other homological properties are obviously important invariants for such infinite groups as are otherwise not easy to investigate. Other groups appearing in similar contexts do have torsion (and thus are of infinite dimension), but often are "virtually" of finite dimension [17], i.e., have a subgroup of finite index with that property. The class of groups of virtually finite dimension includes all finite groups; it seems to be an interesting combination of torsion-freeness and finiteness.

In this lecture we give a survey of some homological aspects of groups fulfilling finiteness assumptions of the above type or something similar. Emphasis is laid on those properties which need (co)homology or even topological constructions for their very formulation; their application leads, in special cases, to ordinary group structure descriptions. It has, of course, been necessary to limit the choice of topics. We consider Poincaré duality for groups and its generalization; high-dimensional cohomology (or Farrell–Tate-cohomology) in the case of virtually finite dimension, and p-periodicity; torsion phenomena detected by characteristic classes of representations; and universal bounds for the order of Euler and Chern classes of rational representations – here we restrict ourselves to the case of finite groups, although some of the arguments should be valid in more generality. Proofs are mostly omitted, except for some sketches of those which have not yet appeared in print. The theory of Euler characteristics of groups is omitted altogether, since it has been described so beautifully in the surveys of Serre [17] and K.S. Brown [7, 8]. It is closely related to some parts of our survey. Under suitable finiteness assumptions on the group the Euler characteristic is defined as an integer, under "virtual" assumptions as a rational number; the denominator displays similarities with the order of characteristic classes of representations, but explicit connections seem not to be available.

We have restricted the references only to those which directly explain or supplement facts referred to in the text, without aiming at listing the vast bibliography.

This survey is dedicated to Saunders MacLane as a mark of gratitude for his friendship, and in recognition of his work and all the stimulation it has provided; it has shown us how to combine geometric and algebraic, abstract and concrete, ideas, and it has given the impulse to reach from the very special to the very general – and vice-versa!

1. Finiteness properties

1.1. The "dimension" of a group G will always mean the cohomology dimension cd G. It is the smallest integer n such that the cohomology groups $H^i(G; A)$ are 0 for all $i > n$ and all coefficient G-modules A. If there is no such integer, then cd $G = \infty$.

We recall that cd $G < \infty$ is equivalent to each of the following statements.

(a) There exists a projective resolution P over $\mathbb{Z}G$ of the trivial G-modulo $\mathbb{Z}$, of finite length

$$P : 0 \to P_m \to P_{m-1} \to \cdots \to P_1 \to P_0 \to \mathbb{Z} \to 0$$

(exact sequence, all P_i projective G-modules).

(b) There exists a free resolution of finite length (as above, but all P_i free G-modules).

(c) There exists a finite-dimensional Eilenberg–MacLane complex $K(G, 1)$ (a CW-complex with fundamental group G whose universal cover $\overline{K(G,1)}$ is contractible).

In (a) and (b) the smallest possible length m is equal to cd G. In (c) the smallest possible dimension of $K(G,1)$ is equal to cd G if cd G is $\neq 2$; it is 2 or perhaps 3 if cd $G = 2$. We recall that $H^i(G; A) \cong H^i(K(G,1); A)$ for all i and all G-modules A; in particular, for any $K(G,1)$, one has cd $G \leq \dim K(G,1)$. Similarly for homology.

1.2. For any subgroup G_1 of G, a resolution P for G can also serve for G_1 (since $\mathbb{Z}G$ is a free $\mathbb{Z}G_1$-module), and hence cd $G_1 \leq$ cd G. A finite cyclic group C_l is of infinite dimension ($H^i(C_l; \mathbb{Z})$ is easily seen to be $\cong \mathbb{Z}/l\mathbb{Z}$ for all even $i > 0$). Thus a group G of finite dimension is *torsion-free;* the converse does not hold. Cohomology dimension seems to be a good and important invariant for torsion-free groups, which in general are not easy to come by, and groups of finite dimension constitute an interesting class arising in many well-known examples (cf. Section 2).

A fundamental theorem of Serre [17] states that if G is torsion-free and G_1 has finite index in G, then cd $G =$ cd G_1 (finite or infinite). If G is a free group $\neq 1$, then cd $G = 1$. By the theorem of Stallings and Swan [18, 19] the converse holds.

1.3. Further finiteness conditions are suggested by those cases where G admits a finite CW-complex $K(G,1)$. If so, G can be finitely presented; and it has a free resolution of the form (b) where all P_i are finitely generated over $\mathbb{Z}G$ (namely, the cellular chain complex of the universal cover $\overline{K(G,1)}$).

If there exists for G a finitely generated free $\mathbb{Z}G$-resolution of finite length, G is said to be of type (FF). This together with finite presentability is in fact equivalent to the existence of a finite $K(G,1)$-complex. If G admits a projective resolution of finite length (a) where all the P_i are finitely generated projective over $\mathbb{Z}G$, then G is said to be of type (FP). Obviously (FF) implies (FP), but it is not known whether the converse holds. Type (FP) together with finite presentability is equivalent to the existence of a $K(G,1)$ which is a (homotopy) retract of a finite CW-complex.

Both type (FP) and type (FF) are easily seen to be inherited by subgroups of finite index. In the other direction, if G_1 is a subgroup of finite index in a torsion-free group G, then G_1 of type (FP) implies G of type (FP); it is not known whether this is true for type (FF) — this is one of the reasons why it is useful to work with type (FP).

1.4. Several effective criteria have been established for deciding whether a group G is of type (FP), provided it is known that G is of finite dimension; see [3, 6]. E.g., G is of type (FP) if and only if cd G is finite and all cohomology groups $H^i(G; A)$ commute with direct limits in the G-module variable A.

If a group G admits a projective resolution, of finite or infinite length, in which all the projective modules P_i are finitely generated over $\mathbb{Z}G$, then of course homology and cohomology groups inherit finiteness properties. E.g., $H_i(G; \mathbb{Z})$ and $H^i(G; \mathbb{Z})$ for all i are finitely generated Abelian groups.

2. Examples. Duality groups

2.1. Let G be a group which operates differentiably on $\mathbb{R}^n$. If the operation is proper and free, then the quotient space $\mathbb{R}^n/G$, having $\mathbb{R}^n$ for its universal cover, is a $K(G, 1)$; it is a differentiable manifold, compact or not, of dimension n, and hence a CW-complex of dimension n. Thus cd $G \leq n$. If $\mathbb{R}^n/G$ is not compact, then cd $G < n$ by well-known cohomology properties of open manifolds.

If $\mathbb{R}^n/G$ is *compact*, it can be "triangulated" as a finite CW-complex, an G is thus of type (FF). Moreover, the (co-)homology of G fulfills Poincaré duality valid for closed manifolds of dimension n:

$$H^i(G; A) \cong H_{n-i}(G; A), \quad \text{or} \tag{1}$$

$$H^i(G; A) \cong H_{n-i}(G; \tilde{\mathbb{Z}} \otimes A) \tag{1'}$$

for all i and all G-modules A. Here (1) holds if $\mathbb{R}^n/G$ is orientable, (1') if non-orientable ($\tilde{\mathbb{Z}}$ is the "orientation module", i.e., the group of integers turned into a G-module by $x \cdot 1 = \pm 1$ according to whether $x \in G$ preserves the orientation of $\mathbb{R}^n$ or not, and G operates diagonally in $\tilde{\mathbb{Z}} \otimes A$). We can use notation (1') also in the orientable case: then $\tilde{\mathbb{Z}}$ is a trivial G-module, and $\tilde{\mathbb{Z}} \otimes A = A$. These duality isomorphisms are natural in any possible sense, since they can be given by the cap-product with a fundamental class $\mu \in H_n(G; \tilde{\mathbb{Z}})$.

A group G satisfying (1') is called a *Poincaré duality group* of dimension n (orientable or non-orientable according to whether $\tilde{\mathbb{Z}}$ is trivial or not), cf. [1, 14]; in short a PD^n-group. Since homology always commutes with direct limits in the coefficient module, the finiteness criterion (see 1.4) tells that such groups are of type (FP). Moreover one immediately infers from (1') that cd $G = n$ and that

$$\begin{aligned} H^i(G; \mathbb{Z}G) &= 0 \quad \text{for } i \neq n, \\ &= \tilde{\mathbb{Z}} \quad \text{for } i = n. \end{aligned} \tag{2}$$

(Thus $\tilde{\mathbb{Z}}$ and n are determined uniquely by the group G). The simple conditions, type (FP) and (2), are sufficient for a group G to be a PD^n-group; cf. in a more general context, Section 2.2 below.

2.2. A very general version of duality similar to (1') is obtained if we replace $\tilde{\mathbb{Z}}$ by a "dualizing complex" D over $\mathbb{Z}G$ ($D_i = 0$ for $i < 0$) and postulate

$$H^i(G; A) = H_{n-i}(G; D \otimes A) \quad \text{for all } i \text{ and } A. \tag{3}$$

Here $D \otimes A$ is a G-complex by diagonal action; homology is understood to be total homology of the double complex $P \otimes (D \otimes A)$, where P is a projective resolution for G. Again (3) implies that G is of type (FP), with cd $G = n$. Moreover, the homology of D is uniquely determined by G; namely, $H_{n-i}(D) = H^i(G; \mathbb{Z}G)$ for all i (this is easily seen from the spectral sequence of double complexes applied to the right-hand side of (3) with $A = \mathbb{Z}G$).

Conversely it can be shown [4, 7] that any group of type (FP) fulfills (3), with dualizing complex $D = P^* = \mathrm{Hom}_G(P, \mathbb{Z}G)$ renumbered as $D_i = P^*_{n-i}$ where P is an (FP)-resolution for G; P^* is a complex over $\mathbb{Z}G$ of type (FP).

There may be various interesting instances of (3), but we restrict here attention to the following special case. We assume that the homology of the dualizing complex D is 0 except for $H_0(D) = C$ which is $\neq 0$ but $\mathbb{Z}$-torsion-free; in other words, that $H^i(G; \mathbb{Z}G)$ is 0 for $i \neq n$ and $H^n(G; \mathbb{Z}G)$ is $\mathbb{Z}$-torsion-free. Then (3) becomes

$$H^i(G; A) \cong H_{n-i}(G; C \otimes A) \quad \text{for all } i \text{ and } A. \tag{4}$$

Groups fulfilling (4) are called duality groups of dimension n, in short D^n-groups, with dualizing module C (the PD^n-groups being those where $C = \mathbb{Z}$ as an Abelian group). Thus the conditions

  (i) G is of type (FP),
  (ii) $H^i(G; \mathbb{Z}G) = 0$ for $i \neq n$,
  (iii) $H^n(G; \mathbb{Z}G)$ is $\mathbb{Z}$-torsion-free,

imply that G is a D^n-group, with dualizing module $C = H^n(G; \mathbb{Z}G)$ (and they are also necessary). The isomorphisms (4) can be described (cf. [2]) by a cap-product $\mu \cap$-with a "fundamental class" $\mu \in H_n(G; C)$, this group being infinite cyclic generated by μ.

2.3. The groups $H^i(G; \mathbb{Z}G)$, which are crucial in that context, are called (generalized) *end-groups* of G; for finitely generated G, $H^1(G; \mathbb{Z}G)$ describes the number of ends of G (the number of points necessary for the endpoint-compactification of the universal cover $\overline{K(G, 1)}$).

For a group admitting a finite $K(G, 1)$ (i.e., of type (FF) and finitely presented) we mention another interpretation of the $H^i(G; \mathbb{Z}G)$, useful in applications. For such a group we may take for $K(G, 1)$ a *compact manifold-with-boundary* M of dimension m. Then $H^i(G; \mathbb{Z}G) = H^i(M; \mathbb{Z}G)$ can be identified with $H^i_{\mathrm{comp}}(\overline{M}; \mathbb{Z})$, where $\overline{M}$ is the universal cover of M and cohomology is meant with compact supports. Relative Poincaré duality for $\overline{M}$ and asphericity of $\overline{M}$ then yield

$$H^i(G; \mathbb{Z}G) = H_{m-i-1}(\partial \overline{M}; \mathbb{Z})$$

(reduced homology if $m - i - 1 = 0$). It can thus be decided from the ordinary integral homology of $\partial \overline{M}$ whether G is a D^n-group; and if so, the dualizing module is $H_{m-n-1}(\partial \overline{M}; \mathbb{Z})$ with the operation of G given by the covering transformations.

As an example of application, let G be a knot-group, i.e., the fundamental group of the complement of a non-trivial knot in S^3. The closed complement M of the knot is known to be aspherical, hence a $K(G, 1)$. It is a compact 3-manifold-with-boundary ($\partial M = $ torus). Since $\partial \overline{M}$ is a disjoint union of planes, $H_j(\partial \overline{M}; \mathbb{Z}) = 0$ except for $j = 0$. Hence G is a duality group ($m = 3$, $m - n - 1 = 0$) of dimension 2. Finitely presented groups of dimension 2 are easy to handle in general: they are of type (FP), and D^2-groups if and only if $H^1(G; \mathbb{Z}G) = 0$ (one end).

2.4. We return to groups G operating in $\mathbb{R}^n$ as in 2.1. If $\mathbb{R}^n/G$ is compact, G is a PD^n-group. If $\mathbb{R}^n/G$ is *not compact*, the group need not be of type (FP). There are, however, many cases where $\mathbb{R}^n/G$ is the interior of a compact manifold-with-boundary M, which can serve as $K(G, 1)$ since it has the same homotopy type as $\mathbb{R}^n/G$. Then G is of type (FF) and finitely presented, and the arguments of 2.3 can be applied to $\partial\bar{M}$.

Such a situation often occurs if G is a torsion-free discrete subgroup of a connected real Lie group L. One knows that G operates properly and freely in the homogeneous space L/K where K is a maximal compact subgroup of L, and that L/K is diffeomorphic to $\mathbb{R}^n$ for some n. E.g., for torsion-free arithmetic subgroups of linear algebraic groups defined over the rationals one takes for L the corresponding groups of real points; according to Borel–Serre [5], $\mathbb{R}^n/G$ (i.e., $G \setminus L/K$) is the interior of a compact M as above, and $\partial\bar{M}$ has precisely one non-vanishing homology group, which is $\mathbb{Z}$-free. Thus torsion-free (rational) arithmetic groups are always duality groups, of a dimension and with a dualizing module which can be described from the Borel–Serre construction.

2.5. If G is a torsion-free discrete subgroup of a real Lie group L, and if G is cocompact in L (i.e., if L/G is compact), then G is a Poincaré duality group, with $G \setminus L/K$ as $K(G, 1)$-manifold. In this way, many examples of PD^n-groups are obtained which are fundamental groups of *compact aspherical manifolds M*.

It is not known whether there are PD^n-groups which are not fundamental groups of aspherical manifolds: it is not even known whether a finitely presented PD^n-group admits a finite $K(G, 1)$ (type (FF)). In the case of dimension 2, partial answers have been obtained: H. Müller [15] has shown that a PD^2-group which has a one-relator presentation is in fact the fundamental group of a closed surface (of genus ≥ 1). For groups of type (FF), Swarup (and H. Müller) have proved that a PD^2-group contains a subgroup of finite index which is a one-relator group[1].

3. Virtual dimension

3.1. A group is said to be *virtually torsion-free* if it contains a subgroup S of finite index which is torsion-free. One similarly defines a group to be virtually of finite dimension, of type (FP), a duality group, etc. For all these concepts, however, it is the virtual presence of torsion that makes all the difference. For example, assume that G contains a subgroup S of finite index with cd $S = n < \infty$; then cd $S' = n$ for any other torsion-free subgroup S' of finite index in G: one considers $S \cap S'$ and applies Serre's theorem concerning the dimension of torsion-free groups and subgroups of finite index. The integer n is called the virtual dimension of G, vcd $G = n$. If G is torsion-free, then vcd $G = $ cd G.

A similar reasoning applies to "type (vFP)"; if there is a subgroup of finite index

---

[1] *Added in proof.* It has been proved [B. Eckmann–H. Müller, to appear in Comm. Math. Helvetici] that all PD^2-groups with positive first Betti number are fundamental groups of closed surfaces.

which is of type (FP), then all torsion-free subgroups of finite index are. Or to G being a vD^n-group; if there is a subgroup of finite index which is a D^n-group, then all torsion-free subgroups of finite index are (here one has to use a finite extension theorem [2] for D^n-groups: if G is torsion-free and G_1 a subgroup of finite index, then G is a D^n-group if and only if G_1 is. The proof simply uses the D^n-criterion in 2.3).

3.2. The groups with vcd $G = 0$ are just the finite groups. All finitely generated linear groups (characteristic 0) are virtually torsion-free [21]. The arithmetic groups as mentioned in Section 2.4 usually do have torsion, but are virtually torsion-free, and hence vD^n-groups for some n. In $G = SL_m(\mathbb{Z})$, for example, the congruence subgroups modulo an integer $q > 2$ are torsion-free and have finite index in G; thus $SL_m(\mathbb{Z})$ is a vD^n-group $(n = \frac{1}{2}m(m-1)$, cf. [5]). Similarly for other arithmetic groups.

3.3. The cohomology groups $H^i(G; A)$, which are 0 for $i > n$ if cd $G = n$, are in general $\neq 0$ if vcd $G = n$ and G has torsion, and we now look closer at these groups.

For any torsion-free subgroup S of finite index k in G, cd $S = n$, the usual restriction-transfer argument shows that $k \cdot H^i(G; A) = 0$ for all $i > n$ and all A. The $H^i(G; A)$, $i > n$, are thus torsion groups; they are annihilated by $y = $ gcd of all finite indices of torsion-free subgroups in G. This integer y is, in turn, a multiple of the lcm of all orders of torsion elements in G. Moreover, y can contain those primes p only which occur in the order of torsion elements in G. Indeed, if G has no p-torsion, there exists a torsion-free subgroup S in G of finite index prime to p (take any torsion-free normal subgroup N of finite index, and let $(G/N)_p$ be a p-Sylow subgroup of G/N; then the preimage S of $(G/N)_p$ in G has the required property).

It follows, in particular, that if the group G of finite virtual dimension n is p-torsion-free, then so are the torsion groups $H^i(G; A)$, $i > n$.

These facts, well-known for finite groups (vcd $G = 0$), suggest a generalization of the *Tate cohomology groups* $\hat{H}^i(G; A)$, $-\infty < i < \infty$, to groups of finite virtual dimension (T. Farrell [13]). These Farrell–Tate groups $\hat{H}^i(G; A)$ have all the torsion properties given above for $H^i(G; A)$, $i > n = $ vcd G; they coincide with the $H^i(G; A)$ for $i > n$. For $i < n$ the situation is somewhat different from the case $n = 0$, where the $\hat{H}^i$, $i < 0$ can be interpreted as homology groups of G. For $n > 0$ no such interpretation is available in general.

For vD^n-groups G, however, one has $\hat{H}^i(G; A) \cong H_{n-i-1}(G; C \otimes A)$ for $i < 0$ where C is the dualizing module of any torsion-free subgroup of finite index; this is in agreement with what happens in the case of finite groups $(n = 0$, $S = 1$, $C = \mathbb{Z})$. In the remaining range $0 \leq i \leq n$ the $\hat{H}^i$ are related to ordinary homology and cohomology by an exact sequence which "measures" the deviation from duality with dualizing module C.

3.4. A question which comes up naturally in that context is to what extent these groups $\hat{H}^i(G; A)$ are *periodic* (with respect to i). This is certainly an exceptional phenomenon; but it may occur more often that for a prime p the p-primary part of the $\hat{H}^i(G; A)$, and thus of the $H^i(G; A)$, $i > n =$ vcd G, is periodic. In that case we say that G is *p-periodic* and write m_p for the smallest p-period. If G is p-periodic, so is every subgroup of G, in particular every finite subgroup. Thus if G contains a finite p-subgroup which is not cyclic (or quaternionic for $p = 2$), G cannot be p-periodic. If G is p-torsion-free, it is trivially p-periodic with $m_p = 1$.

If G is not only of finite virtual dimension, but of type (vFP), there is a simple method for investigating p-periodicity of G (Bürgisser [9]). Let S be a torsion-free normal subgroup of finite index; if the finite group G/S is p-periodic, so is G, and m_p divides the p-period of G/S. The precise value of m_p can sometimes be obtained by looking at suitable finite subgroups of G which are p-periodic.

Among the results of [9], we mention the following example: $SL_m(\mathbb{Z})$ is p-periodic for $\frac{1}{2}m + 1 < p \leq m + 1$, with p-period $m_p = 2(p-1)$ for $p < m + 1$ (for the case $p = m + 1$ see Section 4.5 below). A different method, also appearing in [9], uses characteristic classes of complex and of real representations of G. These classes are dealt with in the next section.

4. Characteristic classes of representations

4.1. With a complex representation $\varrho : G \rightarrow GL_m(\mathbb{C})$ of the group G one associates *Chern classes* $c_k(\varrho) \in H^{2k}(G; \mathbb{Z})$, $k = 0, 1, \ldots, m$; $c_0(\varrho) = 1 \in H^0(G; \mathbb{Z}) = \mathbb{Z}$. One writes $c(\varrho) = 1 + c_1(\varrho) + \cdots + c_m(\varrho) \in H^*(G; \mathbb{Z})$. The following properties will be used:

(i) Representations which are equivalent (over $\mathbb{C}$) have the same Chern classes.

(ii) For any group homomorphism $h : G_1 \rightarrow G$, $c_k(\varrho h) = h^* c_k(\varrho) \in H^{2k}(G_1; \mathbb{Z})$.

(iii) For any two representations ϱ_1, ϱ_2 of G, $c(\varrho_1 \oplus \varrho_2) = c(\varrho_1) \cdot c(\varrho_2)$ (cup-product).

(iv) If $j : \mathbb{Z} \rightarrow \mathbb{Q}$ denotes the embedding, $j_* c_k(\varrho) = 0$ for $k > 0$.

(v) For a finite group G, the first Chern class of 1-dimensional representations establishes an isomorphism $\text{Hom}(G, \mathbb{C}^*) \cong H^2(G; \mathbb{Z})$.

All this is purely algebraic. The definition of $c_k(\varrho)$ and the proofs of the above, however, pass through topology.

We briefly recall a construction of the classes $c_k(\varrho)$. Take a complex $K(G, 1)$, with universal cover $\overline{K(G, 1)}$, and let G operate diagonally on $\overline{K(G, 1)} \times \mathbb{C}^m$ (the operation on $\mathbb{C}^m$ being given by ϱ). The quotient space $\overline{K(G, 1)} \times \mathbb{C}^m / G = E$ is a $\mathbb{C}^m$-bundle over $K(G, 1)$. Its Chern class $c_k(\varrho) \in H^{2k}(K(G, 1); \mathbb{Z}) \cong H^{2k}(G; \mathbb{Z})$ is independent of the choice of $K(G, 1)$. The properties above follow from the general theory of these topological Chern classes, except for (iv) which is due to the fact that the bundle E is flat. In case $H(G; \mathbb{Z})$ is finitely generated, (iv) implies that $c_k(\varrho)$ is a torsion element, $k > 0$.

4.2. Analogous remarks apply to the *Euler class* $e(\varrho)$ of a real representation

$\varrho: G \to GL_m(\mathbb{R})$. It is an element of $H^m(G; \mathbb{Z}(\varrho))$, where $\mathbb{Z}(\varrho)$ is the group of integers $\mathbb{Z}$ turned into a G-module by multiplication with sgn det ϱ (hence a trivial G-module if and only if ϱ has positive determinant only). Properties of $e(\varrho)$ analogous to (i)–(iii) above are:

(i') Representations which are equivalent (over $\mathbb{R}$) have the same Euler class up to sign.

(ii') For any group homomorphism $h : G_1 \to G$, $e(\varrho h) = h^* \varrho(e) \in H^m(G_1; \mathbb{Z}(\varrho h))$.

(iii') For any two representations ϱ_1, ϱ_2 of G, $e(\varrho_1 \oplus \varrho_2) = e(\varrho_1) \cdot e(\varrho_2)$.

The construction of $e(\varrho)$ is by means of the $\mathbb{R}^m$-bundle $\overline{K(G,1) \times \mathbb{R}^m / G}$ over $K(G, 1)$ which inherits an orientation from the $\mathbb{R}^m$ where $GL_m(\mathbb{R})$ operates (whether $\mathbb{Z}(\varrho)$ is trivial or not; cf. [10]). Although this bundle is flat, the analogue of (iv) above is in general not true for $e(\varrho)$.

There is a close relation between the Euler and the Chern classes: If a complex representation $\varrho : G \to GL_m(\mathbb{C})$ is regarded as a real representation $\varrho' : G \to GL_{2m}(\mathbb{R})$ in the canonical way, then the top Chern class $c_m(\varrho)$ is $= e(\varrho') \in H^{2m}(G; \mathbb{Z})$; note that ϱ' has positive determinant. From this it easily follows that if ϱ is a real representation $\varrho : G \to GL_m(\mathbb{R})$ then its complexification $\varrho \otimes \mathbb{C} : G \to GL_m(\mathbb{C})$ has top Chern class

(iv') $c_m(\varrho \otimes \mathbb{C}) = (-1)^{m(m-1)/2} e(\varrho)^2$.

(hence, for $j : \mathbb{Z} \to \mathbb{Q}$, $j_* e(\varrho)^2 = 0$).

We further note that $e(\varrho) = 0$ whenever the real representation ϱ has a real decomposition $\varrho = \varrho_1 \oplus \varrho_2$ with a trivial summand.

4.3. As a first application, we use Chern classes to study the torsion of $H^j(G; A)$, $j > n$, for a group with vcd $G = n$. As noted in 3.3, these cohomology groups are p-torsion-free if G is. We will show that the converse holds in the following more precise sense.

We only need to assume that G is *virtually torsion-free*. Let G have p-torsion, and let $j : C_p \to G$ be the embedding of a cyclic subgroup C_p of order p. We consider a complex representation $\varrho : G \to GL_m(\mathbb{C})$ which is non-trivial on C_p. There exist such representations: Let π be the projection of G onto the finite group G/N, N being a normal torsion-free subgroup of finite index, and take any representation $\bar{\varrho}$ of G/N which is non-trivial on $\pi(C_p) \cong C_p$. Then $\varrho = \bar{\varrho} \pi$ is a representation of G as desired. The restriction ϱj of ϱ to C_p decomposes into m one-dimensional complex representations; let $r > 0$ be the number of non-trivial ones (and hence faithful) among these.

Now the Chern class $c_r(\varrho j) = j^* c_r(\varrho)$ is $\neq 0$ by (iii) and (v) in 4.1. On the other hand $c_r(\varrho) = c_r(\bar{\varrho} \pi) = \pi^* c_r(\bar{\varrho})$ is a torsion element, since $c_r(\bar{\varrho}) \in H^{2r}(G/N; \mathbb{Z})$ is. $c_r(\varrho j)$ being a p-torsion element it follows that the order of $c_r(\varrho) \in H^{2r}(G; \mathbb{Z})$ must contain the prime p. The same holds for $c_{2r}(\varrho \oplus \varrho) = c_r(\varrho)^2$ etc. We thus conclude that if G is virtually torsion-free and has p-torsion, then all $H^{2qr}(G; \mathbb{Z})$, $q = 1, 2, 3, \ldots$ have p-torsion, for some $r > 0$.

In explicit cases, more information on the possible value of r is available, depending on the choice of $\bar{\varrho}$. In some instances, one has representations ϱ of G, non-trivial on C_p, which do not come from a finite factor group G/N; in order to know that $c_r(\varrho)$ is torsion, one has then to use (iv) and finiteness assumptions.

The result above yields the following corollaries.

(1) Let G be virtually torsion-free. If $H^j(G; \mathbb{Z})$ is p-torsion-free for all large even values of j, then G is p-torsion-free.

(2) Let G be virtually torsion-free. If $H^j(G; \mathbb{Z})$ is torsion-free for all large even values of j, then G is torsion-free.

(3) Let vcd G be finite $= n$. Then $H^j(G; \mathbb{Z})$ is p-torsion-free for all large even values of j if and only if G is p-torsion-free.

(4) Let vcd G be finite $= n$. If $H^j(G; \mathbb{Z}) = 0$ for all large even values of n, then cd $G = n$ (and hence $H^j(G; A) = 0$ for all $j > n$ and all A).

4.4. The method of restricting characteristic classes to finite subgroups used before also yields other, more precise, results concerning torsion in the cohomology of G (see also Soulé [19] for far-reaching results).

If ϱ is a complex representation of G which is non-trivial on a cyclic subgroup C_l of arbitrary order l, the restriction ϱh ($h : C_l \hookrightarrow G$ is the embedding) decomposes into complex 1-dimensional representations r of which are non-trivial, $r > 0$. If we assume that these are all faithful, i.e. given on a generator of C_l by a primitive lth root of unity, then $c_r(\varrho h)$ is a generator of $H^{2r}(C_l; \mathbb{Z}) \cong \mathbb{Z}/l\mathbb{Z}$. As before it then follows that the order of the torsion element $c_r(\varrho)$ is a multiple of l.

This applies in particular to arithmetic groups (cf. 2.4; the finiteness conditions hold which guarantee that $c_r(\varrho)$ is torsion for any ϱ since these groups are of type (vFP), cf. [16]). As a typical example we consider $G = \mathrm{SL}_m(\mathbb{Z})$ and its canonical representation $\varrho : G \to \mathrm{GL}_m(\mathbb{C})$ given by the embedding. Since ϱ is in fact a real representation, general properties of Chern classes tell that $2c_k(\varrho) = 0$ if k is odd. We thus fix an even value of k, $0 < k \leq m$, and let l be any integer > 2 with Euler function $\varphi(l)$ dividing k. There exists a subgroup C_l of G such that the restriction of ϱ to C_l decomposes into 1-dimensional representations exactly k of which are faithful, while the others are trivial: Namely, consider the diagonal $m \times m$-matrix in which the $\varphi(l)$ different primitive roots of unit appear $k/\varphi(l)$ times in the diagonal, the other diagonal elements being $= 1$; such a matrix is equivalent to an integral matrix with determinant 1, generating a cyclic group C_l. Thus the order of $c_k(\varrho)$ is a multiple of l, and therefore of $E_k = \mathrm{lcm}\{l \mid \varphi(l) \text{ divides } k\}$. This integer E_k is equal to the denominator of B_k/k written in its lowest terms ($B_k = k$th Bernoulli number, $B_2 = 1/6$, $B_4 = 1/30$, $B_6 = 1/42$, . . .):

(5) The order of the kth Chern class $c_k(\varrho) \in H^{2k}(\mathrm{SL}_m(\mathbb{Z}); \mathbb{Z})$ of the canonical representation $\varrho : \mathrm{SL}_m(\mathbb{Z}) \to \mathrm{GL}_m(\mathbb{C})$, k even, $0 < k \leq m$, is a multiple of the denominator of B_k/k.

The arguments of this section also apply to the Euler class $e(\varrho)$ of real representations ϱ; one may use the top Chern class of $\varrho \otimes \mathbb{C}$ together with (iv') and the structure of the cohomology of cyclic groups C_l. The analogue of (5) is then

(6) The order of the Euler class $e(\varrho_1) \in H^m(\mathrm{SL}_m(\mathbb{Z}); \mathbb{Z})$ of the canonical representation $\varrho_1 : \mathrm{SL}_m(\mathbb{Z}) \to \mathrm{GL}_m(\mathbb{R})$, m even, is a multiple of the denominator of B_m/m.

Remark. It is known, from results of Grothendieck, that the order of $c_k(\varrho)$ in (5) divides the denominator $2E_k$ of $B_k/2k$; and from a topological argument of Sullivan, that the order of $e(\varrho_1)$ in (6) divides $2E_m$. It seems unknown whether the factor 2 is necessary or not. For rational, hence integral, representations of finite groups cf. Section 5 below.

4.5. The same method can also be applied to the question of p-periodicity (cf. 3.4), in particular for various arithmetic groups. We take again the example $G = \mathrm{SL}_m(\mathbb{Z})$, $m \geq 3$, and refer to [9] for other cases and for details.

Let p be a prime with $\frac{1}{2}m + 1 < p \leq m + 1$. The p-Sylow subgroups of G (the maximal finite p-groups in G) are cyclic of order p; let C_p be such a subgroup. The restriction ϱh of $\varrho : G \to \mathrm{GL}_m(\mathbb{C})$ to C_p decomposes into 1-dimensional representations $p - 1$ of which are non-trivial, given on a generator of C_p by the $p - 1$ primitive p-th roots of unity, while the other are trivial (since $2(p - 1) > m$). Hence $c_{p-1}(\varrho h) \in H^{2(p-1)}(C_p; \mathbb{Z}) \cong \mathbb{Z}/p\mathbb{Z}$ is a generator. Now a theorem of Brown–Venkow [7] tells that if an element $a \in H^j(G; \mathbb{Z})$ restricts, on each p-Sylow subgroup S_p of G, to a "maximal generator" (i.e., if $H^j(S_p; \mathbb{Z})$ is cyclic of order $|S_p|$ and generated by the restriction of a), then G is p-periodic; moreover, a p-periodicity isomorphism is given by the cup-product with a, the period m_p thus being a divisor of j. We therefore conclude

(7) $\mathrm{SL}_m(\mathbb{Z})$ is p-periodic for all primes with $\frac{1}{2}m + 1 < p \leq m + 1$, and the p-period m_p is a divisor of $2(p - 1)$.

This result does not go beyond the one obtained in 3.4. However, a better information is provided by applying the method of the present section to the Euler class instead of the Chern class, in the special case $p = m + 1$. Consider, for any prime p the canonical real representation $\varrho : \mathrm{SL}_{p-1}(\mathbb{Z}) \to \mathrm{GL}_{p-1}(\mathbb{R})$, and let C_p be a p-Sylow subgroup of $G = \mathrm{SL}_{p-1}(\mathbb{Z})$. From the properties of the Euler class listed in 4.2 it follows that the restriction of $e(\varrho)$ to C_p is a generator of $H^{p-1}(C_p; \mathbb{Z}) = \mathbb{Z}/p\mathbb{Z}$. Hence the cup-product with $e(\varrho) \in H^{p-1}(G; \mathbb{Z})$ yields p-periodicity of period m_p dividing $p - 1$.

As mentioned in 3.4 the precise value of m_p for $\frac{1}{2}m + 1 < p < m + 1$ is $2(p - 1)$. It is $p - 1$ for $p = m + 1 > 2$. The lower bounds are due to the p-periodicity of suitably chosen finite subgroups, see [9]. In summary

(8) $\mathrm{SL}_m(\mathbb{Z})$ is p-periodic for $\frac{1}{2}m + 1 < p \leq m + 1$, with $m_p = 2(p - 1)$ for $p < m + 1$, $m_p = p - 1$ for $p = m + 1$.

Examples. $\mathrm{SL}_3(\mathbb{Z})$ is 3-periodic with 3-period 4; and $\mathrm{SL}_4(\mathbb{Z})$ is 5-periodic with 5-period 4.

5. Rational representations of finite groups

5.1. We conclude this survey with a brief outline of results on the order of the Euler and Chern classes, in the case where the group G is finite and where the representation is defined over $\mathbb{Q}$ (or some other subfield of $\mathbb{R}$ or $\mathbb{C}$). Certain results may be valid in more generality, e.g., for vcd $G > 0$, but this would require different approaches. We make strong use of the character theory of finite groups; for details we refer to Eckmann–Mislin [10, 11, 12].

We first consider the Euler class $e(\varrho) \in H^m(G; \mathbb{Z}(\varrho))$ of a rational representation $\varrho : G \to \mathrm{GL}_m(\mathbb{Q})$ of the finite group G; of course $e(\varrho)$ stands for $e(\varrho \otimes \mathbb{R})$. For m odd one has $2e(\varrho) = 0$. But we will show that for even m too there is a universal bound for the order of $e(\varrho)$, depending on m only but not on G nor on the representation. Namely, $E_m =$ denominator of B_m/m (cf. 4.3) is such a bound, i.e., the order of $e(\varrho)$ divides E_m. The argument concerning finite cyclic groups which leads to (6) in 4.3 shows that the bound E_m is best possible in that sense. The proof is sketched in 5.2 below; it does not make use of Sullivan's upper bound $2 E_m$ (see Remark in 4.3), and it not only yields the precise bound E_m, but can also be adapted to real fields other than $\mathbb{Q}$.

5.2. Let $\varrho : G \to \mathrm{GL}_m(\mathbb{Q})$ be a rational representation of the finite group G, and let G_p be a p-Sylow subgroup of G. Since the restriction $H^m(G; \mathbb{Z}(\varrho)) \to H^m(G_p; \mathbb{Z}(\varrho))$ is injective on the p-primary component, it will be sufficient to prove our assertion

$$E_m e(\varrho) = 0, \quad m \text{ even}, \tag{1}$$

for p-groups. If G is a p-group, $H^m(G; \mathbb{Z}(\varrho))$ is p-torsion; from $E_m = \mathrm{lcm}\{l \mid \varphi(l)$ divides $m\}$ we infer that (1) simply means

$$
\begin{aligned}
e(\varrho) &= 0 && \text{if } p-1 \text{ does not divide } m, \\
p^{\nu+1} e(\varrho) &= 0 && \text{if } p-1 \text{ divides } m \text{ and } \nu \text{ is the exponent of } p \text{ in } m.
\end{aligned}
\tag{2}
$$

We first assume that ϱ is rationally irreducible, non-trivial. By classical methods of representation theory, one proves that either (A) ϱ is induced from a representation $\sigma : S \to \mathrm{GL}_{m/p}(\mathbb{Q})$ of a subgroup S of index p in G, or (B) ϱ factors through a faithful irreducible rational representation $\bar{\varrho}$ of a factor group $\bar{G} = G/N$ which is "quasi-cyclic"; by this we mean that $\bar{G}$ is cyclic if p is odd, and that $\bar{G}$ is cyclic, quaternionic, a dihedral or a semi-dihedral group if $p = 2$. For each of these quasi-cyclic groups, there exists exactly one faithful irreducible rational representation; if p^β denotes the order of $\bar{G}$, its degree m is $\varphi(p^\beta) = p^{\beta-1}(p-1)$ for odd p and for $p = 2$ (cyclic and quaternionic case), $2^{\beta-2}$ for $p = 2$ (dihedral cases), and it has positive determinant except for $\bar{G} = C_2$.

A simple induction using (A) and (B) now shows that the degree of an irreducible rational representation of a p-group G is either 1 or of the form $p^\nu(p-1)$. Moreover (2) follows immediately in the case (B): $m = p^{\beta-1}(p-1)$ if p is odd and if $p = 2$ and $\bar{G}$

is cyclic or quaternionic; hence (2) means $p^\beta e(\varrho) = 0$ which is trivially true for $\bar\varrho$ and thus for ϱ. And further $m = 2^{\beta-2}$ if $\bar{G}$ is dihedral or semi-dihedral; then (2) means $p^{\beta-1}e(\varrho) = 0$, which is true for $\bar\varrho$ since $H^m(\bar{G}; \mathbb{Z})$ is not cyclic of order $|\bar{G}| = 2^\beta$, and thus is true for ϱ.

To prove (2) in the case (A), we assume by induction that it holds for degrees $< m$. The restriction ϱs of ϱ to the subgroup S decomposes into $\sigma \oplus \sigma'$, σ irreducible of degree m/p. Thus $p^\nu e(\sigma) = 0$, and $p^\nu e(\varrho s) = p^\nu e(\sigma)e(\sigma') = 0$. The transfer from S to G maps $e(\varrho s)$ to $pe(\varrho)$; we thus get $p^{\nu+1}e(\varrho) = p^\nu \operatorname{tr} e(\varrho s) = \operatorname{tr}(p^\nu e(\varrho s)) = 0$.

Let now $\varrho : G \to \mathrm{GL}_m(\mathbb{Q})$ be a reducible representation of the p-group G. It follows that if $p-1$ does not divide m (even) then at least one of the irreducible components of ϱ must be trivial (p is odd), and the product formula (iii) yields $e(\varrho) = 0$. If $p-1$ divides m, the decomposition into irreducible components and the product formula yield $p^{\nu+1}e(\varrho) = 0$, which concludes the proof of (2).

A similar result holds for representations ϱ of finite groups G over a real number field K. In (1) one has to replace E_m by $E_m(K) = \operatorname{lcm}\{l \mid \varphi_K(l) \text{ divides } m\}$ where $\varphi_K(l)$ is the degree of the l-th cyclotomic field over K, cf. [11].

5.3. We now turn to Chern classes $c_k(\varrho)$ of a rational representation $\varrho : G \to \mathrm{GL}_m(\mathbb{Q})$, G finite. Again, for even k, E_k is the (best possible) universal bound for the order of $c_k(\varrho) \in H^{2k}(G; \mathbb{Z})$:

$$E_k c_k(\varrho) = 0, \quad k \text{ even, } 0 < k \le m. \tag{3}$$

As before, the proof of (3) reduces to the case where G is a p-group. But the cases $p = 2$ and p odd have to be treated differently. For $p = 2$, the methods of 5.2 yield the bound E_k, i.e., $2^{\nu+1}c_k(\varrho) = 0$ where ν is the exponent of 2 in k. If p is odd, that procedure seems to lead to complicated computations, and it appears preferable to use a different method, that of "Galois invariance". This has been done in [21], for all p, by comparing the Chern classes $c_k(\varrho)$ with the "algebraic Chern classes" of Grothendieck and applying Galois automorphisms to these.

For finite groups G, such a detour is in fact not necessary. The behaviour of the ordinary Chern classes under an automorphism τ of $\mathbb{C}$ can be described directly, as follows.

Let $\varrho : G \to \mathrm{GL}_m(\mathbb{C})$ be any representation, and let q be the exponent of G. The automorphism τ is given on the qth cyclotomic field over $\mathbb{Q}$ by $\tau(\zeta) = \zeta^r$, where ζ is a qth primitive root of unity and r prime to q; conversely any such automorphism of the qth cyclotomic field can be extended to $\mathbb{C}$. Applying τ to the representation ϱ yields a new representation ϱ^τ. Using character theory and Newton polynomials one proves

$$\varrho^\tau = \psi_r(\varrho), \quad \text{where } \psi_r \text{ is the } r\text{th Adams operation.} \tag{4}$$

The equality is to be understood in the representation ring; but $\psi_r(\varrho)$ is actually a representation. Thus the $\mathbb{C}^m$-bundle over $K(G, 1)$ corresponding to ϱ^τ is obtained from that of ϱ by applying ψ_r. By well-known properties of the Chern classes it follows that

$$c_k(\varrho^\tau) = r^k c_k(\varrho).\tag{5}$$

Now, for a representation ϱ realizable over $\mathbb{Q}$, one has $\varrho^\tau = \varrho$ and thus

$$(r^k - 1)c_k(\varrho) = 0 \quad \text{for all } r \text{ prime to } q = \exp G.\tag{6}$$

If G is a p-group, p odd, $(r^k - 1)c_k(\varrho) = 0$ for all r prime to p implies, by standard number-theoretic arguments, that for $p - 1$ not dividing k one has $c_k(\varrho) = 0$ and otherwise $p^{\nu+1}c_k(\varrho) = 0$ ($\nu =$ exponent of p in k), which completes the proof of (3). For $p = 2$, one gets $2^{\nu+2}c_k(\varrho) = 0$ as in [21], which would yield $2E_k c_k(\varrho) = 0$ instead of (3); this is why the case of 2-groups has to be dealt with separately.

Our procedure can, of course, also be applied to representations realizable over other subfields of $\mathbb{C}$ instead of $\mathbb{Q}$.

References

[1] R. Bieri, Gruppen mit Poincaré-Dualität, Comm. Math. Helvetici 47 (1972) 373–396.

[2] R. Bieri and B. Eckmann, Groups with homological duality generalizing Poincaré duality, Invent. Math. 20 (1973) 103–124.

[3] R. Bieri and B. Eckmann, Finiteness properties of duality groups, Comm. Math. Helvetici 49 (1974) 74–83.

[4] H. Biner, Ph.D. Thesis, ETH Zurich (In preparation).

[5] A. Borel and J.-P. Serre, Corners and arithmetic groups, Comm. Math. Helvetici 48 (1974) 244–297.

[6] K.S. Brown, Homological criteria for finiteness, Comm. Math. Helvetici 50 (1975) 129–135.

[7] K.S. Brown, Groups of virtually finite dimension, Homological Group Theory (Cambridge Univ. Press, 1979) 27–70.

[8] K.S. Brown, Cohomology of infinite groups (Proc. Internat. Congress of Mathematicians, Helsinki 1978).

[9] B. Bürgisser, Ph.D. Thesis, ETH Zurich (1979). For further results see: On the p-periodicity of arithmetic subgroups of general linear groups (to appear in Comm. Math. Helvetici).

[10] B. Eckmann and G. Mislin, Rational representations of finite groups and their Euler class, Math. Ann. 245 (1979) 45–54.

[11] B. Eckmann and G. Mislin, On the Euler class of representations of finite groups over real fields (to appear in Comm. Math. Helvetici).

[12] B. Eckmann and G. Mislin, Chern classes of representations of finite groups (in preparation).

[13] F.T. Farrell, An extension of Tate cohomology to a class of infinite groups, J. Pure Appl. Algebra 10 (1977) 153–161.

[14] F.E.A. Johnson and C.T.C. Wall, On groups satisfying Poincaré duality, Ann. of Math. 69 (1972) 592–598.

[15] Heinz Müller, (1) Théorèmes de décomposition pour les paires de groupes. (2) Groupes et paires de groupes à dualité de Poincaré de dimension 2 (to appear in C.R. Acad. Sci. Paris).

[16] J.-P. Serre, Cohomologie des groupes discrets, C.R. Acad. Sci. Paris 268 (1969) 268–271.

[17] J.-P. Serre, Cohomologie des groupes discrets, Prospects in Math., Ann. of Math. Study 70 (Princeton, 1971) 77–169.

[18] J. Stallings, On torsion-free groups with infinitely many ends, Ann. of Math. 88 (1968) 312–334.

[19] C. Soulé, Thèse Doctorat d'Etat Paris VII (1978).

[20] R. Swan, Groups of cohomological dimension one, J. Algebra 12 (1969) 585–610.

[21] C.B. Thomas, An integral Riemann–Roch formula for flat line bundles, Proc. London Math. Soc. (3)34 (1977) 87–101.

[22] B.A.F. Wehrfritz, Infinite linear groups, Ergebnisse der Math. 76 (Springer, Berlin, 1973).

Poincaré duality groups of dimension two are surface groups

Combinatorial Group Theory and Topology, Annals of Math. Studies (1986), 35–51

1. Introduction

We first explain the terms appearing in the title, and add some general comments.

1.1. A *surface group* G is a group isomorphic to the fundamental group $\pi_1(\Sigma_g)$ of a closed surface Σ_g, orientable or not, of genus $g \geq 1$. Such a group admits a presentation

$$G = \langle x_1, y_1, \ldots, x_g, y_g \mid [x_1, y_1] \ldots [x_g, y_g] = 1 \rangle$$

in the orientable case,

$$G = \langle z_0, z_1, \ldots, z_g \mid z_0^2 z_1^2 \ldots z_g^2 = 1 \rangle$$

in the non-orientable case.

We will also use the concept of a *surface group-pair* $(G; \{S_0, S_1, \ldots, S_m\})$: it consists of the (free) fundamental group of a closed surface of genus $g \geq 0$ with $m+1$ disks removed, $m \geq 0$ (but ≥ 1 if the surface is a sphere), together with the $m+1$ infinite cyclic subgroups generated by the boundary circles of these disks. Surface group-pairs have presentations

$$G = \langle t_1, t_2, \ldots, t_m, x_1, y_1, \ldots, x_g, y_g \rangle,$$
$$S_0 = \langle t_1 t_2 \ldots t_m [x_1, y_1] \ldots [x_g, y_g] \rangle, \quad S_j = \langle t_j \rangle \quad \text{for } j = 1, \ldots, m$$

in the orientable case, $m + g > 0$;

$$G = \langle t_1, t_2, \ldots, t_m, z_0, z_1, \ldots, z_g \rangle$$
$$S_0 = \langle t_1 t_2 \ldots t_m z_0^2 z_1^2 \ldots z_g^2 \rangle, \quad S_j = \langle t_j \rangle \quad \text{for } j = 1, \ldots, m$$

in the non-orientable case, $m \geq 0$, $g \geq 0$.

The "lowest" cases of surface group-pairs are

$$G = \langle t_1 \rangle, \quad S_0 = \langle t_1 \rangle, \quad S_1 = \langle t_1 \rangle$$

and

$$G = \langle z_0 \rangle, \quad S_0 = \langle z_0^2 \rangle.$$

1.2. A *Poincaré duality group G of dimension n*, in short a *PD^n-group*, is a group fulfilling Poincaré duality in (co-)homology for all coefficient ZG-modules A with respect to the formal dimension n and to a certain G-action on the additive group of integers $\tilde{Z}$:

$$H^i(G; A) \cong H_{n-i}(G; \tilde{Z} \otimes A), \quad i \in Z.$$

Here $\tilde{Z} \otimes A$ is a ZG-module by diagonal action, and the isomorphisms are natural in the ZG-modules A. As we will see the dimension n and the ZG-module structure of $\tilde{Z}$ are determined by G (the PD^n-group being orientable or non-orientable according to whether the action is trivial or not).

The definition above is, of course, analogous to the Poincaré duality valid for all closed n-dimensional manifolds X, the coefficients A being $\pi_1(X)$-modules. If X is aspherical then the (co-)homology of X is isomorphic to that of $G = \pi_1(X)$ so that, in that case, G is a PD^n-group. It is not known whether in general the converse is true; i.e., whether a PD^n-group is necessarily isomorphic to the fundamental group of a closed n-dimensional aspherical manifold.

Since the universal covering of the surface Σ_g, $g \geq 1$, is R^2 the surface Σ_g is aspherical. Thus the surface groups in 1.1 above are PD^2-groups. The *Theorem* formulated in the title states that the converse is true, thus solving the problem in the case $n = 2$. The proof of that *Theorem* has been achieved in several steps contained in a series of papers by the present author and various collaborators; these papers were written partly with other objectives in mind. The conference organizers have asked the author to present a survey, as complete as possible, of that proof. We do so and use the opportunity to simplify some of the arguments.

1.3. There are, in fact, different techniques and methods involved in that proof. They belong roughly speaking to the following three areas of ideas:

(I) Homological algebra, homology of groups.
(II) Structure and splitting theorems for groups (Stallings-Bass-Serre and others).
(III) Ranks of projective modules and Euler characteristic (Hattori-Stallings-Bass-Kaplansky).

The papers leading to or containing the steps of the proof are as follows. In the field (I) by Robert Bieri and the author [2], [4], [5]; in (II) by Heinz Müller and the author [12], [14]; in (III) by Peter Linnell and the author [10], [11].

1.4. For compact manifolds-with-boundary (∂-manifolds), in particular for the closed surfaces with disks removed, one has the well-known "relative" Poincaré duality. The surface group-pairs listed in 1.1 above fulfill such a relative Poincaré duality, of dimension 2. To formulate this in a precise way, relative (co-)homology groups for pairs of groups $(G; \{S_0, S_1, \ldots, S_m\})$ have to be considered, cf. Section 2.3 below; this yields the concept of a PD^n-pair of groups. An important step in the proof of the *Theorem* will be to show that all PD^2-pairs of groups are surface group-pairs ("Relative Theorem", Section 3.2).

1.5. We mention here two corollaries of the *Theorem*.

Corollary 1.1 (cf. [12], [11]). *All Poincaré-2-complexes are homotopy equivalent to closed surfaces (of genus ≥ 0).*

Corollary 1.2. *Let G be a torsion-free group containing a surface group S as subgroup of finite index; then G is also a surface group.*

Indeed, a homological argument (see Section 2.2) shows that G is a PD^2-group. This corollary is a special case of the "Nielsen realization conjecture" proved by Kerckhoff [Annals of Math. 117 (1983), 235–265]. The special case above was established by Eckmann-Müller [12] before Kerckhoff's proof, and before our Theorem on PD^2-groups had been completely settled.

1.6. This survey is organized as follows. Section 2 contains general preliminaries on duality and relative duality groups. In Section 3 we assume that the PD^2-group fulfills a certain "splitting" property and show that the *Theorem* can then be reduced to the Relative Theorem; both that reduction and the proof of the Relative Theorem are given on the basis of general splitting arguments explained in Section 4. In Section 5 we show, by quite different methods involving ranks of finitely generated projective modules, that any PD^2-group fulfills the splitting assumption.

2. Duality groups

2.1. The group G is called a *duality group of dimension* $n > 0$ with respect to a dualizing ZG-module C, in short a D^n-group, if one has isomorphisms

$$H^i(G;A) \cong H_{n-i}(G; C \otimes A)$$

for all $i \in Z$ and all ZG-modules A; they are assumed to be natural in A, and $C \otimes A$ is endowed with the diagonal G-action. If $C = Z$ as an Abelian group, one has Poincaré duality (cf. 1.2), i.e., G is a PD^n-group.

From the definition it follows that $H^i(G;A)$ commutes with direct limits in A. This is possible only if G admits a projective resolution $\cdots \to P_i \to P_{i-1} \to \cdots \to P_0 \twoheadrightarrow Z$ with all P_i finitely generated over ZG (by the Bieri Eckmann-Brown finiteness criterion, see [3] and [7]). Furthermore one easily checks that $C \otimes ZG \cong C_0 \otimes ZG$ (where C_0 is the Abelian group underlying C) is an induced module, and thus $H^i(G; ZG) = 0$ for all $i \neq n$; as for $H^n(G; ZG)$, it is isomorphic to $H_0(G; C \otimes ZG) = C$, and this is a (right) ZG-module isomorphism. The dualizing module C is thus determined by G; it is easily seen to be torsion-free as an Abelian group.

The cohomology dimension cd G is clearly $\leq n$, and by the above it is $= n$; hence the integer n is also determined by G, and G admits a finitely generated projective resolution of finite length (equal to n); such groups are said to be of type FP.

Summarizing we see that a D^n-group G fulfills

(1) G is of type FP,
(2) $H^i(G; ZG) = 0$ for $i \neq n$ and $H^n(G; ZG)$ is torsion-free,
(3) cd $G = n$.

It has been proved by Bieri-Eckmann [2] that, conversely, a group fulfilling (1) and (2) is a D^n-group with dualizing module $C = H^n(G; ZG)$.

We note that cd $G = n$ implies that G is torsion-free.

2.2. As an application, let G *be torsion-free and* S *a subgroup of finite index.* By Serre's theorem (see e.g. [8], p. 190) cd $G =$ cd $S = n$. Clearly G admits a finitely generated free resolution over ZG if and only if S does over ZS; hence

G is of type FP if and only if S is. Moreover

$$H^i(S; ZS) \cong H^i(G; \operatorname{Hom}_S(ZG, ZS)) \cong H^i(G; ZS \underset{S}{\otimes} ZG) = H^i(G; ZG),$$

and it follows that G fulfills (1) and (2) above if and only if S does, with the "same" dualizing module C.

Thus G is a D^n-group if and only if S is a D^n-group, and the dualizing modules are isomorphic as Abelian groups. In particular, G *is a PD^n-group if and only if S is.*

1) *Remarks: Dimension 1.* G is a D^1-group if and only if it is finitely generated free. It is a PD^1-group if and only if it is infinite cyclic.

2) *Subgroups of infinite index.* For PD^n-groups G, Strebel [17] has proved, by homological methods, that for a subgroup S of infinite index in G one has $\operatorname{cd} S \leq n - 1$. For $n = 2$ it follows that S is a free group; this is the PD^2-analogue of a fact well-known for surface groups.

2.3. We briefly recall *relative duality* for group pairs (for details see Bieri-Eckmann [4, 5]). A group pair (G, S) consists of a group G and a family $S = \{S_j, j \in I\}$ of subgroups, not necessarily distinct. For any subgroup $S \subset G$ one writes ZG/S for the G-module whose underlying Abelian group is freely generated by the cosets xS, with G-action by left multiplication. Relative (co-)homology is defined by means of the "augmentation kernel"

$$\Delta = \ker \left\{ \bigoplus_j ZG/S_j \xrightarrow{\varepsilon} Z \right\}$$

where $\varepsilon(x S_j) = 1$ for all $x \in G$ and $j \in I$:

$$H_i(G, S; A) = H_{i-1}(G; \Delta \otimes A),$$

$$H^i(G, S; A) = H^{i-1}(G; \operatorname{Hom}(\Delta, A)),$$

A being a G-module, $\otimes$ and Hom equipped with diagonal G-action.

A duality pair of dimension n with dualizing module C is a pair (G, S) fulfilling

$$H^i(G; A) \cong H_{n-i}(G, S; C \otimes A)$$

and

$$H^i(G, S; A) \cong H_{n-i}(G; C \otimes A).$$

In the (orientable) Poincaré duality case, $C = Z$, each of these isomorphisms implies the other one and the first one becomes

$$H^i(G; A) \cong H_{n-i-1}(G; \Delta \otimes A);$$

i.e., (G, S) is a PD^n-pair of groups if and only if G is an D^{n-1}-group with dualizing module Δ. Relative exact sequences show that S must be a finite family of PD^{n-1}-groups $S_0, S_1, \ldots, S_m$.

In particular, if (G, S) is a PD^2-pair then G is finitely generated free and S is a finite family of infinite cyclic groups. The duality isomorphisms yield $H^2(G, S; ZG) = Z$, $H^1(G, S; ZG) = 0$, $H^1(G; ZG) = \Delta$. As shown in [4] the only important homological property is $H^2(G, S; ZG) = Z$; indeed, if G is finitely generated free and S a finite family of infinite cyclic groups it characterizes PD^2-pairs.

3. PD²-groups splitting over a finitely generated subgroup

3.1. A group G is said to split over the subgroup H if either (α) G is a non-trivial amalgamated free product $G = G_1 *_H G_2$, or (β) G is an HNN-extension $G = G_1 *_{H,p}$. Two cases are of special importance: 1) H is finitely generated, 2) H is finite. We recall that Stallings' structure theorem [15], [16] for finitely generated groups tells that 2) holds if and only if $H^1(G; ZG) \neq 0$.

Proposition 3.1. *The assertion of the Theorem holds if the PD^2-group G splits over a finitely generated group L.*

The proof of Proposition 3.1 makes strong use of the "simultaneous splitting theorem" (in short: SST) for groups and subgroups, established by Heinz Müller [14]. The SST is a refinement of the relative version (Swan [18], Swarup [19]) of Stallings' structure theorem; it deals with splittings over finite subgroups H, and since in our case G is torsion-free only $H = 1$ will occur. A short outline of SST and its application in the present context will be given in Section 4 below.

The application of SST to PD^2-groups (α) $G = G_1 *_L G_2$, or (β) $G = G_1 *_{L,p}$, with L finitely generated, is as follows. Since $H^1(G; ZG) = 0$, L is $\neq 1$. The index of L in G is infinite; by Strebel's theorem [17] one has $\operatorname{cd} L \leq 1$, and hence L is (finitely generated) free. We only describe the case (α), the case (β) being similar.

If in $G = G_1 *_L G_2$ the rank of L is > 1 one has, by virtue of SST (see Section 4.4, A)), splittings

$$G_1 = H_1 * H_2, \quad L = L_1 * L_2 \quad \text{with} \quad L_i \subset H_i, \quad i = 1, 2,$$

or

$$G_1 = H * \langle q \rangle = H *_{1,q}, \quad L = L_1 * q L_2^{-1} q \quad \text{with} \quad L_1, L_2 \subset H.$$

The first possibility yields

$$G = H_1 *_{L_1}(H_2 *_{L_2} G_2);$$

if $L_1 \neq H_1$ then G splits over L_1, and if $L_1 = H_1$ then G splits over L_2. Thus G splits over a subgroup whose rank is less than that of L. The second possibility yields

$$G = (H *_{L_1} G_2) *_{L_2, q^{-1}},$$

so G splits over L_2, again of rank less than that of L.

Thus we are always reduced to the case where G splits over an infinite cyclic subgroup C as (α) $G = G_1 *_C G_2$, or (β) $G = G_1 *_{C,p}$.

Since C is a PD^1-group we can apply the general homological arguments of [4], Theorem 8.1 and 8.3. It follows that in the case (α) the group pairs (G_1, C) and (G_2, C) and in the case (β) the group pair $(G_1, \{C, p^{-1} C p\})$ are PD^2-pairs. Now the *Relative Theorem* below tells that these pairs are surface group-pairs (see 1.1) corresponding to closed surfaces with one disk, or two disks respectively, removed.

In (α), $G = G_1 *_C G_2$ is the fundamental group of the closed surface obtained by identifying the boundary circles; in (β), $G = G_1 *_{C,p}$ is the fundamental group of the closed surface obtained by joining the two boundary circles by a tube.

3.2. Relative Theorem. *Any PD^2-pair $(G; \{S_0, \ldots, S_m\})$ is a surface group-pair.*

Again the proof uses SST, in addition to the properties of PD^2-pairs mentioned in 2.3. We proceed by induction on the rank of the finitely generated free group G.

If that rank is 1, i.e., G infinite cyclic $C = \langle c \rangle$, one has $H^1(G; ZG) = Z$ since $G = C$ is a PD^1-group, and $H^1(G; ZG) = \varDelta$ since G is a duality group of dimension 1 with dualizing module $\varDelta$. The exact sequence

$$\varDelta \rightarrowtail \bigoplus_j ZG/S_j \twoheadrightarrow Z$$

then yields $\bigoplus_j ZG/S_j = Z \oplus Z$; this is possible only if $S = \{C, C\}$ or $S = \{\langle c^2 \rangle\}$. Thus the pair (G, S) is either $(C, \{C, C\})$ or $(C = \langle c \rangle, \{\langle c^2 \rangle\})$; i.e., we obtain precisely the lowest orientable case $g = 0$, $m = 1$, or the lowest non-orientable case $g = 0$, $m = 0$ of the presentation list of surface group-pairs in 1.1.

If the rank of G is > 1, SST (see Section 4.4, B)) yields splittings (α) $G = G_1 * G_2$ with $S_0 = \langle g_1 g_2 \rangle$, $1 \neq g_j \in G_j$, while the S_j for $j > 0$ are (conjugate to) subgroups of G_1 and G_2, say $S_1, \ldots, S_k \subset G_1$, $S_{k+1}, \ldots, S_m \subset G_2$; or ($\beta$) $G = G_1 * \langle p \rangle = G_1 *_{1,p}$ with $S_0 = \langle p g_1 p^{-1} g_2 \rangle$, $g_1, g_2 \in G_1$, while $S_1, \ldots, S_m$ are (conjugate to) subgroups of G_1. We will deal with (α) only, the case (β) being similar. We can write

$$G = (G_1 * \langle g_2 \rangle) *_{\langle g_2 \rangle} G_2.$$

Then $S_0 \subset G_1 * \langle g_2 \rangle$. The pair $(G_2; \{\langle g_2 \rangle, S_{k+1}, \ldots, S_m\})$ is a PD^2-pair by Theorem 8.1 of [4]; note that the case $\langle g_2 \rangle = G_2$ needs a special argument, and the pair must then be $(\langle g_2 \rangle, \{\langle g_2 \rangle, \langle g_2 \rangle\})$, cf. [12] p. 517. Similarly the pair $(G_1, \{\langle g_1 \rangle, S_1, \ldots, S_k\})$ is a PD^2-pair. By induction they are surface group-pairs, and one easily checks that so is $(G; \{S_0, S_1, \ldots, S_m\})$.

4. Simultaneous splitting of groups and subgroups

4.1. We consider throughout this section a finitely generated group G which splits over a finite subgroup K as $G = G_1 *_K G_2$ or $G = G_1 *_{K,p}$, i.e., with $H^1(G; ZG) \neq 0$, cf. Section 3.1. Given a finite family $\{S_1, \ldots, S_m\}$ of subgroups of G let $N(G; S_1, \ldots, S_m)$ be the intersection of the kernels of the restriction maps $\mathrm{res}_j: H^1(G; ZG) \to H^1(S_j; ZG)$, $j = 1, \ldots, m$. By the relative version of the structure theorem (Swan [18], Swarup [19]) $N(G; S_1, \ldots, S_m) \neq 0$ if and only if there is a splitting of G such that the S_j are conjugate to subgroups of G_1 or G_2; this will be the case in the following and only such splittings will be considered.

Let T be a further subgroup of G, and assume that T is finitely generated. $\{x_\nu\}$ denoting a set of coset representatives of $G \bmod T$, we consider the restriction map

$$H^1(G; ZG) \xrightarrow{\text{res}} H^1(T; ZG) \cong \bigoplus_v H^1(T; ZT) x_v.$$

The minimal number of non-zero components of $\text{res}(c)$ for all

$$0 \neq c \in N(G; S_1, \dots, S_m)$$

is called the *weight* $n(T)$ of T with respect to G (and to $S_1, \dots, S_m$). Note that $n(T) = 0$ if and only if $N(G; T, S_1, \dots, S_m) \neq 0$; i.e., if there is a splitting of G with $T \subset G_1$ or G_2.

4.2. The simultaneous splitting theorem (*SST*) established by H. Müller [14] (actually for more general G and T) concerns the case $n(T) > 0$. It can be formulated roughly as follows, its full content being in fact more complicated.

There is a tree Γ on which G acts with finite edge-stabilizers and with proper subgroups of G as vertex-stabilizers such that Γ/G has one edge (and that each S_j, $j = 1, \dots, m$ stabilizes a vertex of Γ); and Γ contains a subtree Γ_T invariant under T with Γ_T/T having at most $n(T)$ edges.

Examples. 1) Let G be torsion-free, and $n(T) = 1$. Then one has the following possibilities:

(1) $G = G_1 * G_2$, $T = T_1 * T_2$, $T_1 \subset G_1$, $T_2 \subset G_2$.

(2) $G = G_1 * \langle p \rangle = G_1 *_{1,p}$, $T = T_1 * p T_2 p^{-1}$, $T_1, T_2 \subset G_1$.

(3) $G = \langle p \rangle$, $T = \langle p \rangle$, $S_1 = \dots = S_m = 1$ or $m = 0$.

2) Let G be torsion-free, T infinite cyclic, and $n(T) = 2$. Then one has the following possibilities (cf. [14])

(1) $G = G_1 * G_2$, $T = \langle g_1 g_2 \rangle$, $1 \neq g_i \in G_i$, $i = 1, 2$.

(2) $G = G_1 * \langle p \rangle = G_1 *_{1,p}$, $T = \langle p g_1 p^{-1} g_2 \rangle$, $g_1, g_2 \in G_1$.

(3) $G = \langle p \rangle$, $T = \langle p^2 \rangle$, $S_1 = \dots = S_m = 1$ or $m = 0$.

4.3. We restrict ourselves to some remarks concerning the proof of *SST*. We write Z_2 for $Z/2Z$ and use $Z_2 G$, $Z_2 T$ instead of ZG, ZT; one easily checks that this yields the same weight $n(T)$. Then $H^1(G; Z_2 G)$ can be interpreted as group of all subsets of G which are "almost invariant" under translation without being G or $\emptyset$; almost invariant means invariant except for finite sets. Namely, writing $\overline{Z_2 G}$ for $\text{Hom}(ZG, Z_2) = \prod_{x \in G} Z_2 x$ and εG for $\overline{Z_2 G}/Z_2 G$ (arbitrary modulo finite subsets of G), the exact sequence

$$H^0(G; \overline{ZG}) \to H^0(G; \varepsilon G) \to H^1(G; ZG) \to H^1(G; \overline{Z_2 G}) = 0$$

yields

$$(\varepsilon G)^G/(Z_2 G)^G \cong H^1(G; Z_2 G).$$

The restriction map $H^1(G; Z_2 G) \xrightarrow{\text{res}} \bigoplus_v H^1(T; ZT) x_v$ is then given by $U \mapsto U \cap T x_v$ for all v, where U is a non-trivial almost invariant set. The components $U \cap T x_v$ are almost T-invariant. By the technique of Dunwoody [9] and Swarup [19] these almost invariant sets yield the theorem.

4.4. There remains to compute the weights of those subgroups which occur in the proofs of Proposition 3.1 (subgroup L) and of the "Relative Theorem" in 3.2 (subgroup S_0).

A) We again restrict ourselves to the case (α) in the proof of Proposition 3.1. Thus G is a PD^2-group with $G = G_1 *_L G_2$, L free of rank > 1. We claim that $n(L)$ with respect to $(G_1; \emptyset)$ or to $(G_2; \emptyset)$ is equal to 1. By example 1) in 4.2 this yields the required simultaneous splittings of G_1 and L.

To prove the claim we consider the exact Mayer-Vietoris sequence

$$\cdots \to 0 \to H^1(G_1; ZG) \oplus H^1(G_2; ZG) \xrightarrow{(\mathrm{res}_1 - \mathrm{res}_2)} H^1(L; ZG) \xrightarrow{\delta} H^2(G; ZG) \to \cdots$$

and note that the weight is not 0 since res_1 and res_2 are injective. $H^1(L; ZL)$ is free Abelian of infinite rank since L is free of rank > 1. Thus the restriction of δ to $H^1(L; ZL)$ cannot be injective, $H^2(G; ZG)$ being $= Z$; i.e., the intersection $\mathrm{im}(\mathrm{res}_1, -\mathrm{res}_2) \cap H^1(L; ZL)$ is $\neq 0$. On the other hand, if both $n(L)$ with respect to $(G_1, \emptyset)$ and to $(G_2, \emptyset)$ are > 1, the image $\mathrm{res}_1(c_1) - \mathrm{res}_2(c_2)$ of $0 \neq (c_1, c_2) \in H^1(G_1; ZG) \oplus H^1(G_2; ZG)$ cannot lie in $H^1(L; ZL) \subset H^1(L; ZG)$; this is seen by looking at the lengths of elements with respect to coset representatives of $G \bmod G_1$ and G_2. Thus $n(T)$ with respect to, say, $(G_1, \emptyset)$ is $= 1$, and we obtain the simultaneous splitting.

B) In 3.2, (G, S) is a PD^2-pair, $S = \{S_0, S_1, \ldots, S_m\}$ with $m \geq 0$ and all S_j infinite cyclic, and we have to consider the case where the free group G is of rank > 1. The claim is that $n(S_0)$ with respect to $(G, \{S_1, \ldots, S_m\})$ is $= 2$. By example 2) in 4.2 this yields the required splittings.

The exact relative cohomology sequence of $G \bmod S$ is

$$0 \to H^1(G, S; ZG) \to H^1(G; ZG) \xrightarrow{r} \bigoplus_{j=0}^{m} H^1(S_j; ZG) \xrightarrow{\delta} H^2(G, S; ZG) \to 0,$$

where r denotes the map with components res_j, $j = 0, 1, \ldots, m$. The PD^2-pair properties tell that the first term is 0, the last isomorphic to Z. If S_0 (or any proper subset) is omitted from the family S, the last term becomes 0 and the first term must be $\neq 0$ (cf. [4], Section 11); i.e., the intersection $N = N \cdot (G; S_1, \ldots, S_m)$ of the $\ker \mathrm{res}_j$, $j = 1, \ldots, m$ is non-zero. The weight $n(S_0)$ is the minimal number of components in $H^1(S_0; ZG) = \bigoplus_v H^1(S_0; ZS_0) x_v \cong \bigoplus_v Z x_v$, of $\mathrm{res}_0(c)$

for all $0 \neq c \in N$; note that $\ker \mathrm{res}_0 \cap N = 0$.

Now $r(N) = (\mathrm{res}_0(N), 0, \ldots, 0) = (H^1(S_0; ZG), 0, \ldots, 0) \cap \ker \delta$. One easily checks that δ restricted to any summand $Z x_v$ of $H^1(S_0; ZG)$ is bijective. So obviously the minimum number of components of elements $\neq 0$ in $\mathrm{res}_0(N)$ is 2, which proves the claim.

5. The first Betti number of a PD^2-group

5.1. In order to conclude the proof of the Main *Theorem* we have to show that a PD^2-group, without any further assumption, splits over a finitely generated subgroup. This is guaranteed by the following proposition whose proof will be given in 5.2 and 5.3 below. Recall that PD^n-groups are of type FP so that Betti numbers $\beta_i(G) = \operatorname{rank} H_i(G; Z)$ are defined.

Proposition 5.1. *If G is a PD^2-group then $\beta_1(G) > 0$.*

From this it follows that $H_1(G; Z)$, the abelianized group G, contains at least one infinite cyclic summand C, and thus G admits a factor group $\cong C$. By a result of Bieri-Strebel ([6], Theorem A) this implies that G splits as $G = G_1 *_{L,p}$ over a finitely generated subgroup L (as shown in [6], this holds for any group G of type FP_2, i.e., admitting a projective resolution which is finitely generated in dimensions ≤ 2, and having an infinite cyclic factor group). The splitting is constructed explicitly, the generator p being any element projecting onto a generator of C.

5.2. For the proof of Proposition 5.1 we use the homological Euler characteristic $\chi(G) = \beta_0(G) - \beta_1(G) + \beta_2(G)$ of G. By the Euler-Poincaré formula $\chi(G)$ is equal to the alternating sum of the ranks of the free Abelian groups $Z \underset{G}{\otimes} P_i$ for any FP-resolution over ZG

$$0 \to P_2 \to P_1 \to P_0 \twoheadrightarrow Z.$$

We recall that in the orientable case $\beta_0(G) = \beta_2(G) = 1$ and $\beta_1(G) = \text{even}$; in the non-orientable case $\beta_0(G) = 1$, $\beta_2(G) = 0$ (the proofs are the same as for closed surfaces). Thus in both cases the claim of Proposition 5.1 is $\chi(G) \leq 0$. If G is a non-orientable PD^2-group, let G_0 be the orientable subgroup of index 2. By the multiplicative property of the Euler characteristic (valid for groups of type FP, cf. [8], § IX.6) $\chi(G_0) = 2\chi(G)$; hence $\chi(G_0) \leq 0$ implies $\chi(G) \leq 0$, and we are reduced to the orientable case.

G being an orientable PD^2-group we choose a resolution

$$0 \to P \to ZG^d \xrightarrow{\partial} ZG \twoheadrightarrow Z \tag{1}$$

with P finitely generated projective over ZG. Applying $\operatorname{Hom}_G(-, ZG)$ to (1) we get the sequence

$$0 \leftarrow P^* \xleftarrow{\delta} ZG^d \leftarrow ZG \leftarrow 0 \tag{2}$$

where $P^* = \operatorname{Hom}_G(P, ZG)$ is finitely generated projective. Since $H^i(G; ZG) = 0$ for $i \neq 2$ and $H^2(G; ZG) = Z$ we obtain the exact sequence

$$Z \twoheadleftarrow P^* \xleftarrow{\delta} ZG^d \leftarrow ZG \leftarrow 0, \tag{3}$$

which is another FP-resolution for G; for $\chi(G)$ it yields

$$\chi(G) = \operatorname{rank}\left(Z \underset{G}{\otimes} P^*\right) - d + 1$$

whence

$$\beta_1(G) = 2 - \chi(G) = 1 + d - \operatorname{rank}\left(Z \underset{G}{\otimes} P^*\right). \tag{4}$$

Comparing (1) and (3) we see that $P^*/\delta ZG^d \cong ZG/\partial ZG^d$, and therefore $P^* \oplus \partial ZG^d \cong ZG \oplus \delta ZG^d$. One then has a surjection $ZG^{d+1} \twoheadrightarrow P^* \oplus \partial ZG^d$, and since $\partial Z G^d \neq 0$ a surjection $Z G^{d+1} \twoheadrightarrow P^*$ with non-zero kernel N, i.e.,

$$Z G^{d+1} \cong P^* \oplus N \qquad (5)$$

N is finitely generated projective, and $\text{rank}\left(Z \underset{G}{\otimes} P^*\right) + \text{rank}\left(Z \underset{G}{\otimes} N\right) = d+1$. From (4) we obtain

$$\beta_1(G) = \text{rank}\left(Z \underset{G}{\otimes} N\right). \qquad (6)$$

5.3. It is not known in general whether, for a group G and a non-zero finitely generated projective ZG-module N, the free Abelian group $Z \underset{G}{\otimes} N$ is non-zero. It has, however, proved useful to compare $\text{rank}\left(Z \underset{G}{\otimes} N\right)$ with another rank concept (for f.g. projectives) which we propose to call the *Kaplansky rank* $\varkappa(N)$. It is defined as follows: Let $N \oplus M$ be finitely generated free, and ϕ the idempotent endomorphism of $N \oplus M$ which is 1_N on N and 0 on M. The trace of ϕ (in a basis of $N \oplus M$) is an element of ZG; its coefficient of $1 \in G$ does not depend on the choice of M and of the basis, and this is $\varkappa(N)$. A theorem of Kaplansky [13] states that if $N \neq 0$ then $\varkappa(N) > 0$.

For free modules N one clearly has $\varkappa(N) = \text{rank}_{ZG} N = \text{rank}\left(Z \underset{G}{\otimes} N\right)$; but for projective N the middle term is not defined and one does not known in general whether $\varkappa(N) = \text{rank}\left(Z \underset{G}{\otimes} N\right)$.

In the case of the PD^2-group G and the projective module N in Section 5.2 let us first assume that P and hence P^* is free (e.g., G is the fundamental group of a finite CW-complex). Then (5) immediately implies $\varkappa(N) = \text{rank}\left(Z \underset{G}{\otimes} N\right)$, and by Kaplansky's theorem we get from (6) that $\beta_1(G) > 0$.

In the general case, P^* projective, this does not work. However, an interesting criterion of Bass [1] tells that if, for some projective module, the two ranks are not equal then G contains a subgroup H isomorphic to the additive group $Z\left[\dfrac{1}{p}\right]$ for some prime p. If the index of H in G is finite H would be a PD^2-group − which it is not; if the index is infinite, this would imply, by Strebel's theorem [17], that H is free − which it is not. Thus the two ranks coincide, and we again get from (6) that $\beta_1(G) > 0$.

References

[1] Bass, H.: Euler characteristic and characters of discrete groups. Inventiones Math. *35* (1976), 155−196
[2] Bieri, R.; Eckmann, B.: Groups with homological duality generalizing Poincaré duality. Inventiones Math. *20* (1973), 103−124
[3] Bieri, R.; Eckmann, B.: Finiteness properties of duality groups. Comment. Math. Helv. *49* (1974), 460−478
[4] Bieri, R.; Eckmann, B.: Relative homology and Poincaré duality for group pairs. J. of Pure and Applied Algebra *13* (1978), 277−319

[5] Bieri, R.; Eckmann, B.: Two-dimensional Poincaré duality groups and pairs, in: Homological Group Theory, London Math. Soc. Lecture Notes *36* (1979), 225–230
[6] Bieri, R.; Strebel, R.: Almost finitely presented soluble groups. Comment. Math. Helv. *53* (1978), 258–278
[7] Brown, K. S.: Homological criteria for finiteness. Comment. Math. Helv. *50* (1975), 129–135
[8] Brown, K. S.: Cohomology of Groups. Springer-Verlag, New York 1982
[9] Dunwoody, M. J.: Accessibility and groups of cohomological dimension one. Proc. of the London Math. Soc. *38* (1979), 193–215
[10] Eckmann, B.; Linnell, P.: Groupes à dualité de Poincaré de dimension 2. C.R. Acad. Sci. Paris *295* (1982), Série I, 417–418
[11] Eckmann, B.; Linnell, P.: Poincaré duality groups of dimension two, II. Comment. Math. Helv. *58* (1983), 111–114
[12] Eckmann, B.; Müller, H.: Poincaré duality groups of dimension two. Comment. Math. Helv. *55* (1980), 510–520
[13] Montgomery, S.: Left and right inverses in group algebras. Bull. Amer. Math. Soc. *75* (1969), 539–540
[14] Müller, H.: Decomposition theorems for group pairs. Math. Zeitschrift *176* (1981), 223–246
[15] Stallings, J. R.: On torsion-free groups with infinitely many ends. Ann. of Math. *88* (1968), 312–334
[16] Stallings, J. R.: Group theory and three-dimensional manifolds. Yale Math. Monographs *4*, Yale Univ. Press 1971
[17] Strebel, R.: A remark on subgroups of infinite index in Poincaré duality groups. Comment. Math. Helv. *52* (1977), 317–324
[18] Swan, R. G.: Groups of cohomological dimension one. J. Algebra *12* (1969), 585–601
[19] Swarup, G. A.: Relative version of a theorem of Stallings. J. of Pure and Applied Algebra *11* (1977), 75–82

Continuous solutions of linear equations –
An old problem, its history, and its solution

Expo. Math. 9 (1991), 351–365

Dedicated to the memory of Frank Adams

1. The problem $P[n,r]$

1.1. Our starting point is very elementary. We consider a system of $r < n$ linear equations in n unknowns $x_1,\ldots,x_n$ with real coefficients

$$\sum_{k=1}^{n} a_{jk}x_k = 0, \quad j = 1,\ldots,r. \tag{1}$$

It always has solutions $(x_1,\ldots,x_n) = x \in \mathbb{R}^n$ beyond the trivial one, $x = 0$. Things become quite different if one asks for a solution depending continuously on the coefficients a_{jk}, defined and non-trivial for all those values of the $a_{11},\ldots,a_{rn}$ for which the $r \times n$-matrix (a_{jk}) has maximum rank r; that is, we look for a system of continuous real functions $x_k = f_k(a_{11},\ldots,a_{rn})$, $k = 1,\ldots,n$, defined and without common zeros for all admissible values of the a_{jk} and satisfying the identities

$$\sum_{k=1}^{n} a_{jk}f_k(a_{11},\ldots,a_{rn}) = 0, \quad j = 1,\ldots,r. \tag{2}$$

We will give below examples of pairs of integers r, n where such solutions exist, and pairs where they do not. Non-existence can, of course, be given a positive meaning: namely, that in these cases a system of functions f_k satisfying the above identities (2) must have common zeros (this may be interesting already if the functions are polynomials, etc.).

We denote by $P[n,r]$ our problem "Does there exist for the pair $r < n$ a continuous solution of the system of linear equations (1)?" It will eventually turn out that the answers is "yes" for very few pairs $r < n$ only; and that in these cases the solution always has an interesting significance.

1.2.   Example $r = 1$.

If $r = 1$ and n even $= 2m$ a solution of $P[n, r]$ can be given explicitly: Writing $\alpha_k = a_{1k}$, the system (1) becomes

$$\alpha_1 x_1 + \ldots + \alpha_n x_n = 0.$$

A continuous (linear!) solution is given by

$$x_1 = \alpha_2, \; x_2 = -\alpha_1, \ldots, x_{2m-1} = \alpha_{2m}, \; x_{2m} = -\alpha_{2m-1}.$$

This procedure obviously fails if n is odd. More generally, a *linear* solution $x_k = \sum_{\ell=1}^{n} b_{k\ell} \alpha_\ell, \; k = 1, \ldots, n$ cannot exist: The identity

$$\sum_{k=1}^{n} \alpha_k x_k = \sum_{k,\ell=1}^{n} \alpha_k b_{k\ell} \alpha_\ell = 0$$

means that $(b_{k\ell})$ is skew-symmetric and thus has determinant 0; therefore the x_k have a common zero (for some $(\alpha_1, \ldots, \alpha_n) = a \neq 0$). But even if we allow for the x_k arbitrary continuous functions there is no solution for odd n. This is easily seen, modulo a topological result, from the following geometrical interpretation.

We use vector notation in $\mathbb{R}^n$ and the scalar product $\langle x, y \rangle = \sum_{k=1}^{n} x_k y_k$. The continuous vector function $x(a)$ satisfies $\langle a, x(a) \rangle = 0$ for all $a \neq 0$, in particular for $|a| = 1$; i.e., $x(a)$ is a continuous tangent vector field $\neq 0$ on the unit sphere $|a| = 1$ in $\mathbb{R}^n$. For odd n, i.e. on the even-dimensional sphere S^{n-1} such a field cannot exist. There are many proofs of this old result of algebraic topology. We recall a simple one, using the concept of (topological) degree: If $x(a)$ is a non-zero tangent vector field on S^{n-1}, we consider the great circle determined by a and $x(a)$ and move the point a along the half-circle, in direction of $x(a)$, to the point $-a$. This defines a homotopy from the identity map of S^{n-1} to the antipodal map. The former has degree $+1$, the latter degree $(-1)^n$; since the degree is invariant under homotopy it follows that $1 = (-1)^n$, i.e., n is even.

1.3.   Examples $n = 2$, $n = 3$ and 7

We use vector notation in $\mathbb{R}^n$ and write a and b for the two row-vectors of the $2 \times n$-matrix (a_{jk}), $n = 3$ or 7. We first deal with $n = 3$. A solution of $P[2, 3]$ is a vector function $x(a, b)$ with $\langle x, a \rangle = \langle x, b \rangle = 0$, and $x \neq 0$ for all linearly independent pairs

$a, b \in \mathbb{R}^3$. Such a (bilinear!) function is well-known: namely, the vector product $x = a \times b$; it is defined for all $a, b \in \mathbb{R}^3$ and satisfies $\langle x, a \rangle = \langle x, b \rangle = 0$. Moreover

$$|x|^2 = \text{determinant} \begin{vmatrix} \langle a, a \rangle & \langle a, b \rangle \\ \langle b, a \rangle & \langle b, b \rangle \end{vmatrix} = |a|^2 |b|^2 - \langle a, b \rangle^2,$$

and thus $x \neq 0$ if and only if a, b are linearly independent.

A bilinear "vector product" with the same properties also exists in $\mathbb{R}^7$, thus giving a (bilinear) solution of $P[7, 2]$. It is related to the Cayley number multiplication in $\mathbb{R}^8$ in exactly the same manner the elementary vector product $a \times b$ in $\mathbb{R}^3$ is related to the quaternion multiplication in $\mathbb{R}^4$. We describe now these relation in detail; they will play a decisive role later on.

1.3.1. Quaternions

We define a product $A \cdot B$ for vectors $A, B \in \mathbb{R}^4$ as follows. We write $A \in \mathbb{R}^4 = \mathbb{R} \oplus \mathbb{R}^3$ as $A = \alpha + a$, $\alpha \in \mathbb{R}$, $a \in \mathbb{R}^3$, $B = \beta + b$, and put

$$A \cdot B = \alpha\beta - \langle a, b \rangle + \alpha b + \beta a + a \times b. \tag{3}$$

Using in $\mathbb{R}^4$ the scalar product $\langle A, B \rangle = \alpha\beta + \langle a, b \rangle$ we obtain, in terms of an orthonormal basis consisting of $1 \in \mathbb{R}$, $i, j, k \in \mathbb{R}^3$ the multiplication table $i^2 = j^2 = k^2 = -1$, $i \cdot j = -j \cdot i = k$, $j \cdot k = -k \cdot j = i$, $k \cdot i = -i \cdot k = j$, and 1 is a two-sided unit. This is just the usual multiplication table of quaternions, turning $\mathbb{R}^4$ into an associative algebra with unit. One calls α the real part, a the imaginary part of A, and $\overline{A} = \alpha - a$ the conjugate of A. From (3) or from the multiplication table one notes that $A \cdot B = \overline{B} \cdot \overline{A}$, and that $A \cdot \overline{A} = \alpha^2 + |a|^2 = |A|^2$. The length of $A \cdot B$ is given by $|A \cdot B|^2 = A \cdot B \cdot \overline{A \cdot B} = (A \cdot B) \cdot (\overline{B} \cdot \overline{A}) = A \cdot (B \cdot \overline{B}) \cdot \overline{A} = A \cdot \overline{A} \cdot B \cdot \overline{B}$, whence

$$|A \cdot B|^2 = |A|^2 \cdot |B|^2 \tag{4}$$

known as the "norm product rule".

There is another way of getting (4) from (3):

$$|A \cdot B|^2 = \alpha^2\beta^2 - 2\alpha\beta\langle a, b \rangle + \langle a, b \rangle^2 + \alpha^2|b|^2 + \beta^2|a|^2 + |a \times b|^2 + 2\alpha\beta\langle a, b \rangle;$$

here we have used that $\langle a \times b, a \rangle = \langle a \times b, b \rangle = 0$, and if we further use $|a \times b|^2 = |a|^2|b|^2 - \langle a, b \rangle^2$ there results

$$|A \cdot B|^2 = \alpha^2\beta^2 + \alpha^2|b|^2 + \beta^2|a|^2 + |a|^2|b|^2$$
$$= (\alpha^2 + |a|^2)(\beta^2 + |b|^2) = |A|^2|B|^2.$$

Conversely, if the quaternion algebra in $\mathbb{R}^4$ is defined by the usual basis multiplication table (see above), we can *define* $a \times b$ as the imaginary part of $a \cdot b$, for two purely imaginary quaternions $\in \mathbb{R}^3$:

$$a \times b = a \cdot b + \langle a, b \rangle.$$

The properties of $a \times b$ then follow: Associativity yields

$$(a \cdot b) \cdot b = a \cdot (b \cdot b) = -a \cdot |b|^2$$

which is imaginary; but $(a \cdot b) \cdot b = (-\langle a, b \rangle + a \times b) \cdot b = -\langle a \times b, b \rangle + $ imaginary terms, whence $\langle a \times b, b \rangle = 0$. Similarly $\langle a \times b, a \rangle = 0$. But then, as we have seen above, the norm product rule for quaternions is equivalent to

$$|a \times b|^2 = |a|^2|b|^2 - \langle a, b \rangle^2.$$

1.3.2. Cayley numbers

We recall that in $\mathbb{R}^8$ there exists a bilinear product having all the properties which we have just used for the quaternions: There is a two-sided unit 1; the norm-product rule holds; although the product is not associative, the law $A \cdot (B \cdot B) = (A \cdot B) \cdot B$ (the "alternative law") holds. We do not describe the multiplication table for these "Cayley numbers" or "octaves", but mention that it is most easily obtained by considering vectors of $\mathbb{R}^8$ as pairs of quaternions.

Decomposing $\mathbb{R}^8 = \mathbb{R} \oplus \mathbb{R}^7$, where $1 \in \mathbb{R}$ is the unit, we write A as $\alpha + a$, $\alpha \in \mathbb{R}$, $a \in \mathbb{R}^7$. Then for $a, b \in \mathbb{R}^7$ ($\alpha = \beta = 0$) the definition

$$a \times b = a \cdot b + \langle a, b \rangle$$

yields a bilinear vector product in $\mathbb{R}^7$, with the required properties $\langle a \times b, a \rangle = \langle a \times b, b \rangle = 0$ and $|a \times b|^2 = |a|^2|b|^2 - \langle a, b \rangle^2$. We thus have a (bilinear) solution of $P[7, 2]$.

1.4.   Continuous vector products

Passing from $r = 2$ to any $r < n$, and from bilinearity or multi-linearity to continuity one arrives at the concept of a "continuous vector product of r vectors in $\mathbb{R}^n$". It is a vector function $x(a_1, \ldots, a_r) \in \mathbb{R}^n$ of r vectors $\in \mathbb{R}^n$ satisfying
(i) x is a continuous function of $a_1, \ldots, a_r$.
(ii) $\langle x(a_1, \ldots, a_r), a_j \rangle = 0$, $j = 1, \ldots, r$.

(iii) $|x(a_1,\ldots,a_r)|^2 = $ determinant of $\langle a_j, a_\ell \rangle$, $j, \ell = 1,\ldots,r$.

Condition (iii) implies that $x = 0$ if and only if the vectors $a_1,\ldots,a_r$ are linearly dependent.

This generalized concept of vector product allows for an equivalent formulation of $P[n,r]$: *"Does there exist in $\mathbb{R}^n$ a continuous vector product of r vectors?"* Indeed, identify the r vectors $a_1,\ldots,a_r$ with the row vectors of the $r \times n$-matrix (a_{jk}) in (1); and note that a solution of $P[n,r]$ can be renormalized so that (iii) holds for linearly independent $a_1,\ldots,a_r$, and then extended to *all* matrices (a_{jk}) by $x = 0$ if the vectors are linearly dependent.

1.5.   Example $r = n - 1$, $n \geq 2$

$P[n, n-1]$ has an obvious (multilinear) solution: the components x_k of $x \in \mathbb{R}^n$ are the $(n-1) \times (n-1)$-determinants of the $(n-1) \times n$-matrix (a_{jk}), in appropriate order and with appropriate signs; of course, $x = 0$ if (a_{jk}) has rank $< n - 1$.

In terms of vector products, x above is a (multilinear) vector product of $n-1$ vectors in $\mathbb{R}^n$. The normalization (iii) is a well-known determinant-identity.

1.6.   Example $r = 3$, $n = 8$

Quite generally, to define a continuous vector product of r vectors in $\mathbb{R}^n$ it suffices to do so for orthonormal vectors $a_1,\ldots,a_r$; then one extends to arbitrary linearly independent sets of vectors through Gram-Schmidt orthonormalization and finally imposes the normalization (iii).

So let A, B, C be orthonormal $\in \mathbb{R}^8 = \mathbb{R} \oplus \mathbb{R}^7$, regarded as Cayley numbers, and put

$$x(A, B, C) = A \cdot (A^{-1}B \times A^{-1}C).$$

Here $A^{-1} = \frac{\overline{A}}{|A|^2}$ is the inverse of A, and $\times$ is the vector product of 2 vectors in $\mathbb{R}^7$ (note that $A^{-1}A = 1$, $A^{-1}B$, $A^{-1}C$ are still orthonormal, and hence $A^{-1}B$, $A^{-1}C$ are in $\mathbb{R}^7$).

This is just a simple way to give a solution of $P[8,3]$. More explicitly, in terms of Cayley numbers, a vector product of arbitrary $A, B, C \in \mathbb{R}^8$ is given (cf. [Z]) by

$$x(A, B, C) = -A \cdot (\overline{B} \cdot C) + A\langle B, C \rangle - B\langle C, A \rangle + C\langle A, B \rangle.$$

It is trilinear; (ii) and (iii) can be verified using properties of the Cayley multiplication (caution: it is not associative).

1.7. Summary

We have obtained so far the following list (L) of pairs $r < n$ where an explicit solution of $P[n,r]$ exists.

(L)

$r = 1,$	$n =$ even
$r = n - 1,$	$n \geq 2$
$r = 2,$	$n = 7$
$r = 3,$	$n = 8$

Surprisingly enough, *these turn out to be the only cases of pairs n,r where $P[n,r]$ has a solution*; or equivalently, where there exists in $\mathbb{R}^n$ a continuous vector product of r vectors. The proof will be given in Section 3. It is based on a famous theorem of Frank Adams, combined with arguments of the type (3) and some special computations of algebraic topology.

In all the above cases (L) there exist, as we have seen *multilinear* solutions. We come back to that phenomenon in 5.3. At this stage, however, let us just mention the following fact concerning *bilinear* vector products of *two* vectors in $\mathbb{R}^n$: *They only exist in $\mathbb{R}^3$ and $\mathbb{R}^7$.* Indeed, if such a vector product exists in $\mathbb{R}^n$ then there is in $\mathbb{R}^{n+1}$ a bilinear multiplication with unit and norm product rule; by a classical result of Hurwitz [H] this is possible for $n = 1, 2, 4, 8$ only.

2. Some history

2.1. In 1935 Ważewski [Wa] considered the following problem arising from questions in operator theory: Let U be a ball in $\mathbb{R}^n$ and $(a_{jk}(n))$ a real $r \times n$-matrix, $r < n$, depending continuously on u and of rank r for all $u \in U$; can one add $n - r$ rows, depending continuously on $u \in U$, to $(a_{jk}(n))$ such that the resulting real $n \times n$-matrix is non-singular for all $u \in U$? He proved that this is always possible. Equivalently, his result can be stated in terms of vectors of $\mathbb{R}^n$: There exists a continuous vector function $x(u)$, different from 0 and orthogonal to the row vectors $a_1(u), \ldots, a_r(u)$ of $(a_{jk}(u))$ for all $u \in U$.

2.2. In 1939 and the following years the concepts of "fibration" and "homotopy lifting" were being developed, independently by several authors including myself (see [E], [EF], [HS]). It occurred to me that Ważewski's problem belongs precisely to that framework, and that the same is true for the (more interesting) problem $P[n,r]$. To formulate things let us, for simplicity and without loss of generality, consider only orthogonal matrices (a_{jk}), or orthonormal systems of vectors respectively.

Let $V_{n,r}$ denote the space of all $n \times r$-matrices with orthonormal rows; i.e., of all orthonormal r-frames in $\mathbb{R}^n$. It is a compact manifold (used by Stiefel in his thesis, and called Stiefel manifold). We allow $r = n$ but then require that the determinant be $+1$. For $r < n$ let $p : V_{n,r+1} \to V_{n,r}$ be the map defined by omitting the last row of an $(r+1) \times n$-matrix. The matrix function in Wažewski's problem is a map $f : U \to V_{n,r}$ and one wants a map $g : U \to V_{n,r+1}$ such that $pg = f$ – in the terminology used later a "lifting" g of f. In the problem $P[n,r]$ one considers the identity map $f : V_{n,r} \to V_{n,r}$ and looks for a lifting $g : V_{n,r} \to V_{n,r+1}$, $pg = f$.

The map $p : V_{n,r+1} \to V_{n,r}$ is one of the typical fibrations we were interested in at that time, for geometrical reasons (An important case is $r = 1 : V_{n,1}$ is the sphere S^{n-1} in $\mathbb{R}^n$, and $V_{n,2}$ the space of all its tangent unit vectors; a lifting of the identity is a tangent vector field. Another important class is the projection of a Lie group onto one of its homogeneous spaces). In these cases and similar ones one could check directly that the possibility of lifting a map is a *homotopy property*: If f can be lifted so can any map homotopic to f (and the whole homotopy can be lifted). Later on this was taken as a definition: A map of spaces $p : X \to Y$ is a *fibration* if for an arbitrary space U and a map $f : U \to Y$ which can be lifted to $g : U \to X$ any homotopy of f can be lifted (in the terminology of Serre [S] this was actually required for special spaces U only).

The name "fibration" was chosen from the outset to refer to an intuitive aspect present in all the original examples: The inverse images $p^{-1}(y)$ are all homeomorphic, and called "fibers". If a base-point y_0 is chosen in Y then $p^{-1}(y_0) = F$ is the "typical" fiber.

2.3. We return to our questions. In Wažewski's problem U is a contractible space; hence $f : U \to V_{n,r}$ is homotopic to a constant map, which can trivially be lifted. Thus f can be lifted.

In the problem $P[n,r]$ we want to lift the identity map $f : V_{n,r} \to V_{n,r}$ to $V_{n,r+1}$. In a fibration $p : X \to Y$ a lifting of the identity of Y to X is called a *cross-section* of p. As can be seen from the simplest examples it is important, and in general not easy, to find out whether a fibration admits a cross-section or not (and if not, whether it exists partially, and what are the "singularities").

A strong tool, especially suitable for homotopy problems, were the homotopy groups $\pi_n(X)$, $n = 1, 2, 3, \ldots$. They had already been defined in 1932 by Čech [C], then rediscovered and actually used in a decisive way by Hurewicz [H] in 1935. They generalize the fundamental group of $X (n = 1)$, and their definition, like that of $\pi_1(X)$, uses a base-point $x_0 \in X$; their elements are base-point preserving homotopy classes of sphere-maps $S^m \to X$. Thus only the path-component of X is relevant, and there the structure of $\pi_m(X)$ does not depend on the choice of x_0. We only consider here connected fibrations $p : X \to Y$ with connected fiber $F = p^{-1}(y_0) \subset X$.

In our context one of the results in [E] tells that if $p : X \to Y$ has a cross-section then $\pi_m(X)$ is the direct sum $\pi_m(Y) \oplus \pi_m(F)$. The fiber of $p : V_{n,r+1} \to V_{n,r}$ is the sphere S^{n-r-1}. We assume $r \geq 2$ and $r \leq n - 2$ (the cases $r = 1$ and $r = n - 1$ are easy and have been settled above). Thus a solution of $P[n, r]$ implies

$$\pi_m(V_{n,r+1}) = \pi_m(V_{n,r}) \oplus \pi_m(S^{n-r-1}). \tag{5}$$

In the early years some of the homotopy groups of the $V_{n,r}$ have been computed in [E] and [E$_1$] using the exact homotopy sequence (established, in disguised form, in [E]). They were applied in [E$_1$] to show that one is in contradiction to (5) in the cases $n - r =$ even and $n - r = 3$ or 7:

$P[n, r]$ has a solution (possibly) only if $n - r$ is odd and $\neq 3, 7$.

2.4.   As time progressed, one knew more and more about the $\pi_m(V_{n,r})$ and could eliminate further pairs $r < n$. However one did not succeed in reducing the possibilities for a solution of $P[n, r]$ down to the list (L). So the homotopy group method, although successful in the beginning, was gradually given up.

Then, unexpectedly, came a completely new aspect: Adams' theorem.

3.   Deus ex machina: Adams' theorem

3.1.   A preliminary remark: Reduction

If $P[n, r]$ has a solution so has $P[n - 1, r - 1]$, $r \geq 2$. Indeed, in terms of a vector product of r vectors in $\mathbb{R}^n$, $x(a_1, \ldots, a_r)$ we choose a unit vector $b \in \mathbb{R}^n$ and fix $a_r = b$; for arbitrary $a_1, \ldots, a_{r-1}$ in $\mathbb{R}^{n-1}$ orthogonal to b

$$x'(a_1, \ldots, a_{r-1}) = x(a_1, \ldots, a_{r-1}, b)$$

has all the required properties of a vector product of $r - 1$ vectors in $\mathbb{R}^{n-1}$.

Thus the answer depends essentially on the difference $n - r$; if $P[n, r]$ has a solution so has $P[n - r - 1, 1]$ which is the case (if and) only if $n - r - 1$ is even. Thus a necessary condition is that $n - r$ be odd.

Since $r = 1$ has been settled, we now consider $r = 2$.

3.2.   Let $a \times b$ be a continuous vector product of two vectors in $\mathbb{R}^n$, $n \geq 3$, in the sense of Section 1.4. As we did in the case of quaternions and Cayley numbers, we decompose $\mathbb{R}^{n+1}$ into $\mathbb{R} \oplus \mathbb{R}^n$ and write $A \in \mathbb{R}^{n+1}$ as $A = \alpha + a$, $\alpha \in \mathbb{R}$, $a \in \mathbb{R}^n$, we further use in $\mathbb{R}^{n+1}$ the scalar product $\langle A, B \rangle = \alpha\beta + \langle a, b \rangle$, whence the "norm" $|A|^2 = \alpha^2 + |a|^2$. Then the formula (3)

$$A \cdot B = \alpha\beta - \langle a, b \rangle + \alpha b + \beta a + a \times b$$

defines a continuous product of vectors $A, B \in \mathbb{R}^{n+1}$, and the arguments of 1.3.1 show that the norm product rule $|A \cdot B|^2 = |A|^2 \cdot |B|^2$ holds (and clearly $A = 1 \in \mathbb{R}$ is a two-sided unit):

If there exists in $\mathbb{R}^n$ a continuous vector product of two vectors then $\mathbb{R}^{n+1}$ admits a continuous multiplication with norm product rule and two-sided unit.

3.3.   It is here that Adams' theorem comes into play.

Theorem A   (Adams 1960 [A]). *In $\mathbb{R}^m$ there exists a continuous multiplication with norm product rule and two-sided unit (if and) only if $m = 1, 2, 4, 8$.*

This leaves for a continuous vector product $a \times b$ of two vectors in $\mathbb{R}^n$ the only cases $n = 3$ and $n = 7$. Therefore a solution of $P[n, r]$ can only exist if (apart from trivial case $n - r = 1$) $n - r = 5$.

$n = 7, r = 2$:   The Cayley number vector product, cf. 1.3.2.

$n = 8, r = 3$:   The Cayley number vector product of 3 vectors as described in 1.6.

$n = 9, r = 4$?   Our claim that the list (L) is complete will be proved if we show that $P[9, 4]$ does not have a solution; in other words that $p : V_{9,5} \rightarrow V_{9,4}$ does not admit a cross-section. This can easily be done using the cohomology ring structure of these Stiefel manifolds (see [W]).

In order to give some indication of the proof of Theorem A we have to make use of certain techniques of algebraic topology — which we do without detailed explanation. The trivial case $m = 1$ is always omitted.

3.4.   Adams' original 1960 proof was a real "tour de force", using properties of the cohomology structure of special spaces and primary and secondary cohomology operations. Much simpler proofs were deviced later (see [AA], [E$_2$]), replacing ordinary cohomology by complex K-theory and the Atiyah-Hirzebruch integrality theorem. In either approach it is made clear that the algebraic structure of a topological space — even if that space is a relatively simple cell-complex — has to obey strong restrictions.

A first step, of geometrical nature, is to reduce the claim to a (more general) statement concerning the cohomology ring of a cell-complex. Namely, a multiplication in $\mathbb{R}^m$, $m > 1$, of the required type can be used to construct a "pseudo-projective plane" X of dimension $2m$; i.e., of dimension 2 with respect to the "algebra" $\mathbb{R}^m$: One starts with the affine $\mathbb{R}^m$-plane and adds an $\mathbb{R}^m$-line (in fact a sphere S^m) at infinity, in analogy to the well-known procedure constructing $\mathbb{C}P^2$ from $\mathbb{R}^2 = \mathbb{C}$. The cohomology ring $H^*(X)$ with integer coefficients is then, again in analogy with $H^*(\mathbb{C}P^2)$, the truncated polynomial ring

$$H^*(X) = \mathbb{Z}[a]/(a^3),$$

the algebra generator a lying in $H^m(X)$. Theorem A now follows from

Theorem A'. *A finite cell-complex X with cohomology ring $H^*(X) = \mathbb{Z}[a]/(a^3)$, $a \in H_{n+1}(X)$, exists only for $n+1 = 2, 4, 8$.*

It is clear a priori that $n + 1$ must be even $= 2\ell$, since $H^*(X)$ is anti-commutative as a graded algebra. So the claim is $\ell = 1, 2, 4$. As an appendix (Section 5) we give a sketch of the "simple" K-theoretic proof, using our version in [E$_2$]. But first (Section 4) we emphasize the importance of Theorem A by a series of interesting corollaries.

4.   Other corollaries of Theorem A

4.1.   Division algebras

$\mathbb{R}^{n+1}$ is turned into a division algebra, if a bilinear multiplication $A \cdot B$ is given which has no zero divisors. Note that the norm product rule is not required (but if it holds, there are, of course, no zero divisors). Without loss of generality, we can assume a bilinear multiplication to have a two-sided unit of length 1. If $A \cdot B$ is replaced by $A \cdot B|A||B|/|A \cdot B|$ for $A, B \neq 0$ the new multiplication will satisfy the norm product rule. It will, however, have lost bilinearity, but remain continuous. Thus Theorem A implies

Corollary 1. $\mathbb{R}^{n+1}$ *admits a division algebra structure (if and) only if $n+1 = 1, 2, 4$ or 8.*

4.2.   Parallelizability of spheres

The unit sphere $S^n \subset \mathbb{R}^{n+1}$ is said to be parallelizable, if there exist n tangent vector fields on S^n which are linearly independent at every point $X \in S^n$ ($X =$ unit vector $\in \mathbb{R}^{n+1}$). If this is the case, one can replace these fields by orthonormalized ones. The components of X and those of the n vector fields, with respect to an orthonormal basis of $\mathbb{R}^{n+1}$, taken as rows form an orthogonal $(n+1) \times (n+1)$-matrix $C = (c_{ik}(X))$, the elements $c_{ik}(X)$ being continuous functions of X; we take the components of X as first row $c_{1k} = x_k$. Let y_k, $k = 1, \ldots, n+1$ be the components of $Y \in \mathbb{R}^{n+1}$ and define $Z \in \mathbb{R}^{n+1}$ by its components z_k,

$$z_k = \sum_{j=1}^{n+1} c_{jk}(X) y_j, \qquad k = 1, \ldots, n+1.$$

Writing $X \cdot Y$ for Z, this is a continuous multiplication in $\mathbb{R}^{n+1}$ (for $|X| = 1$, Y arbitrary) with $|X \cdot Y| = |Y|$. We extend it to all $X \in \mathbb{R}^{n+1}$ by $X \cdot Y = |X|((X/|X|) \cdot Y)$

for $X \neq 0$, and $0 \cdot Y = 0$. Then the norm product rule holds, and the vector $E = (1,0,\ldots,0)$, is a right unit. By a suitable choice of the basis one may assume $C(E)$ to be the unit matrix, i.e., E is also a left unit. Thus Theorem A implies

Corollary 2. S^n *is parallelizable only for* $n = 1, 3$, *or* 7.

Remarks.   (a) In these three cases, the classical multiplications in $\mathbb{R}^2$, $\mathbb{R}^4$, and $\mathbb{R}^8$ actually yield a parallelization of S^n, by vector fields which are linear in X. (b) The multiplication $Z = X \cdot Y$ above is linear in Y and continuous in X. Thus Theorem A has not been used in its full generality. Other proofs of Corollary 2, not depending on Theorem A, have been given earlier (see, for example, [K]).

4.3.   Almost-complex structures on spheres

Let M be a complex-analytic manifold; its tangent bundle admits a complex vector bundle structure: Multiplication by $\sqrt{-1}$ of complex vector components in an admissible local complex coordinate system defines, at each point $p \in M$, a linear transformation $J(p)$ of the real tangent space M_p at p, with $J(p)^2 = -\text{identity}$; and $J(p)$ does not depend upon the complex coordinate system used. In fact the field $J(p)$ characterizes the complex-analytic structure given on M. If on a real differentiable manifold M (of even dimension) such a field $J(p)$ of linear transformations with $J(p)^2 = -\text{identity}$ is given, independently of any complex-analytic structure, M is called an almost-complex manifold (and $J(p)$ an almost-complex structure on M).

Corollary 3. S^m *admits an almost-complex structure only for* $m = 2$ *or* 6.

Proof.   If an almost-complex structure J is given on $S^{n-1} \subset \mathbb{R}^n$ a continuous vector product $a \times b$ can be defined in $\mathbb{R}^n$: for $a \in S^{n-1}$ and a unit tangent vector b at a (that is, for a and $b \in \mathbb{R}^n$ with $|a| = |b| = 1$, $\langle a, b \rangle = 0$), we consider the oriented tangent two-plane determined by b and $J(a)b$ and choose $a \times b$ to be the unit vector orthogonal to b in that plane and corresponding to the orientation; the product $a \times b$ is then easily extended to all a and $b \in \mathbb{R}^n$ so as to fulfill the properties of a vector product. Thus $n + 1$ must be $= 2, 4$ or 8, that is, $n = 3$ or 7, and hence the dimension of the sphere is 2 or 6. In these two dimensions, almost-complex structures actually exist.

4.4.   Complex vector products

The definition of a complex continuous vector product is, of course, entirely analogous to the real case; it refers to a positive-definite Hermitean scalar product in $\mathbb{C}^n$.

From a continuous vector product of two vectors in $\mathbb{C}^n$, $n \geq 3$, one obtains by a construction similar to (3) a continuous multiplication in $\mathbb{C}^{n+1}$ with two-sided unit and norm product rule (with respect to the Hermitean length of vectors). This can be interpreted as a multiplication in $\mathbb{R}^{2n+2}$ with unit and norm product rule. Hence $2n + 2 = 2, 4, 8$, that is, $n = 3$:

Corollary 4. *A continuous vector product of two vectors in $\mathbb{C}^n$ exists (if and) only if $n = 3$.*

5.   Proof and Remarks
5.1.   Properties of $K(X)$

We consider the group $K(X)$ for an arbitrary finite polyhedron X; it is the Grothendieck group of stable complex vector bundles over X, or equivalently, the group of homotopy classes of maps of X into ΩU (the loop space of the infinite unitary group). We will use a few standard properties of the contravariant functor K, as follows.

The tensor product of complex vector bundles defines in $K(X)$ a commutative ring structure with unit. There exists a natural transformation ch, called the Chern character of complex vector bundles, of $K(X)$ into $H^*(X; \mathbb{Q})$, the cohomology algebra of X over the rationals $\mathbb{Q}$; it maps $K(X)$ into even-dimensional cohomology only

$$\text{ch } z = a_0 + a_2 + \ldots + a_{2j} + \ldots, \quad a_{2j} \in H^{2j}(X; \mathbb{Q})$$

for $z \in K(X)$, and it is a ring homomorphism. Furthermore there exist natural transformations $\psi_k : K(X) \to K(X)$, the Adams operations, one for each $k = 1, 2, 3, \ldots$ which are related to the Chern character by

$$\text{ch } \psi_k z = a_0 + k a_2 + \ldots + k^j a_{2j} + \ldots$$

where $\text{ch } z = a_0 + a_2 + \ldots + a_{2j} + \ldots$ as above with $a_{2j} \in H^{2j}(X; \mathbb{Q})$. The ψ_k are defined by means of exterior powers Λ_q of complex-vector bundles; in particular $\psi_2 z = z^2 - 2\Lambda_2 z$.

If the cell-complex X is without torsion, i.e., if $H^*(X)$ over the integers has no torsion, $\text{ch} : K(X) \to H^*(X; \mathbb{Q}) = H^*(X) \otimes \mathbb{Q}$ has kernel 0. Moreover, in $\text{ch } z = a_0 + a_2 + \ldots + a_{2j} + \ldots$ the first nonvanishing term, say, a_{2i} is integral, that is, it lies in the subgroup $H^{2i}(X) \subset H^{2i}(X; \mathbb{Q})$ where the integral cohomology is identified with its natural image in $H^{2i}(X; \mathbb{Q})$. Any given element of $H^{2i}(X)$, $i = 0, 1, 2, \ldots$ appears in this way as the first nonvanishing component of $\text{ch } z$ for some $z \in K(X)$ (these last statements constitute the Atiyah-Hirzebruch integrality theorem in its simplest version).

5.2. Proof of Theorem A′

In the case of Theorem A′ we consider a cell-complex X with $H^*(X) = \mathbb{Z}[a]/(a^3)$, $H^*(X;\mathbb{Q}) = \mathbb{Q}[a]/(a^3)$, $a \in H^{2\ell}(X)$, $2\ell = n+1$. There exists an element $z \in K(X)$ with

$$\mathrm{ch}\, z = a + \lambda a^2, \quad \lambda \in \mathbb{Q}.$$

For this z the properties listed in 3.2 yield the following relations:

$$\mathrm{ch}\, z^2 = a^2$$
$$\mathrm{ch}\, \psi_k z = k^\ell a + \lambda k^{2\ell} a^2$$
$$\mathrm{ch}(\psi_k z - k^\ell z) = \lambda k^\ell (k^\ell - 1) a^2 = \lambda k^\ell (k^\ell - 1)\, \mathrm{ch}\, z^2$$

with $\lambda k^\ell (k^\ell - 1) \in \mathbb{Z}$. Since the kernel of ch is 0, $\psi_k z = k^\ell z + \lambda k^\ell (k^\ell - 1) z^2$. In particular, for $k = 2$, we have

$$\psi_2 z = 2^\ell z + \lambda 2^\ell (2^\ell - 1) z^2 = z^2 - 2\Lambda_2 z$$

therefore $\lambda 2^\ell (2^\ell - 1)$ is odd, and thus $\lambda \in \mathbb{Q}$ must be of the form

$$\lambda = \frac{\mathrm{odd}}{\mathrm{odd} \cdot 2^\ell}.$$

For an arbitrary odd integer $k \geq 1$, $(\mathrm{odd}/\mathrm{odd} \cdot 2\ell) \cdot k^\ell (k^\ell - 1)$ is an integer, which means that 2^ℓ divides k^ℓ 1.

Thus from the existence of a cell-complex X satisfying the assumption of Theorem A′, $n+1 = 2\ell$, we conclude that 2^ℓ divides $k^\ell - 1$ for all odd integers $k \geq 1$.

From this purely numerical result one easily draws very precise conclusions. First we note that for $\ell = 1$ the property holds, and we therefore assume $\ell > 1$. Taking $k = 2^{\ell-1} + 1$, the fact that 2^ℓ divides $(2^{\ell-1} + 1)^\ell - 1 = 2^\ell \cdot$ integer $+ \ell 2^{\ell-1}$ implies that ℓ is even, $\ell = 2m$. Taking $k = 2^m + 1$, the fact that 2^{2m} divides $(2^m + 1)^{2m} - 1 = 2^{2m} \cdot$ integer $+ 2m2^m$ implies that 2^{2m} divides $m2^{m+1}$, that is, 2^{m-1} divides m. This is so for $m = 1$ and 2 only, whence $\ell = 2$ or 4. Together with $\ell = 1$, we thus have $n + 1 = 2\ell = 2, 4$ or 8.

5.3. Remarks

We finally remark that Theorem A can be given the following interpretation: With regard to the existence of multiplications in $\mathbb{R}^{n+1}$ with unit and with norm product rule, the *dimensions which are exceptional* (in the sense that such a multiplication exists) *in the continuous case are also exceptional in the bilinear case* (the latter is solved by Hurwitz' classical theorem, that is, by elementary algebra). The proof reproduced above, although it is very simple, does not explain directly why this is so. The situation is similar in various other problems of that type; for example, in the vector product problem (continuous solutions of linear equations) from which we started; or in the problem of finding the maximum number of independent tangent vector fields on a sphere, also solved by Adams [A₁], and in the application thereof to linear families of nonsingular matrices [ALP]. It would be most interesting to find an a priori reason why the existence of "continuous solutions" in all these cases implies the existence of "linear solutions" (or, in any case, simple algebraic solutions). If such a reduction principle from the continuous to the algebraic case could be established, it might clarify the exceptional character of certain low dimensions n of the real n-spaces.

References

[A]   J.F. Adams, On the non-existence of elements of Hopf invariant one, Ann. of Math. 72 (1960), 20-104.

[A₁]   J.F. Adams, Vector fields on spheres, Ann. of Math. 75 (1962), 603-632.

[AA]   J.F. Adams and M.F. Atiyah, K-theory and the Hopf invariant, Quart. J. Math. 17 (1966), 31-38.

[ALP]   J.F. Adams, P. Lax, and R.S. Phillips, On matrices whose real linear combinations are nonsingular, Proc. A.M.S. 16 (1965), 318-322.

[E]   B. Eckmann, Zur Homotopietheorie gefaserter Räume, Comm. Math. Helv. 14 (1941/42), 141-192.

[E₁]   B. Eckmann, Systeme von Richtungsfeldern in Sphären und stetige Lösungen komplexer linearer Gleichungen, Comm. Math. Helv. 15 (1942/43), 1-26.

[E₂]   B. Eckmann, Cohomologie et classes caractéristiques, Cours CIME 1966.

[E₃]   B. Eckmann, Continuous solutions of linear equations - some excepional dimensions in topology. Battelle Rencontres (Benjamin 1967), 516-526.

[EF]   Ch. Ehresmann und J. Feldbau, Sur les propriétés d'homotopie des espaces fibrés, C.R. Acad. Sci. Paris 211 (1941), 945-948.

[HS]   W. Hurewicz and N. Steenrod, Homotopy relations in fibre spaces, Proc. Nat. Acad. Sci. USA 27 (1941), 60-64.

[K]   M. Kervaire, Non-parallelizability of the n-sphere for $n > 7$, Proc. Nat. Acad. Sci. USA (1958), 280-283.

[Wa]   T. Ważewski, Sur les matrices dont les éléments sont des fonctions continues, Comp. Math. II (1935), 63-68.

[W]   G.W. Whitehead, Note on cross-sections in Stiefel manifolds, Comm. Math. Helv. 37 (1962), 239-240.

[Z]   P. Zvengrowski, A 3-fold vector product in $\mathbb{R}^8$, Comm. Math. Helv. 40 (1965/66), 149-152.

Mathematics: Questions and Answers

Am. Math. Mon. 102 (1995), 685–690

The International Congress of Mathematicians, which takes place every 4 years and which is being held this time in Zürich opens today. It brings together some 3000 mathematicians, active in research and university teaching, from all over the world. Not only in view of the (temporary) inundation of the city with this particular species of scientist but also at other times is the question often asked, what do mathematicians really do? In what follows I will try to give some small insight into the nature and processes of this science.

Fermat's Theorem

In June 1993 a sensational report went round the mathematical world; by electronic mail it reached even faraway universities, academies and colleges with lightning speed. The famous 350-year-old Fermat Theorem had been proved. To the surprise of most mathematicians this report was also published in many non-specialist media, thereby reaching a broad public. The *New York Times* devoted a front page article to it and Andrew Wiles (Princeton), who had announced the proof [1], along with those who had prepared the way, especially Gerhard Frey (Essen) and Kenneth Ribet (Berkeley), became as famous overnight as the stars of the arts and sport. The problem, unsolved for 350 years, seemed to exercise as strong a fascination for laymen as for specialists.

For once it was neither the powerful technico-scientific applications nor the attractive coloured computer pictures and graphics which excited a large public, but

---

The original text of this article, in German, appeared in the Swiss newspaper *Neue Zürcher Zeitung* on August 3, 1994, to mark the opening of the International Congress of Mathematicians. It was suggested to the author, Beno Eckmann, that an English version would be welcomed and would reach a wider public. The translation was undertaken by Peter Hilton.

[1] At the time of writing (8/3/94-translator) it appears to experts that there is a gap in the complicated chain of inferences constructed by A. Wiles. This does not imply false reasoning, but rather that the argument must be supplemented. The great achievement of Wiles is only marginally affected by this.
Added by translator (2/21/95) – It has now been announced by Andrew Wiles and R. L. Taylor (Cambridge), and verified by colleagues, that the gap has been filled.

the actual mathematical process – and this with an unprecedented intensity. It was therefore only to be expected that, over the ensuing days, we mathematicians were bombarded with questions from all sides. Did we take the opportunity to make people, near and far, more familiar with our science? For the questions, on the whole, went to the heart of the matter.

The basic underlying question in Fermat's Theorem should be explained for the sake of completeness. The equation $x^2 + y^2 = z^2$ has many solutions in positive integers, for example, $x = 3$, $y = 4$, $z = 5$. On the other hand, the equation $x^3 + y^3 = z^3$ has no solutions in integers, and Fermat asserted, in 1635, that he could prove that the equation $x^n + y^n = z^n$ for $n > 2$, has no solutions in positive integers. We may doubt whether he really had a proof. The problem seems simple and innocuous; it has aroused the interest of many amateur and professional mathematicians over the centuries, and many false proofs have been offered.

On the other hand the assertion has been proved for very many values of the exponent n; up to last year, this included all values of n up to 4,000,000. It is noteworthy that the very profound methods which were developed to do this have had a decisive influence on modern mathematics; the theory of the so-called *algebraic numbers*, from which so many general ideas stem, arose primarily from these efforts. And now we know, provided Wiles' arguments are found to be watertight, that the assertion is valid for *all n*.

All this leaves the layman somewhat perplexed so simple a statement and yet so difficult to prove? And do professional mathematicians occupy themselves with such things and get paid for doing so? The further questions, below, give us the opportunity, in this respect, to correct some misapprehensions.

An invisible part of our culture

What have we gained? What are the consequences for our world of the proved result?

Here we must in all honesty reply: we have gained nothing. That the theorem has been proved has no consequences, even for number theory itself. But – does one pose such a question in the face of a masterwork of art or an impressive achievement in sport? Mathematics is, like the arts, a part of our *cultural tradition*, and has always, in ancient and modern times, obtained its justification from this fact. But, in contrast to the arts and sport, mathematics has no general public. Its assertions, as we must recognize, are immediately accessible only to a small circle; and the newer, the deeper, the more abstract the result, the narrower the circle will be. Thus mathematics can scarcely rely on making a resounding splash in the media, apart from very exceptional cases such as Fermat's Theorem.

But, of course, that is just one side of the story. On the other side stand the innumerable applications of mathematics which one meets everywhere. Mathematics has become an indispensable tool in the technico-scientific world of today, whether it is concerned with various kinds of calculation, with physics, chemistry, biology, medicine, meteorology, telecommunications etc. Even in simpler matters do all of us, knowingly or unknowingly, apply mathematical reasoning, when we speak of probability, extrapolation, analysis and interpretation of graphs, coding, averages and such like.

One does not, however, reflect that all the mathematical concepts, methods and results which are applied are *abstractions*, which had to be thought up. And even the solution of apparently 'frivolous' and useless problems à la Fermat – and many others originating in simple, practical questions – demand the elaboration of theoretical structures of great generality. The universal applicability of mathematics, which, as a rule, is neither intended nor foreseeable, seems to depend on those conceptions; a few examples to illustrate this will be cited below.

The relationship between these two very different aspects of mathematics is not easily comprehended. The instrument we employ for recognizing, describing, understanding and expressing by means of theoretical construction is mathematics, its language, its mode of thought, its results; that is, a structure of thought which is abstract and which is not primarily erected for this purpose. The applications bear witness to the power of mathematics, but are not its real motivation. The springs of "mathematization" seem to be of a very different kind. If we try to describe them, we need words like curiosity, thirst for knowledge, the impulse towards play.

A game then, pretentious and difficult, as all good games should be? In a certain sense, yes. But one knows that ultimately it has significance and effect and that places the motivation close to that of the artist. And, as in the arts, the criteria of value and rightness are not easily made precise. They include intensity, beauty and unity of the expression, the opening of new horizons, and insights which stem from a profound struggle to understand the problem. Even this remains inevitably restricted to the circle of the 'initiated'. Thus is our art invisible to a wider public.

Mathematical proof

Why prove something which is known to be correct in 4,000,000 cases, and more besides? Wouldn't one regard this, in any other endeavour, itself as "proof"?

Here we must again go further back and, above all, insist that all those mathematical concepts, which are daily and hourly in action, find no place in the *real* world we observe. The apparently simplest things like a straight line, 3-dimensional space, whole numbers, probability are creations of the human spirit, to say nothing of real or complex numbers, groups, vector spaces, integrals etc. Whether all these exist outside our thoughts or not, i.e., whether it is a matter of discovering or inventing, is also a bone of contention among mathematicians – but irrelevant here.

Certainly these ideas arise originally from our observations and experience, mainly in the domain of geometry and physics on the one hand and numbers and counting on the other. But first must come the complete abstraction, the release from reality, to form from that experience a *mathematical object*. This is only defined by its combinatorial properties, which vary from case to case and which satisfy certain axioms; essential here is the structure of *mutual relations*. In the framework thus established we apprehend, guided by intuition and experiment, relationships, results, theorems. Whether they are correct one can only determine by a strictly logical analysis of the proof – otherwise one does not know whether they are valid. Experience shows that intuition may lead us astray. So long as we have no proof of Fermat's Theorem, we cannot be sure that integer solutions do not exist for large values of the exponent n.

Concerning the multiplicity of applications of mathematical structures and results, this obviously stems from their *universality*, their independence from concrete objects. Whether it concerns the forecast of an eclipse of the sun or the moon, the mathematical design of a bridge, the formulation of cosmological theories, the schemata of the physics of elementary particles, or the analysis of computer tomograms, there are always abstract, mathematical tools behind it, far removed from any reality. It would be very dangerous to apply them if one were not sure of their validity.

No Nobel Prize

Will Andrew Wiles receive the Nobel Prize?

There is no Nobel Prize for mathematicians; this doesn't seem to be well-known, but it gives rise to speculation. Many explanations circulate, stories about conflicts between Nobel and a prominent mathematician of the time, and much more besides; as the President of the Nobel Committee once expressed it, none of these stories can be true. We don't know the reason, we can only conjecture: mathematics was simply *forgotten*. As so often happens, it was seen – as a tool, which is simply to hand and which we apply; the mathematician's task is merely to carry out the necessary calculations. Even today when we generally recognize the significance of mathematics, people know very little of its true nature and inner beauty – because the research takes place within a narrower circle and is invisible from the outside. The non-mathematician sees only the tip of the iceberg. What is beneath? There lies this difficult and scarcely intelligible process of creating mathematical ideas and structures out of the vague experience and intuition of our environment, putting them to work and recognizing their connections; and even struggling with totally *unexpected consequences* of our own thinking. These are consequences which can give rise to far-reaching applications, from which further problems arise which call for new solutions or demand more new ideas.

An example which especially well illustrates how mathematical thought emerges from the depths to break surface is the discovery of electromagnetic waves, certainly one of the most important events in the history of science and modern mankind. The credit should be given to the physicists James Clark Maxwell (1831–1879) and Heinrich Hertz (1857–1894); but it rests heavily on mathematical theories which had been developed much earlier for other reasons (analysis, the wave equation), and which showed that the Maxwell-Heaviside equations lead inevitably to *waves* – and this was experimentally verified by Hertz.

Similarly much else came in unexpected ways to be applied to the physical world: group theory, developed by Galois to study the solution of algebraic equations, has been applied to the elucidation of atomic spectra; Boolean algebra, which stems from mathematical logic, is applied to electric circuit theory; the Radon transform has been applied to computer tomography; category theory to the design of automata and formal languages; differential geometry, topology and algebra to the new theoretical physics. Always there were completely different reasons for creating and formulating the mathematical concepts – or perhaps no other reason but the inner beauty of the conceptual construction?

What about the computer?

Can one not simply leave the difficult considerations involved in the Fermat proof to the computer?

This question is often asked, with some justification. For it is known not only to those involved, but also to the outsiders, that this is the era of the computer, which has immeasurably increased the possibilities for applying mathematical thought to our world. Moreover, not only applied, but also *pure* mathematicians, are using the computer in the most intensive way, to experiment, to verify conjectures, to render complicated geometric situations intelligible, and to push through difficult algebraic manipulations. But none of this replaces strict conceptual proof; on the contrary, it, in fact, depends on its logical foundations.

Now in an article which appeared last year in the *Scientific American* the "Death of Proof" was announced[2]. The text was very well documented and contained quotations from well-known mathematicians. Classical proofs within a conceptual framework were to be replaced by visualization and verification, naturally on a computer, the Fermat proof by Wiles was characterized as a "splendid anachronism". The article released a flood of indignant protests, even from mathematicians quoted in the article. All were agreed that the actual situation had been completely misunderstood. Semistrict arguments lead to semitruths which are correct only with a certain probability, or even false (and for whose uncertain validity huge amounts of computer time must be financed).

One could ignore this if a danger did not present itself whose consequences could be worse than one thinks. On the basis of such thinking a worldwide, fundamental restructuring of mathematics education could be proposed, which would replace everything by *interdisciplinary games* on the computer. It appears that already textbooks and software in this direction have been prepared, and here too certain reformers are following the same trend. Thus would the growing generations believe what they see on the screen, without knowing that "nothing has been proved". And the experience of the inner beauty of mathematical thought would be withheld from them. Mathematics must be used according to its true nature, abstract, valid within a strict context, universal, and, precisely for that reason, eminently practical.

---

[2] John Horgan, The death of proof, *Scientific American*, October 1993.

So do the words of Hermann Weyl[3], uttered 50 years ago, take on a new urgency: "We do not claim for mathematics the prerogative of a Queen of Science; there are other fields which are of the same or even higher importance in education. But mathematics sets the standard of objective truth for all intellectual endeavours; science and technology bear witness to its practical usefulness. Besides language and music it is one of the primary manifestations of the free creative power of the human mind, and it is the universal organ for world-understanding through theoretical construction. Mathematics must therefore remain an essential element of the knowledge and abilities we have to teach, of the culture we have to transmit, to the next generation."

---

[3] From the first page of the *Collected Works of Hermann Weyl*, edited by K. Chandrasekharan (Springer Verlag, 1968).

Hurwitz-Radon matrices revisited:
From effective solution of the Hurwitz matrix equations to Bott periodicity

CRM Proceedings and Lecture Notes 6 (1994), 23–35

The matrix equations $A_j^2 = -E$, $A_j A_k + A_k A_j = 0$, $j, k = 1, \ldots, s$, $j \neq k$ were discussed independently by Hurwitz and Radon around 1920*. Solutions have played, during the years, an important rôle in many parts of mathematics and mathematical physics: as coefficient matrices of elliptic differential operators, in the topology of vector fields and fiber bundles, in combinatorial analysis, in the composition problem for quadratic forms—this was the original motivation and starting point, and in that problem the (complex) matrices have to be *orthogonal*. A closely related variant is to solve the equations by *unitary* matrices.

In this survey we describe a procedure for constructing effectively the solutions, first in the unitary and from there in the orthogonal case. The essential tool is a "product" which reduces everything to small values of s; this product exhibits at the same time a very close relation with the well-known Bott periodicity in homotopy and K-theory.

I have tried to keep large parts of the presentation (except for those concerning complex and real K-theory) accessible to non-specialists, in order to emphasize the elementary aspects of the problem and the solution. I hope that this is somewhat in the spirit of Peter Hilton's beautiful contributions to the teaching of mathematics at various levels.

1. Orthogonal and unitary H-R matrices

1.1. Complex $n \times n$ matrices $A_1, A_2, \ldots, A_s$ are called *Hurwitz-Radon matrices* if they fulfill the conditions

$$
\text{(I)} \qquad
\begin{cases}
(1) & A_j^2 = -E \\
(2) & A_j A_k + A_k A_j = 0 \\
(3) & A_j A_j^T = E
\end{cases}
$$

for $j, k = 1, 2, \ldots, s$, $j \neq k$.

NOTATIONS. A^T is the transposed of the matrix A. We write E (or E_n if the *size* n is to be emphasized) for the $n \times n$ unit matrix. Hurwitz-Radon will be abbreviated by H-R.

1991 *Mathematics Subject Classification.* Primary: 15A24, 15A63, 55R45; Secondary: 20C15.
This is the final form of the paper.

*Hurwitz died in 1919. His paper [H] appeared in 1923. Radon's work was submitted in 1922 and published also in 1923 [R].

(3) means that the A_j are *orthogonal*. In view of (1) it is equivalent to $A_j^T = -A_j$, i.e., the A_j are skew-symmetric.

1.2. An equivalent formulation of (I) is obtained as follows. Consider $s + 1$ matrices $A_0, A_1, \ldots, A_s$ and require that their linear combination

$$x_0 A_0 + x_1 A_1 + \cdots + x_0 A_0$$

be an *orthogonal* matrix for all (real or complex) values of the x_j "except for a factor $\sum_0^s x_j^2$"; i.e.,

$$\left(\sum_0^s x_j A_j \right) \left(\sum_0^s x_k A_k^T \right) = \sum_0^s x_j^2 E.$$

This is the case if and only if

(II)
$$\begin{cases} A_j A_j^T = E \\ A_j A_k^T + A_k A_j^T = 0 \end{cases}$$

for $j, k = 0, 1, \ldots, s$, $j \neq k$. To find all solutions it is clearly enough to consider those with $A_0 = E$. Substituting this for A_k we get $A_j + A_j^T = 0$ for $j = 1, 2, \ldots, s$, and then (II) for $j, k = 1, 2, \ldots, s$ is equivalent to (I).

1.3. The problem considered (independently) by Hurwitz [**H**] and Radon [**R**] concerns the "composition of quadratic forms"

(III)
$$\left(\sum_0^s x_j^2 \right) \left(\sum_1^n y_\ell^2 \right) = \sum_1^n z_\ell^2$$

where the z_ℓ are complex bilinear forms of the $x_0, \ldots, x_s$ and $y_1, \ldots, y_n$. They determined for given n the maximum number $s + 1$ for which such bilinear forms exist. If $n = \text{odd} \cdot 16^\alpha \cdot 2^\beta$, $\beta = 0, 1, 2, 3$, the maximal $s + 1$ is $\rho(n) = 8\alpha + 2^\beta$, the "Radon number". The bilinear forms can be written in matrix notation as

$$\begin{pmatrix} z_1 \\ \vdots \\ z_n \end{pmatrix} = \left(\sum_0^s x_j A_j \right) \begin{pmatrix} y_1 \\ \vdots \\ y_n \end{pmatrix}.$$

Then (III) simply tells that $\sum_0^s x_j A_j$ is orthogonal "except for the factor $\sum_0^s x_j^2$".

The H-R problem thus consist in solving (I), for given n, by a maximum number s of matrices $A_1, \ldots, A_s$. Or, in other words, given s to find the minimum n for which such $n \times n$ matrices exist. In this note we will present a method for not only determing the minimum n but also to construct effectively *all* solutions of (I). We will do this in detail for the *unitary* variant (I') of (I), see below 1.4. The original *orthogonal* problem (I) will then only be sketched; it is more complicated but is dealt with by essentially the same procedure.

It turns out that this effective procedure leads to an interesting relation with the homotopy groups of the infinite unitary, or orthogonal respectively, group (Bott periodicity).

1.4. The *unitary* matrix problem (I') is obtained by simply replacing (3) by

$$(3') \qquad\qquad\qquad A_j \bar{A}_j^T = E,$$

i.e., one asks for unitary matrices fulfilling (1) and (2). Condition (3') is then equivalent to $\bar{A}_j^T = -A$.

Again, if we put $A_0 = E$, the problem (I') is equivalent to

$$x_0 A_0 + x_1 A_1 + \cdots + x_s A_s$$

being a *unitary* matrix for all (real) values of the x_j "except for the factor $\sum_0^s x_j^2$"; i.e.,

$$\left(\sum_0^s x_j A_j\right)\left(\sum_0^s x_k \bar{A}_k^T\right) = \left(\sum_0^s x_j^2\right) E.$$

2. Irreducible unitary solutions

2.1. We deal in the following, until Section 4, with *unitary* H-R matrices exclusively; i.e., with unitary matrices fulfilling (1) and (2). Given two solutions for the same s of size n and m respectively

$$\sigma = \{A_1, \ldots, A_s\}, \quad \sigma' = \{A_1', \ldots, A_s'\}$$

their *sum* $\sigma + \sigma'$ is defined by

$$\sigma + \sigma' = \left\{\begin{pmatrix} A_1 & \\ & B_1 \end{pmatrix}, \ldots, \begin{pmatrix} A_s & \\ & B_s \end{pmatrix}\right\}$$

(empty entries always mean 0 or 0-matrices). $\sigma + \sigma'$ is easily checked to be again a solution, of size $n + m$. If M is an invertible matrix then

$$M\sigma M^{-1} = \{M A_1 M^{-1}, \ldots, M A_s M^{-1}\}$$

is again a solution of (1) and (2) of size n, and two solutions σ and τ are called *equivalent* if there is an M such that $\tau = M\sigma M^{-1}$.

2.2. A unitary solution is called *irreducible* if it is not equivalent to a sum of two solutions. Minimal size n solutions, for given s, must of course be irreducible. It turns out that there are very few irreducible solutions, and that only *one* size, depending on s, occurs:

PROPOSITION 2.1. *If s is even there is only one irreducible solution, up to equivalence, of size $n = 2^{s/2}$. If s is odd there are two irreducible solutions, up to equivalence, of size $n = 2^{(s-1)/2}$.*

COROLLARY 2.2. *The maximum number s of unitary H-R matrices of size $n = \text{odd} 2^t$ is $2t + 1$.*

The proof of Proposition 2.1 is contained in the author's paper [E] published 50 year ago. It is an application of some basic facts of the theory of *group representations*. A reader familiar with that theory will have noticed that the terminology above is taken from group representations (except that the "size" is called there "degree"). We recall in 2.3 some of the arguments of [E] although these are not needed explicitly in the sequel of the present note.

2.3. Let G_s be the group given by the presentation

$$\langle a_1,\ldots,a_s,\varepsilon \mid \varepsilon^2 = 1, a_j^2 = \varepsilon, a_j a_k = \varepsilon a_k a_j \rangle$$

where $j, k = 1,\ldots,s$, $j \neq k$. A unitary solution of (1), (2) of size n yields a representation of G_s of degree n by assigning the matrix A_j to the generator a_j and $-E$ to ε. And vice-versa: any unitary representation of G_s with $\varepsilon \mapsto -E$ yields a solution.

G_s is a finite group of order 2^{s+1}. The commutator subgroup consists, for $s > 1$, of 1 and ε only; G_1 is Abelian (cyclic of order 4). The center of G_s consists of 1 and ε if s is even; and of 1, ε and two further elements z, εz if s is odd ($z = a_1 a_2 \cdots a_s$).

G_1 has two irreducible representations, of degree 1, with $\varepsilon \mapsto (-1)$; namely $A_1 = (i)$ or $(-i)$. For $s > 1$ representations of degree 1 cannot be used since they assign the matrix (1) to ε; their number is 2^s. Irreducible representations of degree > 1 automatically assign $-E$ to ε; their number is 1 for even s and 2 for odd s, and one has for their degrees d_s, d_s' ($=$ powers of 2)

$$2^{s+1} = 2^s + d_s^2 \qquad \text{for even } s,$$
$$2^{s+1} = 2^s + d_s^2 + d_s'^2 \quad \text{for odd } s$$

which immediately yields $d_s = 2^{s/2}$ or $d_s = d_s' = 2^{(s-1)/2}$ respectively.

By a general result any representation of a finite group is equivalent to a unitary one, and equivalent to a sum of irreducible representations. Thus the term "irreducible solution" in 2.2 is in agreement with the usual meaning of irreducible representation.

2.4. Since any unitary solution of (1), (2) is equivalent to a sum of irreducible solutions, which we will denote by ρ_s for even s and by ρ_s, ρ_s' for odd s, it is convenient to consider the free Abelian group D_s generated by (the equivalence class of) ρ_s or of ρ_s, ρ_s' respectively. Thus

$$D_s \cong \mathbb{Z} \qquad \text{if } s \text{ is even,}$$
$$D_s \cong \mathbb{Z} \oplus \mathbb{Z} \quad \text{if } s \text{ is odd.}$$

Positive integers k are to be interpreted, for even s, as sums of k solutions ρ_s; and positive pairs $(k, k') \in \mathbb{Z} \oplus \mathbb{Z}$ as sums of k solutions ρ_s and k' solutions ρ_s', for odd s. The zeros and the negative integers have a symbolic meaning only ("Grothendieck groups").

2.5. Solutions for small values of s.

$$s = 1 \quad \rho_1 : A_1 = (i) \quad \text{or} \quad \rho_1' : A_1 = (-i)$$

$$s = 2 \quad \rho_2 : A_1 = \begin{pmatrix} i & \\ & -i \end{pmatrix}, \quad A_2 = \begin{pmatrix} & 1 \\ -1 & \end{pmatrix}$$

$$s = 3 \quad \rho_3 : A_1 = \begin{pmatrix} i & \\ & -i \end{pmatrix}, \quad A_2 = \begin{pmatrix} & 1 \\ -1 & \end{pmatrix}, \quad A_3 = \begin{pmatrix} & i \\ i & \end{pmatrix}$$

or

$$\rho_3' : A_1 = \begin{pmatrix} -i & \\ & i \end{pmatrix}, \quad A_2 = \begin{pmatrix} & -1 \\ 1 & \end{pmatrix}, \quad A_3 = \begin{pmatrix} & -i \\ -i & \end{pmatrix}$$

$$s = 4 \quad \rho_4 : A_1 = \begin{pmatrix} i & & & \\ & -i & & \\ & & -i & \\ & & & i \end{pmatrix}, \quad A_2 = \begin{pmatrix} & & & 1 \\ -1 & & & \\ & & & -1 \\ & 1 & & \end{pmatrix}$$

$$A_3 = \begin{pmatrix} & i & & \\ i & & & \\ & & & -i \\ & & -i & \end{pmatrix}, \quad A_4 = \begin{pmatrix} & & & 1 \\ & & 1 & \\ -1 & & & \\ & -1 & & \end{pmatrix}$$

It is convenient to include in the list of the irreducible solutions the case $s = 0$: The size is 1, and ρ_0 is the empty set of matrices.

3. How to construct unitary solutions

3.1. We describe in this section a procedure for constructing explicitly all unitary solutions of (1), (2). "All" means, of course, up to equivalence; our notations will often not distinguish between equivalence classes and their representatives. In Section 5 the procedure will be restated in a more formal framework.

As stated in the previous section we only have to produce the irreducible solutions ρ_s, s even, and ρ_s, ρ'_s, s odd.

An almost trivial observation shows that one can forget about the ρ_s for even s. Namely, let $\rho_{s+1} = \{A_1, A_2, \ldots, A_s, A_{s+1}\}$; if s is even the size of ρ_{s+1} is $2^{((s+1)-1)/2} = 2^{s/2}$. If A_{s+1} is omitted $\{A_1, A_2, \ldots, A_s\}$ is a solution of size $2^{s/2}$, whence irreducible. Since there is only *one* such it must be ρ_s. Of course one might omit, instead of A_{s+1}, any other A_j to yield the same ρ_s up to equivalence.

PROPOSITION 3.1. *If s is even then omitting a matrix from ρ_{s+1} or ρ'_{s+1} yields the irreducible solution ρ_s.*

3.2. What happens if the same procedure is applied to ρ_{s+1} for odd s? The size of ρ_{s+1} is $2^{(s+1)/2} = 2 \cdot 2^{(s-1)/2}$. Therefore omitting A_{s+1} from ρ_{s+1} yields the sum of *two* irreducible solutions. What are they? For reasons of symmetry one will guess that one obtains $\rho_s + \rho'_s$.

Here is a proof. We consider $\{A_1, \ldots, A_s\}$ from ρ_{s+1}. Since $s + 1$ is even $\{A_1, A_2, \ldots, A_{s+1}\}$ is equivalent to $\{-A_1, A_2, \ldots, A_{s+1}\}$. The matrices $A_1 A_2 \ldots A_s$ and $-A_1 A_2 \ldots A_s$ are similar and therefore have the same *trace*:

$$\text{trace}(A_1 A_2 \ldots A_s) = -\text{trace}(A_1 A_2 \ldots A_s) = 0.$$

Let further ρ_s be $\{B_1, B_2, \ldots, B_s\}$. The matrix $B_1 B_2 \ldots B_s$ commutes with all B_j and thus is a non-zero multiple of E, cE (here again a result of representation theory is used: A matrix which commutes with all matrices of an irreducible representation is a multiple of E). This also shows that one can take $\rho'_s = \{-B_1, \ldots, -B_s\}$; for the product is $-cE$; and since $\text{trace}(cE) \neq \text{trace}(-cE)$ these two solutions cannot be equivalent.

It is clear now that $\{A_1, \ldots, A_s\}$ must be $\rho_s + \rho'_s$ in order to make the trace of the product equal to 0.

PROPOSITION 3.2. *If s is odd and $\rho_{s+1} = \{A_1, \ldots, A_s, A_{s+1}\}$ then omitting A_{s+1} yields $\rho_s + \rho'_s$.*

PROPOSITION 3.3. *If s is odd then ρ'_s is obtained from ρ_s by changing the signs of all matrices (or of an odd number of matrices).*

3.3. We now describe our crucial tool, a *product of solutions*. We first recall that $A_1, A_2, \ldots, A_s$ are unitary H-R matrices of size n if and only if

$$f(x) = \sum_0^s x_j A_j$$

(with $A_0 = E_n$) is a unitary matrix for all $x = (x_0, \ldots, x_s) \in \mathbb{R}^{s+1}$ with $\sum_0^s x_j^2 = 1$.

Given two solutions $\sigma_s = \{A_1, \ldots, A_s\}$, $\sigma_t = \{B_1, \ldots, B_t\}$ of size n and m respectively we put

$$g(y) = \sum_0^s y_k B_k,$$

$B_0 = E_m$, $y = (y_0, \ldots, y_t) \in \mathbb{R}^{t+1}$ and set

$$(4) \qquad F(x, y) = \begin{pmatrix} f(x) \otimes E_m & E_n \otimes g(y) \\ -E_n \otimes \overline{g(y)}^T & \overline{f(x)}^T \otimes E_m \end{pmatrix}.$$

This is a matrix of size $2nm$. The entry $f(x) \otimes E_m$, e.g., means $\sum_0^s x_j A_j \otimes E_m$ where $A_j \otimes E_m$ is the matrix of the tensor product of the linear transformations corresponding to A_j and E_m. One easily checks that

$$F(x, y) \overline{F(x, y)}^T = \left(\sum_0^s x_j^2 + \sum_0^t y_k^2 \right) E_{2nm}.$$

Thus if $\sum_0^s x_j^2 + \sum_0^t y_k^2 = 1$ then $F(x, y)$ is a unitary matrix. The coefficient matrix of x_0 is E_{2nm}; therefore the coefficient matrices of $x_1, \ldots, x_s, y_0, y_1, \ldots, y_t$ are $s + t + 1$ unitary H-R matrices of size $2nm$ and constitute a solution $\sigma_{s+t+1} = \{C_1, \ldots, C_s, C_{s+1}, \ldots, C_{s+t+1}\}$.

Explicitly:

$$(5) \qquad \begin{cases} C_j = \begin{pmatrix} A_j \otimes E_m & \\ & -A_j \otimes E_m \end{pmatrix} & \text{for } j = 1, \ldots, s, \\[2ex] C_j = \begin{pmatrix} & E_{nm} \\ -E_{nm} & \end{pmatrix} & \text{for } j = s + 1, \\[2ex] C_j = \begin{pmatrix} & E_n \otimes B_k \\ E_n \otimes B_k & \end{pmatrix} & \begin{array}{l} \text{for } j = s+2, \ldots, s+t+1, \\ \text{with } k = j - s - 1. \end{array} \end{cases}$$

We call σ_{s+t+1} the *product* of σ_s and σ_t, written $\sigma_s \cdot \sigma_t$.

3.4. If we take, in particular, the product $\rho_s \cdot \rho_t$ with t odd then its size is as follows:

$$2 \cdot 2^{(s-1)/2} 2^{(t-1)/2} = 2^{(s+t)/2} \qquad \text{if } s \text{ is odd}$$
$$2 \cdot 2^{s/2} 2^{(t-1)/2} = 2^{(s+t+1)/2} \qquad \text{if } s \text{ is even};$$

i.e., in both cases $\rho_s \cdot \rho_t = \rho_{s+t+1}$.

As remarked in 3.1 above the case s even, $s+t+1$ even, is not important, while s and t odd explicitly produce new solutions. In particular, $\rho_s \cdot \rho_1 = \rho_{s+2}$ (or ρ'_{s+2}) so that all ρ_s, s odd, can be constructed inductively from ρ_1.

THEOREM 3.4. *All irreducible unitary solutions ρ_{s+2} for odd s are obtained inductively from $\rho_1 = \{(i)\}$ by the product $\rho_s \cdot \rho_1$ (and ρ'_{s+2} by changing the signs of all matrices).*

In explicit form, let $\rho_s = \{A_1, \ldots, A_s\}$ of size $n = 2^{(s-1)/2}$. Then ρ_{s+2} is given by the $s + 2$ matrices

$$(6) \qquad \begin{pmatrix} A_j & \\ & -A_j \end{pmatrix}, j = 1, \ldots, s; \quad \begin{pmatrix} & E_n \\ -E_n & \end{pmatrix}; \quad \begin{pmatrix} & iE_n \\ iE_n & \end{pmatrix}.$$

Note that the examples in 2.4 are exactly of the above form.

COROLLARY 3.5. *All unitary H-R matrices can be written with entries 0, ± 1, $\pm i$ alone.*

REMARK 3.6.. The short expression for (6) corresponding to (4), with $A_0 = E$ and $f(x) = \sum_0^s x_j A_j$, $g(y) = y_0 + iy_1$ is

$$F(x, z) = \begin{pmatrix} f(x) & zE_n \\ -\bar{z}E_n & \overline{f(x)}^T \end{pmatrix}$$

where we have written z for $y_0 + iy_1$.

3.5. In 2.4 we have introduced the notation D_s for the free Abelian group generated by the (equivalence classes of) irreducible solutions ρ_s, or ρ_s, ρ'_s respectively. D_s is $\cong \mathbb{Z}$ for even s, $\cong \mathbb{Z} \oplus \mathbb{Z}$ for odd s; only positive integers have a "concrete" meaning.

Omitting in a solution $\sigma_{s+1} \in D_{s+1}$ the last matrix A_{s+1} is compatible with the sum of solutions. It thus defines a homomorphism $h \colon D_{s+1} \to D_s$ for all $s \geq 0$. (In defining h the rôle of A_{s+1} can also be played by another matrix A_j). We write $\mathcal{E}_s$ for the cokernel $D_s / h(D_{s+1})$ of h and call it the *reduced unitary* H-R *group*. Since by Proposition 3.1 $h(\rho_{s+1}) = \rho_s$ for even s the map h is surjective, whence $\mathcal{E}_s = 0$ for even s. For odd s, $h(\rho_{s+1}) = \rho_s + \rho'_s$, by Proposition 3.2; the elements of $\mathcal{E}_s$ are classes in D_s modulo the "diagonal" $\mathbb{Z}(\rho_s + \rho'_s)$. Thus $\mathcal{E}_s \cong \mathbb{Z}$ (choose ρ_s and $\rho_s + \rho'_s$ as a basis of D_s instead of ρ_s, ρ'_s).

For odd s the elements of $\mathcal{E}_s$ can be described as follows. For positive $k \in \mathbb{Z}$ they are represented by sums of solutions ρ_s, the element $0 \in \mathbb{Z}$ by $\rho_s + \rho'_s$; and for negative k by sums of ρ'_s. In this way all elements of $\mathcal{E}_s$ have a "concrete" meaning in terms of solutions.

THEOREM 3.7. *The reduced unitary H-R group $\mathcal{E}_s$, $s \geq 0$, is 0 for even s; and is $\mathbb{Z}$ for odd s, generated by ρ_s or by $\rho'_s = -\rho_s$.*

4. How to construct orthogonal H-R matrices

4.1. The discussion of orthogonal solutions of the H-R matrix problem (1), (2) is quite parallel to the unitary case, though a little more complicated.

Clearly, if a unitary solution σ_s is equivalent to a *real* one then we get an orthogonal solution. This is actually the only way to obtain orthogonal solutions: By a theorem of Frobenius-Schur in representation theory a unitary representation is equivalent to a complex orthogonal one (if and) only if it is equivalent to a real one. If σ_s is *not* equivalent to a real solution we have to combine it with its conjugate-complex $\bar{\sigma}_s$, as $\sigma_s + \bar{\sigma}_s$; it is then easy to find a transformation T such that $T(\sigma_s + \bar{\sigma}_s)T^{-1}$ is real.

The following classification of the real—i.e., orthogonal—irreducible representations τ_s is taken from [E]. It is based on determining for which s the unitary irreducible solution ρ_s is equivalent to a real solution; and if not, equivalent to $\bar{\rho}_s$ or not. A different procedure can be found in [E2]; it applies an elegant approach of T. Y. Lam and T. Smith [LS] deviced in a more general framework.

4.2. The statements concerning the τ_s depend on the value of s modulo 8.

If $s \equiv 0, 6, 7$ then ρ_s is (equivalent to a) *real* solution, thus $\tau_s = \rho_s$; and the size is $2^{s/2}$ for $s \equiv 0, 6$, $2^{(s-1)/2}$ for $s \equiv 7$ (where one has another orthogonal solution $\tau'_s = \rho'_s$).

If $s \equiv 2, 3, 4$ then ρ_s is *not real*, and equivalent to $\bar{\rho}_s$, thus $\tau_s = \rho_s + \bar{\rho}_s$, and the size is $2^{(s+2)/2}$ for $s \equiv 2, 4$, $2^{(s+1)/2}$ for $s \equiv 3$ (where one has another solution $\tau'_s = \rho'_s + \bar{\rho}'_s$).

If $s \equiv 1, 5$ then ρ_s is *not real*, and *not* equivalent to $\bar{\rho}_s$ (whence $\bar{\rho}_s = \rho'_s$), thus $\tau_s = \rho_s + \bar{\rho}_s = \rho_s + \rho'_s$; and the size is $2^{(s+1)/2}$. There is only *one* irreducible orthogonal solution.

4.3. As in the unitary case, we consider the free Abelian groups generated by the irreducible orthogonal solutions τ_s, or τ_s, τ'_s and denote them by $D_s^{\mathcal{O}}$. They are $\cong \mathbb{Z}$ for $s \equiv 0, 1, 2, 4, 5, 6$ modulo 8, and $\cong \mathbb{Z} \oplus \mathbb{Z}$ for $s \equiv 3, 7$ modulo 8.

The *reduced orthogonal* H-R groups $D_s^{\mathcal{O}}/h(D_{s+1}^{\mathcal{O}})$ are denoted by $\mathcal{E}_s^{\mathcal{O}}$. They are easily computed from the list in 4.2 and the sizes of the τ_s, for $s \equiv 0, 1, \ldots, 7$ modulo 8:

$$
\begin{aligned}
s \equiv 0: \quad & h(\tau_1) = h(\rho_1 + \bar{\rho}_1) = 2\rho_0 = 2\tau_0 \\
s \equiv 1: \quad & h(\tau_2) = h(\rho_2 + \bar{\rho}_2) = \rho_1 + \rho'_1 + \bar{\rho}_1 + \bar{\rho}'_1 = 2\tau_1 \\
s \equiv 2: \quad & h(\tau_3) = h(\rho_3 + \bar{\rho}_3) = \rho_2 + \bar{\rho}_2 = \tau_2 \\
s \equiv 3: \quad & h(\tau_4) = h(\rho_4 + \bar{\rho}_4) = \rho_3 + \rho'_3 + \bar{\rho}_3 + \bar{\rho}'_3 = \tau_3 + \tau'_3 \\
s \equiv 4: \quad & h(\tau_5) = h(\rho_5 + \bar{\rho}_5) = \rho_4 + \bar{\rho}_4 = \tau_4 \\
s \equiv 5: \quad & h(\tau_6) = h(\rho_6) = \rho_5 + \rho'_5 = \tau_5 \\
s \equiv 6: \quad & h(\tau_7) = h(\rho_7) = \rho_6 = \tau_6 \\
s \equiv 7: \quad & h(\tau_8) = h(\rho_8) = \rho_7 + \rho'_7 = \tau_7 + \tau'_7.
\end{aligned}
$$

This yields

THEOREM 4.1.. *The reduced orthogonal H-R groups $\mathcal{E}_s^{\mathcal{O}}$ are $\mathbb{Z}/2$ for $s \equiv 0$ and 1 modulo 8; 0 for $s \equiv 2, 4, 5, 6$; and $\mathbb{Z}$ for $s \equiv 3, 7$.*

4.4. The explicit construction of "all" orthogonal H-R matrices is again based on the product defined in 3.3. It yields a product $D_s^{\mathcal{O}} \times D_t^{\mathcal{O}} \to D_{s+t+1}^{\mathcal{O}}$.

We first note that

$$\tau_s \cdot \tau_7 = \tau_{s+8}.$$

Indeed, as can be seen from the list in 4.2, the size of τ_{s+8} is $16n$ where n is the size of τ_s; and this is precisely the size of $\tau_s \cdot \tau_7$. We thus get all τ_s from those for $0 \leq s \leq 7$ by multiplying with powers of τ_7.

We can forget about $\tau_2, \tau_4, \tau_5, \tau_6$; omitting the last matrix in τ_3 yields τ_2, the last, the two last, and the three last matrices in τ_7 yields τ_6, τ_5, τ_4 respectively. τ_0 is the empty set of matrices of size 1, and $\tau_1 = \left\{ \left(\begin{smallmatrix} & 1 \\ -1 & \end{smallmatrix} \right) \right\}$; note that this is $= \tau_0^2$. And

$\tau_8 = \tau_7 \cdot \tau_0 = \{\begin{pmatrix} A_j \\ -A_j \end{pmatrix}, j = 1, \ldots, 7, \begin{pmatrix} & E_8 \\ -E_8 & \end{pmatrix}\}$ where the A_j are the matrices of τ_7.

As for τ_3 and τ_7, they are described by the well-known multiplication tables of the quaternions and the Cayley octaves respectively, as follows.

Let $\tau_3 = \{A_1, A_2, A_3\}$, $A_0 = E_4$, and $f(x) = \sum_0^3 x_j A_j$. Then

$$f(x) = \begin{pmatrix} x_0 & x_1 & x_2 & x_3 \\ -x_1 & x_0 & x_3 & -x_2 \\ -x_2 & -x_3 & x_0 & x_1 \\ -x_3 & x_2 & -x_1 & x_0 \end{pmatrix},$$

and $f(x)y$, where $y = (y_0, y_1, y_2, y_3)$, is the quaternion product of x and $y \in \mathbb{R}^4$.

Similarly, for $\tau_7 = \{A_1, \ldots, A_7\}$, $A_0 = E_8$,

$$f(x) = \begin{pmatrix} x_0 & x_1 & x_2 & x_3 & x_4 & x_5 & x_6 & x_7 \\ -x_1 & x_0 & x_3 & -x_2 & x_5 & -x_4 & -x_7 & x_6 \\ -x_2 & -x_3 & x_0 & x_1 & x_6 & x_7 & -x_4 & -x_5 \\ -x_3 & x_2 & -x_1 & x_0 & x_7 & -x_6 & x_5 & -x_4 \\ -x_4 & -x_5 & -x_6 & -x_7 & x_0 & x_1 & x_2 & x_3 \\ -x_5 & x_4 & -x_7 & x_6 & -x_1 & x_0 & -x_3 & x_2 \\ -x_6 & x_7 & x_4 & -x_5 & -x_2 & x_3 & x_0 & -x_1 \\ -x_7 & -x_6 & x_5 & x_4 & -x_3 & -x_2 & x_1 & x_0 \end{pmatrix}$$

The coefficient matrices of the x_j have, in both cases, entries $0, \pm 1$ only.

THEOREM 4.2. *All irreducible orthogonal solutions ρ_s are obtained from $\tau_0 = \emptyset$, $\tau_1 = \begin{pmatrix} & 1 \\ -1 & \end{pmatrix}$, and τ_3, τ_7 as above by multiplying with powers of τ_7 (and omitting matrices for $s \equiv 2, 4, 5, 6$ modulo 8).*

COROLLARY 4.3. *All orthogonal H-R matrices can be written with entries $0, \pm 1$ alone.*

REMARK 4.4. $\tau_0^2 = \tau_1$ and $\tau_3^2 = 4\tau_7$.

The second assertion follows from $\tau_3 = \rho_3 + \bar{\rho}_3 = 2\rho_3$ and $\tau_7 = \rho_7$.

REMARK 4.5. From the sizes of the minimal, i.e., irreducible orthogonal solutions τ_s for the values of s modulo 8 (see 4.2 above) one easily obtains the "Radon number" mentioned in 1.3, as follows.

We write $n = \text{odd} \cdot 2^{4\alpha+\beta}$, $\beta = 0, 1, 2, 3$, and look for the largest s such that the corresponding size is $2^{4\alpha+\beta}$. Since solutions for $s \equiv 2, 4, 5, 6$ are obtained be *omitting* matrices we need only consider the other s.

$\beta = 0$:	$s = 8\alpha$	yields	$2^{s/2} = 2^{4\alpha}$
$\beta = 1$:	$s = 8\alpha + 1$	yields	$2^{(s+1)/2} = 2^{4\alpha+1}$
$\beta = 2$:	$s = 8\alpha + 3$	yields	$2^{(s+1)/2} = 2^{4\alpha+2}$
$\beta = 3$:	$s = 8\alpha + 7$	yields	$2^{(s-1)/2} = 2^{4\alpha+3}$.

In summary: The maximum number of orthogonal H-R matrices of size $n = \text{odd} \cdot 2^{4\alpha+\beta}$ is $8\alpha + 2^\beta - 1$.

5. The reduced H-R rings and Bott periodicity

5.1. In both the unitary and the orthogonal case properties of the *product* $\sigma_s \cdot \sigma_t = \sigma_{s+t+1}$ can be expressed in terms of the groups D_s and $\mathcal{E}_s$ (or $D_s^{\mathcal{O}}$ and $\mathcal{E}_s^{\mathcal{O}}$). It is easy to see that it is distributive relative to the sum $\sigma_s + \sigma_s'$ of solutions. It is a little harder to check that it is associative.

With regard to $h \colon D_{s+1} \to D_s$, it is immediate from the matrix list (5) of $\sigma_{s+1} \cdot \sigma_t$ or of $\sigma_s \cdot \sigma_{t+1}$ that

$$h(\sigma_{s+1}) \cdot \sigma_t = h(\sigma_{s+1} \cdot \sigma_t),$$
$$\sigma_s \cdot h(\sigma_{t+1}) = h(\sigma_s \cdot \sigma_{t+1}).$$

The product of representatives thus yields a product $\mathcal{E}_s \times \mathcal{E}_t \to \mathcal{E}_{s+t+1}$, or $\mathcal{E}_s^{\mathcal{O}} \times \mathcal{E}_t^{\mathcal{O}} \to \mathcal{E}_{s+t+1}^{\mathcal{O}}$ respectively. It is *graded* if $\mathcal{E}_s$ is given the degree $s + 1$.

The direct sum $\mathcal{E}_* = \bigoplus_{s=-1}^{\infty} \mathcal{E}_s$, where we have added a term $\mathcal{E}_{-1} = \mathbb{Z}$ generated by the unit for the product, is a graded ring. We call it the *reduced unitary H-R ring*. Similarly we have a graded ring $\mathcal{E}_*^{\mathcal{O}} = \bigoplus_{s=-1}^{\infty} \mathcal{E}_s^{\mathcal{O}}$, the *reduced orthogonal H-R ring*. The discussion in Sections 3 and 4 gives the complete structure of these rings, as follows.

5.2. In the unitary case: $\mathcal{E}_s = 0$ for even s, and for odd s, $\mathcal{E}_s \cong \mathbb{Z}$ generated by ρ_s, where $\rho_{2k-1} = \rho_1^k$, $k = 1, 2, \ldots$.

THEOREM 5.1. *The reduced unitary H-R ring $\mathcal{E}_*$ is the polynomial ring $\mathbb{Z}[\rho_1]$ generated by $\rho_1 = \{(i)\}$.*

In the orthogonal case: $\mathcal{E}_s^{\mathcal{O}} = 0$ for $s \equiv 2, 4, 5, 6$ modulo 8. $\mathcal{E}_0^{\mathcal{O}} \cong \mathbb{Z}/2$ generated by τ_0, $\mathcal{E}_1^{\mathcal{O}} \cong \mathbb{Z}/2$ generated by $\tau_1 = \{\left(\begin{smallmatrix} & 1 \\ -1 & \end{smallmatrix}\right)\}$. $\mathcal{E}_3$ and $\mathcal{E}_7$ are $\cong \mathbb{Z}$ generated by τ_3 and τ_7 (see 4.4), and the product with τ_7 is an isomorphism $\mathcal{E}_s^{\mathcal{O}} \cong \mathcal{E}_{s+8}^{\mathcal{O}}$ for all s. In particular, τ_7 generates a polynomial subring $\mathbb{Z}[\tau_7]$ of $\mathcal{E}_*^{\mathcal{O}}$.

The ring $\mathcal{E}_*^{\mathcal{O}}$ is commutative. For products involving $\mathcal{E}_s^{\mathcal{O}}$, $s \equiv 2, 4, 5, 6$ modulo 8, there is nothing to prove. $\tau_0 \tau_1 = \tau_1 \tau_0 = \tau_0^3 = 0$ in $\mathcal{E}_2^{\mathcal{O}}$; $\tau_0 \tau_3 = \tau_3 \tau_0 = 0$; $\tau_0 \tau_7 = \tau_7 \tau_0 = \tau_8$ in $\mathcal{E}_8^{\mathcal{O}} = \mathbb{Z}/2$. Similarly $\tau_1 \tau_3 = \tau_3 \tau_1$ in $\mathcal{E}_4^{\mathcal{O}}$; $\tau_1 \tau_7 = \tau_7 \tau_1 = \tau_9$ in $\mathcal{E}_9^{\mathcal{O}} = \mathbb{Z}/2$. Finally $\tau_3 \tau_7 = 4\tau_3^3 = \tau_7 \tau_3$. We summarize:

THEOREM 5.2. *The reduced orthogonal H-R ring $\mathcal{E}_*^{\mathcal{O}}$ is the commutative ring, graded by $s + 1$ for $\mathcal{E}_s^{\mathcal{O}}$, generated by three elements τ_0, τ_3, τ_7 with relations $2\tau_0 = 0$, $\tau_0^3 = 0$, $\tau_3^2 = 4\tau_7$.*

ADDENDA 5.3. $\tau_0^2 = \tau_1$; τ_7 generates the polynomial ring $\mathbb{Z}[\tau_7] \subset \mathcal{E}_*^{\mathcal{O}}$; the product with τ_7 is an isomorphism $\mathcal{E}_s^{\mathcal{O}} \cong \mathcal{E}_{s+8}^{\mathcal{O}}$; $\mathcal{E}_s = 0$ for $s \equiv 2, 4, 5, 6$ modulo 8.

5.3. All this is so reminiscent of the *Bott periodicity theorems* (see [K], for example) that there has to be an explicit relationship.

In its original form the Bott periodicity is about the homotopy groups of the infinite unitary group $U = \lim U(n)$ and the infinite orthogonal group $\mathcal{O} = \lim \mathcal{O}(n)$. The limit is understood with respect to the imbeddings $U(n) \to U(n+1)$, $\mathcal{O}(n) \to \mathcal{O}(n+1)$ by adding a 1 to the diagonal, $n = 1, 2, 3, \ldots$. The homotopy groups $\pi_s(U)$ are the "stable" homotopy groups of $U(n)$; if f is odd, $\pi_s\big(U(n)\big) \cong \pi_s\big(U((s+1)/2)\big)$ for $n \geq (s+1)/2$, and if s is even, $\cong \pi_s\big(U((s+2)/2)\big)$ for $n \geq (s+2)/2$. Similarly for $\pi_s(\mathcal{O})$. Bott proved in 1956 ff (see [K]) that the $\pi_s(U)$ are periodic with period 2, the $\pi_s(\mathcal{O})$ with period 8; and that $\pi_s(U) = \mathbb{Z}$ for odd s, $= 0$ for even s; and that $\pi_s(\mathcal{O}) \cong \mathbb{Z}/2$ for $s \equiv 0, 1$ modulo 8, $= 0$ for $s \equiv 2, 4, 5, 6$, $\cong \mathbb{Z}$ for $s \equiv 3, 7$.

Thus $\mathcal{E}_s \cong \pi_s(U)$ *and* $\mathcal{E}_s^{\mathcal{O}} \cong \pi_s(\mathcal{O})$ *for all* $s \geq 0$. Of course one wants to find a *map* which provides these isomorphisms. The natural choice is as follows.

Given a solution $\sigma_s = \{A_1, A_2, \ldots, A_s\}$, unitary or orthogonal, consider as earlier $f(x) = \sum_0^s x_j A_j$ with $A_0 = E$, $x = (x_0, \ldots, x_s) \in \mathbb{R}^{s+1}$, $\sum_0^s x_j^2 = 1$. Then f is a map of $S^s \subset \mathbb{R}^{s+1}$ into some $U(n)$ or $\mathcal{O}(n)$. Passing to U or $\mathcal{O}$ and to the homotopy class of f we get a map $\Phi: D_s \to \pi_s(U)$ or $\Psi: D_s \to \pi_s(\mathcal{O})$ respectively. If $\sigma_s = h(\sigma_{s+1})$ then f can be extended to $S^{s+1} \subset \mathbb{R}^{s+2}$ and is thus nullhomotopic. The maps $\mathcal{E}_s \to \pi_s(U)$ and $\mathcal{E}_s^{\mathcal{O}} \to \pi_s(\mathcal{O})$ thus obtained are again written Φ and Ψ.

Φ and Ψ are easily seen to be (additive) homomorphisms: A sum $\sigma_s + \sigma_s'$ becomes in U, e.g.,

$$\begin{pmatrix} f(x) & & & & & \\ & f'(x) & & & & \\ & & 1 & & & \\ & & & \ddots & & \\ & & & & 1 & \\ & & & & & \ddots \end{pmatrix}$$

which is homotopic to the matrix product of

$$\begin{pmatrix} f(x) & & \\ & 1 & \\ & & \ddots \end{pmatrix} \quad \text{and} \quad \begin{pmatrix} f'(x) & & \\ & 1 & \\ & & \ddots \end{pmatrix}$$

But the homotopy group addition, in the case of a group space, is by *multiplying* the values in the group.—That they are *isomorphisms* is easily checked for small values of s:

Case U. $s = 1 : \rho_1 - (i)$; $f(x) = (x_0 + ix_1) \in U(1)$ with $x_0^2 + x_1^2 = 1$. This is a generator a_1 of $\pi_1(U(1)) = \pi_1(U) = \mathbb{Z}$.

$$s = 3 : \rho_3 = \left\{ \begin{pmatrix} i & \\ & -i \end{pmatrix}, \begin{pmatrix} & 1 \\ -1 & \end{pmatrix}, \begin{pmatrix} & i \\ i & \end{pmatrix} \right\}; \; f(x) = \begin{pmatrix} x_0 + ix_1 & x_2 + ix_3 \\ -x_2 + ix_3 & x_0 - ix_1 \end{pmatrix}$$

with $x_0^2 + x_1^2 + x_2^2 + x_3^2 = 1$. $f: S^3 \to SU(2) = S^3$ is a generator of $\pi_3(SU(3)) = \pi_3(U(3)) = \pi_3(U) = \mathbb{Z}$.

Case $\mathcal{O}$. $s = 0 : \tau_0 = \varnothing$; $f(x_0) = (x_0)$ with $x_0^2 = 1$, $x_0 = \pm 1$. This is a generator b_0 of $\pi_0(\mathcal{O}(1)) = \pi_0(\mathcal{O}) = \mathbb{Z}/2$.

$$s = 1 : \tau_1 = \left\{ \begin{pmatrix} & 1 \\ -1 & \end{pmatrix} \right\}; \; f(x) = \begin{pmatrix} x_0 & x_1 \\ -x_1 & x_0 \end{pmatrix} \quad \text{with} \quad x_0^2 + x_1^2 = 1.$$

This is a generator of $\pi_1(SO(2)) = \mathbb{Z}$; under $SO(2) \to SO(3) \to \mathcal{O}$ it becomes the generator b_1 of $\pi_1(SO(3)) = \pi_1(\mathcal{O}) = \mathbb{Z}/2$.

$s = 3 : \tau_3$ has been described in 4.4 by a matrix $f(x)$, $x \in S^3 \subset \mathbb{R}^4$. The map $f: S^3 \to SO(4)$ becomes, under $SO(4) \to SO(5)$ a generator b_3 of $\pi_3(SO(5)) \cong \pi_3(\mathcal{O}) = \mathbb{Z}$.

For higher values of s it is not easy to establish directly the isomorphisms. This is due to the fact, in the case U for example, that ρ_s has the size $2^{(s-1)/2}$; one then obtains a map $f: S^s \to U(2^{(s-1)/2})$ while the usual generators of $\pi_s(U)$ are in $\pi_s(U((s+1)/2))$, and $2^{(s-1)/2}$ is $> (s+1)/2$ for $s > 3$.

5.4. Using the ring structures the isomorphism proof becomes simple, as follows.

$\pi_*(U) = \bigoplus_{s=-1}^{\infty} \pi_s(U)$ (with a term $s = -1$ being $\mathbb{Z}$) is turned into a ring by exactly the same product formula (4) used for $\sigma_s \cdot \sigma_t$, which makes sense for continuous maps $f(x)$, $g(y)$; similarly for the case $\mathcal{O}$. Then trivially Φ and Ψ are ring homomorphisms

$$\Phi \colon \mathcal{E}_* \to \pi_*(U),$$

$$\Psi \colon \mathcal{E}_*^{\mathcal{O}} \to \pi_*(\mathcal{O}).$$

We have to determine the ring structure of $\pi_*(U)$ and $\pi_*(\mathcal{O})$. To this end we pass to (complex and real) *K-theory* and recall that $\pi_s(U) \cong \tilde{K}_{\mathbb{C}}(S^{s+1})$ and $\pi_s(\mathcal{O}) \cong \tilde{K}_{\mathbb{R}}(S^{s+1})$. As shown in [**E2**] the product (4) for elements of the homotopy groups $\pi_s(U)$, or $\pi_s(\mathcal{O})$ respectively, corresponds under these isomorphisms to the exterior tensor product of stable vector bundles, i.e., of elements of $\tilde{K}_{\mathbb{C}}(S^{s+1})$ or of $\tilde{K}_{\mathbb{R}}(S^{s+1})$. With regard to this product the complete K-theoretic formulation of Bott periodicity yields the ring structure of $\bigoplus_{-1}^{\infty} K_{\mathbb{C}}(S^{s+1}) = \pi_*(U)$, and similarly for $K_{\mathbb{R}}$ and $\pi_*(\mathcal{O})$. Namely (see [**K**]):

$\pi_*(U)$ is the polynomial ring $\mathbb{Z}[a_1]$ in the generator $a_1 \in \pi_1(U)$. And $\pi_*(\mathcal{O})$ is the (graded) commutative ring generated by the generators $b_0 \in \pi_0(\mathcal{O})$, $b_3 \in \pi_3(\mathcal{O})$, and $b_7 \in \pi_7(\mathcal{O})$ with only relations $2b_0 = 0$, $b_0^3 = 0$, $b_3^2 = 4b_7$.

Thus all we have to prove is that $\Phi(\rho_1) = a_1$ and $\Psi(\tau_0) = b_0$, $\Psi(\tau_3) = b_3$. But this has been done above.

THEOREM 5.7. *The maps* $\Phi \colon \mathcal{E}_* \to \pi_*(U)$ *and* $\Psi \colon \mathcal{E}_*^{\mathcal{O}} \to \pi_*(\mathcal{O})$ *are ring isomorphisms.*

COROLLARY 5.8. *For all* s, *$\Phi(\rho_s)$ is a generator of $\pi_s(U)$ and $\Psi(\tau_s)$ is a generator of $\pi_s(\mathcal{O})$. In other words, these generators of $\pi_s(U)$ and $\pi_s(\mathcal{O})$ are given by maps $S^s \to U$ or $S^s \to \mathcal{O}$ which are linear with respect to the coordinates of $S^s \subset \mathbb{R}^{s+1}$, the coefficient matrices being E and the minimal size H-R solutions.*

COROLLARY 5.9. *(a) Any continuous map $S^s \to U$ is homotopic to a linear map $f \colon S^s \to U(n) \to U$, for suitable n, of the form $f(x) = \sum_0^s x_j A_j$, $x = (x_0, x_1, \ldots, x_s) \in S^s \subset \mathbb{R}^{s+1}$ with $A_0 = E$ and where $A_1, \ldots, A_s$ are H-R matrices of size n.*

(b) If a linear map $f \colon S^s \to U(n)$ is nullhomotopic in U then it is linearly nullhomotopic, i.e., restricted from a linear map $S^{s+1} \to U(n)$.

Similarly for maps $S^s \to \mathcal{O}$.

PROBLEM. Prove Corollary 5.9 directly, i.e., without using Bott periodicity. This would reduce the proof of the Theorems of Bott to the algebra of H-R matrices as described in this survey.

References

[E]   B. Eckmann, *Gruppentheoretischer Beweis des Satzes von Hurwitz-Radon über die Komposition quadratischer Formen*, Comment. Math. Helvetici **15** (1942/43), 358–366.

[E2] _____, *Hurwitz-Radon matrices and periodicity modulo 8*, L'Ens. Math. **35** (1989), 77–91.

[H]   A. Hurwitz, *Über die Komposition der quadratischen Formen*, Math. Ann. **88** (1923), 1–25.

[LS] T. Y. Lam and T. Smith.

[K]   M. Karoubi, *K-Theory*, Springer-Verlag, Berlin-Heidelberg-New York, 1978.

[R]   J. Radon, *Lineare Scharen orthogonaler Matrizen*, Abh. Math. Sem. Hamburg (1922), 1–14.

Birth of fibre spaces, and homotopy

Expo. Math. 17 (1999), 23–34

Home is where one starts from. As one grows older
the world becomes stranger, the pattern more complicated
of dead and living
T.S. Eliot

In this article I will describe an episode in the history of mathematics, very much restricted in content and in time: the origins of the homotopy theory of fibre spaces, roughly from 1935 to 1950. My account, following as closely as possible my oral presentation, will describe the episode more or less as I experienced it myself and will thus have a somewhat personal flavour.

Two warnings: 1) The beginnings of the theory of vector bundles (with structure group, "fibre bundles", "sphere bundles"), which date roughly from the same period, will not be treated. I will confine myself to results and problems of homotopical nature.

2) In speaking of "us" I am thinking on the one hand of the group of students of Heinz Hopf, and on the other hand of the three authors or groups of authors who independently developed the subject, communications having been interrupted by the war: Ehresmann-Feldbau [166], Hurewicz-Steenrod [260], and myself [148]. As to references I take advantage of the opportunity to refer to the bibliography of the monumental work by Dieudonné "A History of Algebraic and Differential Topology 1900-1960".

The beginnings were simple. We understood that one could combine the ideas of Hurewicz [256] on the homotopy groups with the notion of a fibre space suggested by the Hopf fibrations [245]. There emerged a mass of new results and interesting problems. One can say – and I will speak of that later – that many subsequent developments in topology, algebra, and several other disciplines originated during this episode; a network of disciplines and relations between them of increasing complexity was created thus contributing to the unity of mathematics itself.

1. Hopf fibrations and generalizations

1.1 The term "fibration" in the sense of this article appears for the first time in 1935 in Hopf's memoir [245] "Über die Abbildungen von Sphären auf Sphären niedrigerer Dimension". In an appendix one finds the "Hopf fibrations" of spheres by spheres

---

* Lecture delivered (in French) at the Colloque "Matériaux pour l'histoire des mathématiques au XXième siècle" in honor of Jean Dieudonné, Nice 6-8 January 1996. Translated by Peter Hilton

a) $p: S^{2n+1} \longrightarrow \mathbb{C}P^n$, fibre S^1
b) $p: S^{4n+3} \longrightarrow \mathbb{H}P^n$, fibre S^3
c) $p: S^{15} \longrightarrow \mathbb{O}P^1$, fibre S^7

They are obtained by representing the sphere by coordinates $z_0, z_1, ..., z_n$, which are, respectively, complex numbers, quaternions, and octonions (Cayley numbers), and then passing to homogeneous coordinates $z_0 : z_1 : ... : z_n$. Thus $\mathbb{C}P^n$ is complex projective space, $\mathbb{H}P^n$ is quaternionic projective space, and $\mathbb{O}P^1$ is the octonionic projective line – in the last case we can only have $n = 1$ because the multiplication of octonions is not associative. The case $n = 1$ gives the fibrations $S^3 \longrightarrow S^2$, $S^7 \longrightarrow S^4$, $S^{15} \longrightarrow S^8$. The projection p is a continuous function and the counterimage $p^{-1}(x)$ of a point x is, respectively, S^1, S^3, S^7. Thus the spheres in question are partitioned, in a very particular way, into spheres S^1, S^3, S^7.

1.2 Without giving a definition of the term, Hopf called these partitions fibrations (this expression had been used previously by Seifert [420] in a special case concerning three-dimensional manifolds, where he allowed the possibility of "exceptional" fibres; this concept remains interesting to this day). But soon after we noticed that one was dealing with a situation that we met in geometry in many other cases: one had a continuous map $p : E \longrightarrow B$ where all the counterimages $p^{-1}(b) = F_b, b \in B$ are homeomorphic, and where each point $b \in B$ has a neighborhood U such that $p^{-1}(U)$ is homeomorphic, via p, to $U \times F_b$. One says that E is a (locally trivial) fibre space, with base B, projection p, and fibres F_b, homeomorphic to a fibre-type F.

1.3 Typical examples. 1) E is the space of unit tangent vectors to a differentiable manifold B of dimension n (equipped with a Riemannian metric), F_b is the set of vectors at b , $F = S^{n-1}$.

2) E is a Lie group, F a closed subgroup, B is the corresponding homogeneous space.

3) $E = V_{n,m}$, $m \leq n$, the space of frames of m orthonormal vectors in $\mathbb{R}^n$, $B = V_{n,m-1}$, p consists of omitting the last vector, and $F = S^{n-m}$. There is the unitary analogue in $\mathbb{C}^n$, and other analogues obtained by replacing $m - 1$ by $m - k$.

4) A special case of 2) and 3) is given by the unitary groups: $E = U(n)$, $F = U(n-1)$ and $B = S^{2n-1}$. There is, of course, an analogous construction for the orthogonal groups. In all these cases the projection p is the obvious map.

2. Homotopy groups

It was also in 1935, and in 1936, that there appeared the Notes of Hurewicz [256] "Beiträge zur Topologie der Deformationen". It was very quickly clear that one was dealing here with something very different from what had been done previously in algebraic topology (called "combinatorial" at that time) – with the exception of the fundamental group of which the homotopy groups $\pi_n(X), n = 1, 2, 3...$ were, to be sure, a generalization. They had been invented back in 1932 by Čech [121] but were redefined and applied by Hurewicz with quite unexpected results. Moreover they seemed to us, a little later, wonderfully adapted to the study of homotopical questions related to fibrations.

2.1. First, quickly, the definitions. One considers pointed spaces X, that is, spaces furnished with a base-point x_0. Maps and homotopies are pointed, that is, respect the base-point. The elements of $\pi_n(X)$ are homotopy classes of maps $S^n \longrightarrow X$, the base-point of S^n being s_0. If h is a standard map of the unit cube I^n to S^n which maps the interior of I^n homeomorphically to $S^n - s_0$ and which sends the frontier $\dot{I}^n$ of I^n to s_0, one may use h to identify $\pi_n(X)$ with the set of homotopy classes $I^n, \dot{I}^n \longrightarrow X, x_o$. In I^n one chooses a distinguished direction and decomposes I_n as $I \times I^{n-1}$. The group operation is then defined by $f + g$ as follows: $I, \{0 \le t \le 1\}$ is separated into I_1 and I_2, $\{0 \le t \le 1/2\}$ and $\{1/2 \le t \le 1\}$. One then compresses f onto $I_1 \times I^{n-1}$ and g onto $I_2 \times I^{n-1}$ and obtains $f + g$. For $n = 1$ this is indeed the rule of addition (non-commutative in general) of the fundamental group $\pi_1(X)$. The definition is homotopy invariant, and the group axioms are verified exactly as for $\pi_1(X)$. In particular, the neutral element is the homotopy class of the constant map (to the base-point). Quite generally, whatever space one maps into X, a map in this class is said to be "nullhomotopic". To simplify the presentation I will permit myself not always to distinguish between a map f and its homotopy class.

A map $h : X \longrightarrow Y$ induces, by composition $S^n \longrightarrow X \longrightarrow Y$, a homomorphisme $h_* : \pi_n(X) \longrightarrow \pi_n(Y)$. One sees easily that in the case of a covering map $\overline{X} \longrightarrow X$ the homotopy groups $\pi_n(X)$ and $\pi_n(\overline{X})$ are isomorphic for $n \ge 2$; for example $\pi_n(S^1) = \pi_n(\mathbb{R}) = 0$. Other elementary properties: $\pi_i(S^n) = 0$ for $i < n$ (by simplicial approximation one lands in $S^n - s_0$ which is contractible), and $\pi_i(X \times Y) = \pi_i(X) \times \pi_i(Y)$. If X is simply connected then pointed homotopy is equivalent to free homotopy.

2.2 If X is an H-space, that is, a space furnished with a continuous multiplication, denoted $x.x'$, with neutral element e up to homotopy, one readily verifies that

$$(f_1 + f_2).(g_1 + g_2) = f_1.g_1 + f_2.g_2,$$

always up to homotopy. Hence e being the constant map

$$(f + e).(e + g) = f + g = f.g,$$

$$(e + g).(f + e) = f + g = g.f.$$

It follows that $f + g$ is given by multiplication X, and that $f + g = g + f$. Thus for a topological group G, $\pi_1(G)$ is abelian – as was well-known previously. But in general, for arbitrary X, maps $I^n, \dot{I}^n \longrightarrow X, x_0$ can be identified with maps $I^{n-1}, \dot{I}^{n-1} \longrightarrow \Omega X, x_0$, where ΩX is the space of maps $I, \dot{I} \longrightarrow X, x_0$ (the space of loops on X at x_0). One thus has $\pi_n(X) = \pi_{n-1}(\Omega X)$ for $n \ge 2$. But ΩX is an H-space by composition of loops.

The homotopy groups $\pi_n(X)$ are thus abelian for all X and all $n \ge 2$. It seems that for this reason the groups were believed not to be interesting when they were first presented by Čech in 1932.

2.3. *Two "miracles":*

1) We were surprised to note that one could give a very simple and transparent proof of the fact that

$$\pi_n(S^n) = \mathbb{Z},$$

the isomorphism being given by associating with a map $f : S^n \longrightarrow S^n$ its degree. Indeed, one sees by the methods of simplicial approximation that the degree is homotopy invariant and that one has

a) the function degree $\pi_n(S^n) \longrightarrow \mathbb{Z}$ is a homomorphisme,

b) $\pi_n(S^n)$ is generated by the identity map (degree = 1).

In other words, one retrieves Hopf's theorem [244] which says that two maps $S^n \longrightarrow S^n$ of the same degree are homotopic.

With the help of the exact sequence of a fibration, which will be discussed in the next section, one concludes that

$$\pi_3(S^2) = \mathbb{Z},$$

generated by the Hopf fibration $S^3 \longrightarrow S^2$. Thus, in particular, there exist infinitely many non-homotopic maps $S^3 \longrightarrow S^2$. This fact was established in 1931 by Hopf [243] in his famous work where he introduced the "Hopf invariant"; the new proof was quite different, both more simple and more precise: In the exact sequence

$$... \longrightarrow \pi_3(S^1) \longrightarrow \pi_3(S^3) \xrightarrow{p_*} \pi_3(S^2) \longrightarrow \pi_2(S^1) \longrightarrow ...$$

one sees that $\pi_3(S^3) = \mathbb{Z}$ is isomorphic to $\pi_3(S^2)$ via p_* where $p : S^3 \longrightarrow S^2$ is the Hopf fibration; and since $\pi_3(S^3)$ is generated by the identity, $\pi_3(S^2)$ is generated by p_* of the identity, that is, simply by p.

3. The exact sequence of a fibration

What one called later the exact sequence of a fibration $p : E \longrightarrow B$ with fibre F was first formulated without arrows, simply as a series of isomorphisms relating the homotopy groups of E, B, and F. I allow myself to use here from the outset the notation of exact sequences, which is much more convenient [in fact the history of exact sequences is quite complex and curious; I will not go into this subject].

3.1 One has induced homomorphisms

$$\pi_n(F) \xrightarrow{i_*} \pi_n(E) \xrightarrow{p_*} \pi_n(B),$$

where p is the fibration and $i : F \longrightarrow E$ the inclusion in E of the fibre F over the base-point of B; the base-point of E is chosen in F. To define a homomorphism $\Delta : \pi_n(B) \longrightarrow \pi_{n-1}(F)$, and to demonstrate that the long sequence thus obtained

$$... \longrightarrow \pi_n(F) \xrightarrow{i_*} \pi_n(E) \xrightarrow{p_*} \pi_n(B) \xrightarrow{\Delta} \pi_{n-1}(F) \longrightarrow ...$$

is exact (at each stage image = kernel) one uses the *lifting homotopy* property:

Lemma. Let $f' : X \longrightarrow E$ and $f = pf' : X \longrightarrow B$. Then any homotopy H of f can be lifted to a homotopy H' of f' such that $H = pH'$.

In fact the only spaces X which come into the proof of exactness are the spheres S^n and the cubes I^n. Let us grant this Lemma; I return in **3.3** to its proof. The Lemma is suggested by the classical case of a covering map (discrete fibre), where the lift exists and is even unique. Let us note an immediate consequence.

If $f = pf'$ and if f is homotopic to g then there exists $g' : X \longrightarrow E$, homotopic to f', such that $g = pg'$. In other words: "Every map homotopic to a projection is itself a projection. In particular, every nullhomotopic map is a projection".

3.2 An element of $\pi_n(B)$ can be represented by a map f of I^n into B such that the frontier of I^n, which is a sphere S^{n-1}, is sent to the base-point. Since I^n is contractible, f is a projection $f = pf'$ where f' maps I^n into E and its frontier S^{n-1} into F. All such homotopy classes of maps of I^n into E, with a law of addition analogous to that of $\pi_n(E)$, form the *relative group* $\pi_n(E \bmod F)$ of E modulo F. It is mapped by p_* into $\pi_n(B)$, and what just has been said shows that p_* surjective; the injectivity follows immediately from the Lemma. Thus one has

$$\pi_n(E \bmod F) = \pi_n(B).$$

The homomorphism Δ is then defined by passing from $f : I_n \longrightarrow B$ to f' as above, followed by the restriction of f' to S^{n-1}.

The fact that the sequence is exact may be easily verified at each of the three stages. For example, if for $f : S^n \longrightarrow E$ one has $p_* f = 0$ then the nullhomotopy of pf may be lifted, giving a homotopy from f to a map $S^n \longrightarrow F$, so that f is in the image of i_*. I leave the reader to study what happens in low dimensions.

3.3 All this seemed clear to us, in particular the Lemma. But of course there was something to be proved, under a hypothesis requiring verification in the interesting examples.

The hypothesis that I chose was that of a *retraction*: One supposes that every point $b \in B$ has a neighborhood $U(b)$ such that there exist a retraction $R(x, b)$ of $p^{-1}(U(b))$ onto F_b depending on b :

$$R(x, b) \in F_b \text{ for all } x \in p^{-1}(U(b)), \; = x \text{ if } x \in F_b.$$

In the concrete examples mentioned in **1.3** E, B and F are differentiable manifolds; one can furnish them with a Riemannian metric and easily construct such a retraction by means of geodesics orthogonal to each fibre.

If a retraction $R(x, b)$ is given one chooses $y \in F_b$ and sets for each $b' \in U(b)$

$$t(b') = R(y, b').$$

Then t is a map of $U(b)$ in E which is a homeomorphism of $U(b)$ onto a set $V(y)$ transversal to the fibres $F_{b'}$. This "lifting of a neighborhood around an arbitrary point y" allows one to lift a homotopy bit by bit and thus to demonstrate the Lemma.

Remark: To establish the Lemma and the exact homotopy sequence it suffices to consider a map $p : E \longrightarrow B$, to call the inverse images $p^{-1}(b)$ "fibres" , and to suppose the existence of a retraction $R(x, b)$; it is not necessary to have a locally trivial fibration in the sense of **1.2**. Serre [429] went even further and merely supposed the existence of liftings for maps $I^n \longrightarrow B$.

4. Results

Once the Lemma and the exact homotopy sequence were established, results fell from the skies! The most immediate results were simply based on what was already familiar, by elementary reasoning, on the homotopy groups of spheres: $\pi_i(S^1) = 0$ for $i \geq 2$, $\pi_i(S^n) = 0$ for $i < n$, and $\pi_n(S^n) = \mathbb{Z}$ for all $n \geq 1$.

4.1 The Hopf fibrations $S^3 \longrightarrow S^2$ with fibre S^1, $S^7 \longrightarrow S^4$ with fibre S^3, and $S^{15} \longrightarrow S^8$ with fibre S^7, give

$$\pi_3(S^2) = \mathbb{Z}, \pi_7(S^4) \supset \mathbb{Z}, \text{ and } \pi_{15}(S^8) \supset \mathbb{Z},$$

the generator of $\mathbb{Z}$ being the projection.

4.2 The fibrations $S^{2k+1} \longrightarrow \mathbb{C}P^k$, $k \geq 1$ with fibre S^1, give

$$\pi_2(\mathbb{C}P^k) = \mathbb{Z}, \text{ and } \pi_i(\mathbb{C}P^k) = \pi_i(S^{2k+1}), i > 2.$$

4.3 The fibrations $U(n) \longrightarrow S^{2n-1}$ with fibre $U(n-1)$ give

$$\pi_s((U(n)) = \pi_s(U(\frac{s+1}{2})), \ n \geq \frac{s+1}{2} \text{ if } s \text{ is odd,}$$

$$\pi_s(U(n)) = \pi_s(U(\frac{s+2}{2})), \ n \geq \frac{s+2}{2} \text{ if } s \text{ is even.}$$

This is what was later called the stability of $\pi_s(U(n))$. For small values of s the stable groups are easily obtained: $\pi_1(U(n)) = \pi_1((U(1)) = \mathbb{Z}$, generated by the identity $S^1 \longrightarrow U(1)$. For $s = 2$ one has

$$\pi_2 U(n)) = \pi_2(U(2)) = \pi_2(SU(2)) = \pi_2(S^3) = 0,$$

and for $s = 3$ by analogous reasoning

$$\pi_3(U(n)) = \pi_3(S^3) = \mathbb{Z}, \ n \geq 2.$$

For $s > 3$ things proved much more difficult. By examining very closely the homomorphism Δ in the homotopy exact sequence of the fibration in question I succeeded in showing [149] that $\pi_4 = 0$, et $\pi_5 = \mathbb{Z}$.

4.4 The Lemma may clearly be applied not only to homotopy groups but as well to all sorts of homotopy properties for maps of X into the base B of a fibration $E \longrightarrow B$. Let us consider three examples:

a) $X = I^n$. Then every map $f : I^n \longrightarrow B$ is a projection pf'. This result, almost trivial, in the case of the fibration 3) in **1.3** seems to be the first consequence of the Lemma for a problem in analysis (Theorem of Wazewski, cited in [148]): to complete a real orthogonal $n \times (m - 1)$-matrix with $m < n$, which is a continuous function in I^n, by a supplementary line.

b) $X = B$ and $f = $ identity. A lifting of f is called a *section* of the fibration; vector fields or tangent frames of vector fields to a manifold are special cases. From the exact sequence one deduces a necessary condition for the existence of a section: The groups $\pi_n(E)$ decompose into the direct sum of $\pi_n(B)$ and $\pi_n(F)$. One thus establishes interesting cases of non-existence of sections. In [E1] I showed, by means of more complicated arguments adapted to this case, that *spheres of dimension $4k+1$, $k \geq 1$ do not admit two orthogonal unit tangent vector fields.*

c) One calls a map $f : X \longrightarrow Y$ *essential* if every map homotopic to f is surjective, that is, has an image covering all of Y. In view of the Lemma one sees that if the identity of E is essential then so is the projection $E \longrightarrow B$. All the fibrations in **1.3** are examples since E is a closed manifold and the identity map has degree 1.

To be sure, not only results fell from the skies but also problems. The examples above show this clearly. Let us simply mention that the determination of the stable homotopy groups of the unitary groups, for $s > 5$, and of the maximum number of orthonormal tangent vector fields on a sphere remained open.

5. Homotopy equivalence, aspherical spaces

5.1 One cannot mention the Notes of Hurewicz [256] without speaking of aspherical spaces. The spaces considered were cell-complexes (at that time, simplicial); X is said to be *aspherical* if $\pi_n(X) = 0$ for all $n \geq 2$. For such spaces X and Y Hurewicz showed by induction on the n-skeleton of X, $n = 1, 2, ...$ that

a) If $f, g : X \longrightarrow Y$ induce the same homomorphism $f_* = g_* : \pi_1(X) \longrightarrow \pi_1(Y)$ of fundamental groups then f and g are homotopic.

b) Every homomorphism $h : \pi_1(X) \longrightarrow \pi_1(Y)$ may be realized by a map $f : X \longrightarrow Y$ such that $f_* = h$. Whence

c) If $\pi_1(X)$ and $\pi_1(Y)$ isomorphic, there exist $f : X \longrightarrow Y$ and $g : Y \longrightarrow X$ such that gf and fg are homotopic to the identity of X and Y respectively. One thus is led to the notion of homotopy equivalence.

In general, for spaces X and Y (not necessarily aspherical) a map $f : X \longrightarrow Y$ such that there exists $g : Y \longrightarrow X$ as in c) above is called a *homotopy equivalence*. One says in this case that X and Y are of the same homotopy type (notation $X \sim Y$). Plainly this is a generalization, and a very important one, of the notion of homeomorphism. A homotopy equivalence $f : X \longrightarrow Y$ induces isomorphisms $f_* : \pi_n X \longrightarrow \pi_n Y$ for all $n \geq 1$; the same is true for homology groups.

For an *aspherical space X* one thus has the following results.

The homotopy type of X is completely determined by its fundamental group. In particular, all homology groups of X are determined by the fundamental group of X.

And *if the fundamental group of an aspherical space is trivial then $X \sim$ point, that is, X is contractible.*

5.2 Analogous arguments apply to spaces aspherical in dimensions $n < N$ for some $N \geq 2$. It suffices to modify slightly a) and b) above. Thus we are now concerned with maps of the N-skeleta of X and Y; and in a) f is not necessarily homotopic to g, but to a map coinciding with g on the $(N-1)$-skeleton of X. One thereby proves that the homology groups $H_i(X)$, $i < N$ are determined by $\pi_1(X)$. It is not so, in general, for

$H_N(X)$; but the quotient group H'_N of H_N modulo the subgroup of "spherical" elements (represented by cycles which are images if spheres) is determined by $\pi_1(X)$.

More generally, if $\pi_i(X) = 0$ for $i < k$ and $k < i < N$, then $H_i(X)$, $i < N$, and $H'_N(X)$ are determined by $\pi_k(X)$.

5.3 The résumé above does not correspond precisely to what is to be found in the Notes of Hurewicz on this subject. On the one hand the Notes go considerably further (Hurewicz homomorphism, etc.), and on the other I have gone beyond them a little in 5.2 – but at that time it was already obvious that the ideas could be applied in this more general fashion. Thus I would like not only to emphasize the importance of Hurewicz' ideas but also lay the ground work for a typical, concrete application to fibrations.

6. Fibrations of spheres by tori

6.1 Let us consider [ESW] a fibration $S^n \longrightarrow B$ with fibre a torus T^s of dimension s, where $n > 2$ and $s \geq 1$. One sees easily that the case $n = 2$, $s = 1$ is impossible. The fibration is supposed sufficiently regular for the Lemma and its consequences (Section **3**) to apply. The exact homotopy sequence gives

$$\pi_1(B) = 0, \ \pi_2(B) = \pi_1(T^s) = \mathbb{Z}^s,$$

$$\pi_i(B) = 0 \text{ for } 2 < i < n$$

One wishes to compare B with a known space Y having the same homotopy groups as those above. One chooses for Y the topological product of s copies of $\mathbb{C}P^m$ with m sufficiently large; it suffices (see **4.2**) to use the homotopy sequence of the fibration a) in **1.1** to infer that the values of π_i are the desired ones, and even that $\pi_i = 0$ a little beyond n.

In the light of **5.2** one thus has for the homology groups

$$H_i(B) = H_i(Y) \text{ for } i < n,$$

$$H'_n(B) = H'_n(Y) = H_n(Y)$$

The dimension of B being $n-s$ it follows that $H_n(Y) = 0$, so n is odd ($\mathbb{C}P^m$, and hence Y has non-zero homology if and only if the dimension is even). Since then $H_{n-1}(Y) \neq 0$, s must be $= 1$.

The sphere S^n can be fibred by tori T^s only if n is odd and $s = 1$. And in that case one has the Hopf fibration.

6.2 One thus recovers the well-known result that if some sphere S^n is a Lie group, then its rank (the dimension of the maximal abelian subgroups) must be $= 1$. But the geometric method takes us further [S].

If S^n, $n > 1$, is a Lie group, hence of rank 1, then the one-parameter subgroups, homeomorphic to S^1, are all conjugate to each other. As they are determined by the tangent at the neutral element, the collection of them can be identified with real projective space $\mathbb{R}P^{n-1}$; on the other hand it may be identified with the classes of S^n modulo the normaliser $N(S^1)$ of a subgroup S^1. This normaliser consists of a finite number of copies of S^1. Thus one has a fibration $S^n \longrightarrow \mathbb{R}P^{n-1}$ which gives rise to an exact sequence where π_i of the base is $\pi_i(S^{n-1})$ and π_i of the fibre is $\pi_i(S^1)$. In particular, since

$$\pi_2(S^1) \longrightarrow \pi_2(S^n) \longrightarrow \pi_2(S^{n-1}) \longrightarrow \pi_1(S^1) \longrightarrow \pi_1(S^n)$$

is exact it follows that $\pi_2(S^{n-1}) = \pi_1(S^1) = \mathbb{Z}$, whence $n - 1 = 2$, $n = 3$:
   The only spheres which can be Lie groups are S^1 and S^3.

7. And then?

One can trace the influence of the episode, described in this lecture, on almost all mathematical disciplines right up to the present day (this is probably true of most ideas which were new a long time ago). To study the mutual relations between the various trends, schools, and methods would be a fascinating but very difficult task. Have there not been ideas and methods which were very popular and important for a certain type of problem, which suddenly disappeared – only to be reborn later in a different context? "The pattern becomes more and more complicated of dead and living."

I restrict myself to mentioning here a list of developments directly related to what I have said above, and which occurred immediately after or even during that episode. It will only be possible to make brief allusions.

7.1 *Homotopy groups of spheres.* The results of **4.1** concern the case of $\pi_n(S^m)$ with $n = 2m - 1$ where there is a term $\mathbb{Z}$. Other similar cases were known. But in 1950 Serre [429] proved some sensational results about $\pi_n(S^m)$ for $n > m$: These groups are always finite except for the case of m even and $n = 2m - 1$ where the group is a direct sum of $\mathbb{Z}$ and a finite group. But besides, the topic of $\pi_n(S^m)$ is too vast to be described here more generally ("suspension" of Freudenthal [201], stable homotopy groups of spheres).

7.2. *Eilenberg-MacLane spaces $K(G, n)$.* Eilenberg and MacLane [179], [181] studied spaces with $\pi_i = 0$ for $i \neq n$. Following **5.2** the homotopy type of such a space is determined by n and $\pi_n = G$, abelian if $n \geq 1$. These spaces play a universal role in homology, cohomology, and all operations. Their existence, for given G and n was established by J.H.C.Whitehead [500].

7.3 *Group homology and cohomology, homological algebra.* The homology of an aspherical space being determined by its fundamental group – which is arbitrary – algebraic methods were swiftly developed for calculating the homology as a function of G. Thus was the (co-)homology of groups born, and with it the much more general field of homological algebra.

7.4. *Homotopy type.* J.H.C.Whitehead had already shown around 1940 that a map $f : X \longrightarrow Y$ which induces isomorphisms of all homotopy groups is a homotopy equivalence. He simplified the proof later by inventing the notion of CW-complex [493, pp.95-105].

7.5. *Categories and functors.* Eilenberg and MacLane [178], [180] realised that the general ideas behind such notions as homotopy equivalence, "natural" isomorphisms, etc.

have a much deeper significance. At first their theory of categories, functors, and natural equivalences seemed just to be a precise language; but later one abstracted from them a mathematical structure both fundamental and useful.

7.6 *Parallelisable spheres.* A differentiable manifold of dimension n (furnished with a Riemannian metric) is said to be parallelisable if it admits a continuous field of orthonormal tangent n-frames. In view of **4.4** the spheres S^{4k+1}, $k \geq 1$ are certainly not parallelisable. Kervaire [272] and Bott-Milnor [82] showed in 1958 that S^n is parallelisable (if and) only if $n = 1, 3$, or 7. This result could later be easily derived from the celebrated theorem of Adams [2], 1960. In 1966 Adams [A] determined the exact number k such that S^n admits a tangent k-frame.

7.7 *Stable homotopy groups of the unitary groups.* The very partial results of **4.3** have been gradually improved. The problem was completely solved in 1956 by Bott [77], using subtle methods of differential geometry: $\pi_n(U(m)) = \mathbb{Z}$ for n odd, $m \geq \frac{n+1}{2}$, and $= 0$ for n even and $m \geq \frac{n+2}{2}$. This is "Bott periodicity" (there is an analogous but more complicated result for the orthogonal groups). This geometric solution has had enormous consequences (topological K-theory, general cohomological functors).

7.8. Let me add a personal remark concerning this last topic. I have been considering, since the earliest calculations, *linear* maps f of $S^n = \{x \in \mathbb{R}^{n+1}, x = (x_o, x_1, ..., x_n, |x| = 1\}$ into $U(m)$, that is maps of the form $f(x) = \sum x_j A_j$ where the A_j are $m \times m$-matrices. It follows that the A_j are unitary "Hurwitz-Radon" matrices [H], [R] and [E2]; and conversely, every system of $n + 1$ unitary Hurwitz-Radon matrices determines a linear map of S^n into $U(m)$. I conjectured that in the stable range every homotopy class contains a linear map, and that if a linear map is nullhomotopic then it is so in a linear fashion, that is by a supplementary Hurwitz-Radon matrix which provides a linear map of S^{n+1} into $U(m)$. In the light of what one knows about these matrices this would have yielded Bott's theorem. But it is only following the work of Bott, using his result on the multiplicative structure of K-theory, that I have been able to demonstrate the truth of this conjecture (see [E3]). There still remains the problem of demonstrating it directly, thus reducing Bott periodicity to the purely algebraic discussion of Hurwitz-Radon matrices.

References

The numbers in brackets [ ] refer to the bibliography in the work of Jean Dieudonné "A History of Algebraic and Differential Topology 1900-1960".

Some additions
[A] J.F.Adams, Vector fields on spheres, Ann. of Math. 75 (1962), 603-632
[E1] B.Eckmann, Systeme von Richtungsfeldern auf Sphären und stetige Lösungen komplexer linearer Gleichungen, Comment. Math. Helv. 15 (1942/43), 1-26
[E2] B.Eckmann, Gruppentheoretischer Beweis des Satzes von Hurwitz-Radon über die Komposition quadratischer Formen, Comment. Math. Helv. 15 (1942/43), 358-366

[E3] B.Eckmann, Hurwitz-Radon matrices revisited, CRM Proceedings and Lecture Notes, Vol.6 (AMS 1994), 23-35

[ESW] B.Eckmann, H. Samelson, G.W.Whitehead, On fibering spheres by toruses, Bull. Amer. Math. Soc. 55 (1949), 433-438

[S] H.Samelson, Über die Sphären, die als Gruppenmannigfaltigkeiten auftreten, Comment. Math. Helv. 13 (1940), 144-155

4-Manifolds, group invariants, and ℓ_2-Betti numbers

L'Enseignement Mathématique 43 (1997), 271–279

It has been known for some time that closed 4-manifolds provide, via the fundamental group and the Euler characteristic, interesting invariants for finitely presented groups. In this short survey we describe these and more refined invariants (using also the signature of the manifold), and explain some of their significance. The invariants are not easily calculated in general, but quite good information is obtained using l_2-Betti numbers.

The topic has been developed by several authors, more or less independently. We mention Hausmann-Weinberger [H-W], Kotschick [K], Lück [L], and myself [E1], [E2]. The paper [K] contains a wealth of information on the invariants and further important references; the application of l_2-Betti numbers appears in [E2] and in [L].

1. A BASIC CONSTRUCTION

1.1. We will always denote by M a connected orientable closed 4-manifold (compact without boundary) admitting a cell decomposition. The fundamental group $G = \pi_1(M)$ is finitely presented. Indeed, homotopy classes of loops can be represented by edge-polygons and null-homotopies of these by using 2 cells. Conversely, any finitely presented group G is the fundamental group of a closed 4-manifold. If

$$G = \langle g_1, \ldots, g_m \mid r_1, \ldots, r_n \rangle$$

is a presentation of G, there is a standard procedure for constructing such a manifold: One first puts $M' = S^1 \times S^3 + \cdots + S^1 \times S^3$, connected sum, one copy for each generator g_i of G. Then $\pi_1(M')$ is a free group on generators $g_1, \ldots, g_m$. A relator, say r_1, is a word in the g_i and can be represented by a loop S^1 in M'.

A tubular neighbourhood $S^1 \times B^3$ of S^1, where B^k is the k-dimensional ball, has boundary $S^1 \times S^2$. Replacing the interior by $B^2 \times S^2$ with the same boundary yields a new 4-manifold where the element corresponding to r_1 has been killed; and similarly for the other r_i. Let M_0 be the 4-manifold thus obtained, fulfilling $\pi_1(M_0) = G$. The idea of that construction can already be found in the old book [S-T]. Much later the procedure, in a more general context, has been called "elementary surgery".

1.2. We recall that the (good old) Euler characteristic $\chi(X)$ of a finite cell complex X is the alternating sum

$$\chi(X) = \sum (-1)^i \alpha_i ,$$

where α_i is the number of i-cells. It is easily computed for M_0 above: For M' it is $2 - 2m$ since it is $= 0$ for $S^1 \times S^3$ and since it decreases by 2 in a connected sum. Under the surgery process above it increases by 2 [use the fact that for the union of two complexes X and Y with intersection Z the characteristic is $\chi(X) + \chi(Y) - \chi(Z)$; and that $\chi(B^2 \times S^2) = 2$]. Whence

$$\chi(M_0) = 2 - 2m + 2n = 2 - 2(m - n) .$$

The difference $m - n$ is called the deficiency of the presentation of G.

1.3. On the other hand the characteristic can be expressed by the Betti numbers of the cell complex X as $\sum (-1)^i \beta_i(X)$ where $\beta_i(X) = \dim_{\mathbf{R}} H_i(X; \mathbf{R})$ (and is therefore a topological invariant). Moreover the β_i of a manifold fulfill Poincaré duality, i.e. they are equal in complementary dimensions. Thus $\chi(M) = 2 - 2\beta_1(M) + \beta_2(M)$. We recall that homology in dimension 1 depends on the fundamental group G only; β_1 is the $\mathbf{Q}$-rank of G Abelianised and we write $\beta_1(G)$ for $\beta_1(M)$. Comparing with $\chi(M_0)$ above we see that the deficiency of the presentation is $\leq \beta_1(G)$. Thus there is a maximum for the deficiency of all presentations of G, called the deficiency $\mathrm{def}(G)$ of G. [For this simple side result there are, of course, much easier arguments.]

2.   The Hausmann-Weinberger invariant

2.1. As seen above, the Euler characteristic of a 4-manifold M with given finitely presented fundamental group G is bounded below by $2 - 2\beta_1(G)$. The minimum of $\chi(M)$ for all such M has been considered by Hausmann-Weinberger [H-W] and denoted by $q(G)$. Using M_0 above we have the inequalities

$$2 - 2\beta_1(G) \leq q(G) \leq 2 - 2\,\mathrm{def}(G) .$$

2.2. EXAMPLES.

1) In [H-W] it is shown, by a simple argument, that $q(\mathbf{Z}^n) \geq 0$ for all $n \geq 1$. We return to that case later on. Here we just recall that $q(\mathbf{Z}) = q(\mathbf{Z}^2) = q(\mathbf{Z}^4) = 0$, as is easily seen by taking an appropriate M with $\chi(M) = 0$. However for $\mathbf{Z}^3$ one only gets $0 \leq q \leq 2$, the deficiency being 0.

2) For the surface group Σ_g, $g \geq 2$, i.e. the fundamental group of the closed orientable surface of genus g, one has $\mathrm{def}(\Sigma_g) = 2g - 1$ and $\beta_1 = 2g$. Thus

$$2 - 4g \leq q(\Sigma_g) \leq 4 - 4g.$$

3) For any knot group G (the fundamental group of the complement of a classical knot in S^3) the deficiency is 1 and $\beta_1 = 1$ whence $q(G) = 0$.

4) Let G be a 2-knot-group, i.e. the fundamental group of the complement of two-dimensional knot S^2 in S^4. As for classical knots $\beta_1(G) = 1$. Surgery along the imbedded sphere S^2 produces a 4-manifold M with fundamental group G, and with $\beta_2(M) = 0$, whence $\chi M = 0$. Thus again $q(G) = 0$.

2.3.   There is a topological ingredient available in 4-manifolds which has not been used, namely the signature. This has suggested a more refined group invariant associated with 4-manifolds, see the next section.

3.   THE $(\chi + \sigma)$-INVARIANT

3.1.   We recall that the cohomology group $H^2(M; \mathbf{R})$ is a real quadratic space, the quadratic form being given by the cup-product evaluated on the fundamental cycle of M. It is non-degenerate, and the space splits into a positive-definite and a negative-definite subspace of dimensions β_2^+ and β_2^- respectively. The difference $\beta_2^+ - \beta_2^- = \sigma(M)$ is the signature of M. Its sign clearly depends on the orientation of M and we assume the orientation chosen in such a way that $\sigma(M) \leq 0$, i.e., $\beta_2^+ \leq \beta_2^-$. Since $\beta_2 = \beta_2^+ + \beta_2^-$ the sum $\chi(M) + \sigma(M)$ is equal to $2 - 2\beta_1(G) + 2\beta_2^+(M)$, where as always $G = \pi_1(M)$. Since that sum is bounded below by $2 - 2\beta_1(G)$ depending on G only one can define an invariant $p(G)$ to be the minimum of $\chi(M) + \sigma(M)$ for all M with fundamental group G and oriented in such a way that $\sigma(M) \leq 0$. Obviously $p(G) \leq q(G)$. An equivalent way to define $p(G)$ is to take, independently of orientations, the minimum of $\chi(M) - |\sigma(M)|$.

Putting together all above inequalities we get

$$2 - 2\beta_1(G) \leq p(G) \leq q(G) \leq 2 - 2\,\mathrm{def}(G).$$

3.2. It seems difficult in general to compute the value of $p(G)$ and $q(G)$, and their group-theoretic meaning is not known. We first show how one can proceed in special cases where information on $H^2(G)$, i.e. H^2 of the Eilenberg-MacLane space $K(G,1)$ is available. We then show (Section 3.3) that it is quite interesting for applications to know that the two invariants are non-negative. (This is clearly the case if $\beta_1(G) \leq 1$, in particular if G is finite).

Any 4-manifold M with $\pi_1(M) = G$ can be imbedded in a $K(G,1)$ by adding cells of dimension $2, 3, \ldots$ in order to kill the homotopy groups in dimensions ≥ 2. This yields an injective map $H^2(G; \mathbf{R}) \longrightarrow H^2(M; \mathbf{R})$. If in $H^2(G; \mathbf{R})$ the cup-product happens to be trivial then $H^2(M; \mathbf{R})$ contains an isotropic subspace of dimension $\beta_2(G)$. In that case $\beta_2^+(M)$ must be $\geq \beta_2(G)$ so that

$$p(G) \geq 2 - 2\beta_1(G) + 2\beta_2(G).$$

This applies to examples in 2.2:

For the group $G = \mathbf{Z}^3$ the 3-dimensional torus is a $K(G,1)$ and the cup-product in H^2 is trivial. Since $\beta_1(G) = \beta_2(G) = 3$ we get $p(\mathbf{Z}^3) \geq 2$ whence $p(\mathbf{Z}^3) = q(\mathbf{Z}^3) = 2$.

For $G = \Sigma_g$, $g \geq 2$, the surface of genus g is a $K(G,1)$, and $\beta_1(G) = 2g$, $\beta_2(G) = 1$. Thus $p(G) \geq 4 - 4g$ whence $p(\Sigma_g) = q(\Sigma_g) = 4 - 4g$. So here the invariants are negative. Another such case is the free group F_m on $m \geq 2$ generators where one easily finds $p(F_m) = q(F_m) = 2 - 2m$.

3.3. There are several instances where the sign of the invariants yields significant information on the 4-manifolds or the groups involved. We mention three of them.

I) *Deficiency.* From the inequality in 2.1 one immediately notes that if $q(G) \geq 0$ then $\mathrm{def}(G) \leq 1$. We will return to this fact later on.

II) *Complex surfaces.* We assume that our 4-manifold M is a complex surface (complex dimension 2). Then it is known that $\chi + \sigma$ of M can be expressed in different ways: We write c_2 for the second Chern class $c_2(M)$ evaluated on M, c_1^2 for the cup-square of the first Chern class evaluated on M. Then $\chi(M) = c_2$ and $\sigma(M) = 1/3(c_1^2 - 2c_2)$ [since the signature is $1/3$ of the first Pontrjagin number, which in the complex case can be expressed by the Chern classes as above]. Thus

$$\chi(M) + \sigma(M) = c_2 + 1/3(c_1^2 - 2c_2) = 1/3(c_1^2 + c_2).$$

This is 4 times the holomorphic Euler characteristic $1 - g_1 + g_2$ of M by the Riemann-Roch theorem.

PROPOSITION 1.   *Let M be a complex surface, and assume that its funda-mental group G fulfills $p(G) \geq 0$. Then the holomorphic Euler characteristic of M is ≥ 0.*

By the Kodaira-Enriques classification it follows that M cannot be ruled over a curve of genus ≥ 2.

REMARK.   The formulae above leading to the holomorphic Euler charac-teristic refer to the orientation of the complex surface dictated by the complex structure. Thus the argument is valid only if in *that* orientation $\sigma(M) \leq 0$. If however $\sigma(M) > 0$ then $p(G) \geq 0$ implies that $2 - 2\beta_1(G) + 2\beta_2^+{}_{\text{wrong}}(M) \geq 0$ where $\beta_2^+{}_{\text{wrong}}$ refers to the "wrong" orientation and is $= \beta_2^-(M)$. Now $\beta_2^+(M) > \beta_2^-(M)$ by assumption. Thus the result remains true; the holomor-phic characteristic is > 0.

III) *Donaldson Theory.* Finitely presented groups G with $p(G) \geq 0$ and $\beta_1(G) \geq 4$ do not qualify for the Theorems A,B, and C of Donaldson [D] relating to non-simply connected topological manifolds. Indeed in these theorems the signature is assumed to be negative with $\beta_2^+ = 0$, 1 or 2. However $p(G) \geq 0$ means $2 - 2\beta_1(G) + 2\beta_2^+(M) \geq 0$, i.e. $\beta_2^+(M) \geq \beta_1(G) - 1$.

4.   DEUS EX MACHINA: l_2-COHOMOLOGY

4.1.   We recall in a few words the (cellular) definition of l_2-cohomology and l_2-Betti numbers, in the case of a 4-manifold M but things apply to any finite cell-complex.

Some definitions: For any countable group G let l_2G be the Hilbert space of square-integrable real functions on G, with G operating on the left, and NG the algebra of bounded G-equivariant linear operators on l_2G. A Hilbert-G-module H is a Hilbert space with isometric left G-action which admits an isometric G-equivariant imbedding into some l_2G^m (direct sum of m copies of l_2G). The projection operator ϕ of l_2G^m with image H is given by a matrix (ϕ_{kl}), $\phi_{kl} \in NG$. The "trace" $\sum \langle \phi_{kk}(1), 1 \rangle$ is the von Neumann dimension $\dim_G H$; it is a real number ≥ 0, and $= 0$ if and only if $H = 0$.

Let $\tilde{M}$ be the universal cover of M with the cell-decomposition corre-sponding to that chosen in M. The square-integrable real i-cochains of $\tilde{M}$ constitute a Hilbert space $C_{(2)}^i(\tilde{M})$ with isometric G-action. It decomposes into the direct sum of α_i copies of l_2G, $i = 0, \ldots, 4$. As before α_i denotes the number of i-cells of M; G is the fundamental group of M acting by permutation of the cells of $\tilde{M}$. The $C_{(2)}^i$ with the induced coboundary operators form a Hilbert-G-module chain complex. The cohomology H^i of that complex is easily identified with $H^i(M; l_2 G)$, cohomology with local coefficients (see, e.g. [E2]). The *reduced* cohomology group $\overline{H}^i$ (i.e. cocycles modulo the closure of coboundaries) of that complex can be imbedded in $C_{(2)}^i$ as a G-invariant subspace and is therefore a Hilbert-G-module. Its von Neumann dimension $\dim_G \overline{H}^i(\tilde{M})$ is the i-th l_2-Betti number $\overline{\beta}_i(M)$. It is a topological, even a homotopy, invariant of M.

4.2. Since $\dim_G C_{(2)}^i = \alpha_i$ and since the von Neumann dimension behaves like a rank, the usual Euler-Poincaré argument shows that the l_2-Betti numbers compute the Euler characteristic exactly as the ordinary Betti numbers do:

$$\chi(M) = \sum (-1)^i \overline{\beta}_i(M).$$

Moreover the $\overline{\beta}_i$ of a closed manifold fulfill Poincaré duality. Thus

$$\chi(M) = 2\overline{\beta}_0 - 2\overline{\beta}_1 + \overline{\beta}_2.$$

According to Atiyah's l_2-signature theorem [A], $\sigma(M)$ can also be expressed by appropriate l_2-Betti numbers: $\overline{H}^2(\tilde{M})$ splits into two complementary G-invariant subspaces with von Neumann dimensions $\overline{\beta}_2^+(M)$ and $\overline{\beta}_2^-(M)$, and $\sigma(M)$ is their difference. Thus, as with ordinary Betti numbers, one has

$$\chi(M) + \sigma(M) = 2\overline{\beta}_0(G) - 2\overline{\beta}_1(G) + 2\overline{\beta}_2^+(M).$$

We now assume G to be infinite. Then $\overline{\beta}_0(G) = 0$. Indeed a 0-cocycle f in $\tilde{M}$ is a constant and if $\tilde{M}$ is an infinite complex f can be l_2 only if it is $= 0$.

THEOREM 2. *If for a finitely presented group G the first l_2-Betti number $\overline{\beta}_1(G)$ is 0 then the invariants $p(G)$ and $q(G)$ are non-negative.*

COROLLARY 3. *If $\overline{\beta}_1(G) = 0$ then $\operatorname{def}(G) \leq 1$.*

COROLLARY 4. *If $G = \pi_1(\text{complex surface } M)$ with $\overline{\beta}_1(G) = 0$ then the holomorphic Euler characteristic of M is non-negative.*

4.3. There are many groups for which it is known that $\overline{\beta}_1(G) = 0$. A good list is given in [B-V]. We mention here three big and interesting classes of groups with that property.

1) All finitely generated amenable groups [C-G]. We recall that this class includes the virtually solvable groups, thus in particular the finitely generated Abelian groups (whence $\mathbf{Z}^n$, example 1) in 2.2). [Actually for an amenable group G with $K(G,1)$ of finite type, i.e. there is a $K(G,1)$ with finite m-skeleta, all l_2-Betti numbers are 0.]

THEOREM 5. *If G is a finitely presented amenable group then $p(G)$ and $q(G)$ are non-negative.*

2) [L1] All finitely presented groups G containing an infinite finitely generated normal subgroup N such that there is in G/N an element of infinite order. For these "Lück groups" one has the same conclusions as in the amenable case. — In [L1] the subgroup N is assumed to be finitely presented. Lück has shown later [L2] that the weaker assumption above is sufficient.

3) The statement of Theorem 5 also holds more generally for a finitely presented group G which contains a finitely generated normal subgroup N such that G/N is infinite and amenable [E2]. The proof is somewhat different: It makes use not of the universal cover but of the cover belonging to N. The amenable group G/N operates on that cover and one can use the l_2-Betti numbers relative to G/N. — A simple example is given by a group with finitely generated commutator subgroup and infinite Abelianisation.

4.4. REMARKS.

1) We note that for finitely presented infinite amenable groups, and also for groups as in 4.3, 3) above, the deficiency is ≤ 1. This can also be proved without 4-manifolds: It suffices to consider a $K(G,1)$ with 2-skeleton corresponding to a presentation of G.

2) It is well-known that a group with deficiency ≥ 2 cannot be amenable since it contains free subgroups of rank ≥ 2; see [B-P], where a stronger result is proved.

3) There is a class of groups for which $\overline{\beta}_1$ is positive: The groups G with infinitely many ends (i.e. with $H^1(G;\mathbf{Z}G)$ of infinite rank; here one takes ordinary cohomology with local coefficients). A nice proof for this can be found in [B-V]. Another approach is to use Stallings' structure theorem from which it follows that these groups contain free subgroups of rank ≥ 2 and thus are non-amenable. For non-amenable groups the Guichardet amenability criterion [G] tells that $\overline{H}^1(G;l_2G) = H^1(G;l_2G)$. The coefficient

map $H^1(G;\mathbf{Z}G) \longrightarrow H^1(G;l_2G)$ induced by the imbedding $\mathbf{Z}G \longrightarrow l_2G$ is easily seen to be injective. Since we have assumed $H^1(G;\mathbf{Z}G) \neq 0$ the result follows.

5.   The vanishing of $q(G)$

5.1.   Here we mention in a few words what happens when for a finitely presented group G the invariant $q(G)$ is 0. For the details and more comments we refer to the paper [E2]. We thus consider a 4-manifold M with $\pi_1(M) = G$ and $\chi(M) = 0$.

Since we restrict attention to groups with $\overline{\beta}_1(G) = 0$ the vanishing of $\chi(M)$ implies $\overline{\beta}_2(M) = 0$, whence $\overline{H}^2(\widetilde{M}) = 0$. As shown in [E2] by a spectral sequence argument it follows that $H^2(M;\mathbf{Z}G)$ is isomorphic to $H^2(G;\mathbf{Z}G)$, ordinary cohomology with local coefficients $\mathbf{Z}G$. By Poincaré duality $H^2(M;\mathbf{Z}G) = H_2(M;\mathbf{Z}G)$ which can be identified with $H_2(\widetilde{M};\mathbf{Z})$. Since $\widetilde{M}$ is simply connected, $H_2(\widetilde{M};\mathbf{Z})$ is isomorphic to the second homotopy group $\pi_2(\widetilde{M}) = \pi_2(M)$.

What about $H_3(\widetilde{M};\mathbf{Z})$? It can be identified with $H_3(M;\mathbf{Z}G)$ which, by Poincaré duality, is $\cong H^1(M;\mathbf{Z}G) = H^1(G;\mathbf{Z}G)$. This group, the "endpoint-group" of G, is known to be either 0 or $\mathbf{Z}$ or of infinite rank. As mentioned in 4.4, remark 3) the latter case is excluded by our assumption $\overline{\beta}_1(G) = 0$. The case $H^1(G;\mathbf{Z}G) = \mathbf{Z}$ is exceptional: it means that G is virtually infinite cyclic, and we exclude this. Then $H_3(\widetilde{M};\mathbf{Z}) = 0$.

5.2.   We now add the assumption that $H^2(G;\mathbf{Z}G) = 0$. This is a property shared by many groups (e.g. duality groups). Then the homology groups $H_i(\widetilde{M};\mathbf{Z})$ are $= 0$ for $i = 1,2,3,4$ ($i = 4$ because $\widetilde{M}$ is an open manifold). Thus all homotopy groups of $\widetilde{M}$ are $= 0$, $\widetilde{M}$ is contractible, M is a $K(G,1)$, and the group G fulfills Poincaré duality.

THEOREM 6.   *Let G be an infinite, finitely presented group, not virtually infinite cyclic, fulfilling $\overline{\beta}_1(G) = 0$ and $H^2(G;\mathbf{Z}G) = 0$, and let M be a manifold with fundamental group G. If the Euler characteristic $\chi(M) = 0$, then M is an Eilenberg-MacLane space for G and G is a Poincaré duality group of dimension 4.*

We recall that for knot groups and 2-knot groups $q(G) = 0$, see examples 3) and 4) in 2.2. Theorem 6 can only be applied to 2-knot groups which are not classical knot groups since the latter have cohomological dimension 2.

[A]   ATIYAH, M. Elliptic operators, discrete groups and von Neumann algebras. *Astérisque 32* (1976), 43–72.

[B-P]   BAUMSLAG, G. and S. J. PRIDE. Groups with two more generators than relators. *J. London Math. Soc. 17* (1978), 425–426.

[B-V]   BEKKA, M. E. B. and A. VALETTE. Group cohomology, harmonic functions and the first L^2-Betti number. Preprint; to appear in *Potential Analysis*.

[C-G]   CHEEGER, J. and M. GROMOV. L^2-cohomology and group cohomology. *Topology 25* (1986), 189–215.

[D]   DONALDSON, S.K. The orientation of Yang-Mills moduli spaces and 4-manifold topology. *J. Differential Geometry 26* (1987), 397–428.

[E1]   ECKMANN, B. Amenable groups and Euler characteristic. *Comment. Math. Helvetici 67* (1992), 383–393.

[E2]   —— Manifolds of even dimension with amenable fundamental group. *Comment. Math. Helvetici 69* (1995), 501–511.

[G]   GUICHARDET, A. *Cohomologie des groupes topologiques et des algèbres de Lie.* Cedic–F. Nathan, 1980.

[H-W]   HAUSMANN, J.-C. et S. WEINBERGER. Caractéristique d'Euler et groupes fondamentaux des variétés de dimension 4. *Comment. Math. Helvetici 60* (1985), 139–144.

[K]   KOTSCHICK, D. Four-manifold invariants of finitely presentable groups; in *Topology, Geometry and Field Theory*. World Scientific.

[L1]   LÜCK, W. L^2-Betti numbers of mapping tori and groups. *Topology 33* (1994), 203–214.

[L2]   —— Hilbert modules and modules over finite von Neumann algebras and applications to L^2-invariants. Preprint, Mainz 1995; to appear in *Math. Annalen*.

[S-T]   SEIFERT, H. und W. THRELFALL. *Lehrbuch der Topologie.* Teubner, 1934.

The Euler characteristic –
a few highlights in its long history

There are, throughout mathematics, countless variations and ramifications of the concept of Euler characteristic. We choose here a very few ones which are highlights in its development; they can be characterized by the names Euler, Poincaré, Atiyah, Gromov.

1. The polyhedron formula.

1.1. To start from the very beginning we consider the old **Euler Polyhedron Formula** for a convex polyhedron in 3-space with α_0 vertices, α_1 edges, and α_2 faces (2-cells, polygons):

$$\alpha_0 + \alpha_2 = \alpha_1 + 2.$$

It was published, with two closely related proofs, in 1758. Historians tell us that it was known and proved by Descartes already, but not published. Euler's proofs are often considered to be "false" or incomplete. They can, however, easily be completed so as to yield a correct proof [S].

Here is a simple proof suitable for students with a little knowledge of spherical geometry (spherical triangles or polygons on the unit sphere), namely, that the sum of the angles of an n-gon is equal to $(n - 2)\pi + \text{area}$. The proof then goes as follows: Project the polygon from an interior point to the unit sphere around that point. You get a spherical polyhedron with the same number of vertices, edges, and faces which are n_ν-gons, $\nu = 1, 2, .., \alpha_2$. The sum of all angles of these faces is

$$\sum_\nu (n_\nu - 2)\pi + \text{area of the sphere} = \sum_\nu n_\nu \pi - \alpha_2 2\pi + 4\pi.$$

But that sum of angles is $= \alpha_0 2\pi$, and $\sum_\nu n_\nu = 2\alpha_1$, whence

$$2\alpha_1 \pi - \alpha_2 2\pi + 4\pi = \alpha_0 2\pi,$$

---

* English version of a lecture delivered at the annual meeting of the Swiss Mathematical Society October 1999. The author had been asked to give, at the last meeting of the century, a survey talk on some development; it had to be addressed to a general audience of non-experts and is thus largely elementary.

i.e.
$$\alpha_0 - \alpha_1 + \alpha_2 = 2.$$

1.2. Remark. The number 2 on the right-hand side is the area of the unit sphere divided by 2π. If instead of the sphere an arbitrary orientable closed surface in 3-space is considered, subdivided into geodesic polygons, then the same formula holds where the right-hand side is the total Gauss curvature divided by 2π (Gauss-Bonnet formula). This is the starting point for a large number of ramifications not mentioned in this talk. They could be the topic of several separate lectures (characteristic classes, singularities of vector fields, etc).

1.3. We now consider an arbitrary finite *cell-complex* X. According to his preference the reader may think of a simplicial complex, or of a CW-complex with arbitrary or with "regular" cells (homeomorphic images of balls of dimension $i = 0, 1, ..., n$). The number of i-cells is written α_i. The **Euler characteristic** of X is the integer
$$\chi(X) = \sum_i (-1)^i \alpha_i.$$

If X is the 2-sphere with a given cell-decomposition then, according to the above, $\chi(X) = 2$, whence *independent of the cell-decomposition*. At the beginning of the 20th century Poincaré discovered that this is a general phenomenon, and much more, that $\chi(X)$ is a topological invariant of the underlying space X (although this was proved much later).

2. Euler-Poincaré and Betti numbers

2.1. The above facts are due to the **Euler-Poincaré formula**
$$\sum_i (-1)^i \alpha_i = \sum_i (-1)^i \beta_i(X)$$

where the $\beta_i(X)$ are the *Betti numbers* of the underlying space X – we use the same notation as for the cell complex. These are topological, even homotopy invariants of the space X; we will come back to that point below.

While the left-hand side of the formula is the Euler characteristic of the cell complex the right-hand side is called the Euler-Poincaré characteristic of the space X. There are, throughout mathematics, several variations with the same name where instead of the Betti numbers other invariants occur.

The formula can be considered as the very source of Homological Algebra – actually not the formula itself, but the way it has been proved in the twenties by Heinz Hopf under the influence of Emmy Noether. We will give below a somewhat different proof, using elementary linear algebra of vector spaces. Our procedure will be used later on for certain generalizations.

2.2. What are the Betti numbers? They are well-known in differential geometry: If X is a closed Riemannian manifold then β_i is the dimension of the

$\mathbb{R}$-vector space of real harmonic differential i-forms on X; these are those i-forms on which the Laplace-Beltrami operator Δ vanishes. We use the analogous "combinatorial" definition in an arbitrary finite cell-complex X where one has a combinatorial Laplace operator Δ relative to the given cell-decomposition of X.

To this end we associate with the cell-decomposition of X $\mathbb{R}$-vector spaces C_i, or $C_i(X)$ if the space X need to be mentioned: Namely, for $i = 0, 1, ..., n$ let C_i be the vector space with basis the i-cells σ_i^ν, $\nu = 1, 2, ..., \alpha_i$ and $C_i = 0$ if $i < 0$ or if $i > n$. We use in C_i the scalar product $<, >$ for which the σ_i^ν form an orthonormal basis. Elements $c_i \in C_i$, i.e. linear combinations of the cells, are called i-chains or just chains (index i omitted if it is clear from the context). Note that $< c, \sigma_i^\nu >$ is the coefficient of σ_i^ν in c.

The usual cellular boundary defines for all i a linear map $\partial : C_i \longrightarrow C_{i-1}$, with $\partial\partial = 0$. The "coboundary" $\delta : C_{i-1} \longrightarrow C_i$ is the adjoint map $\delta = \partial^*$, defined by

$$< \delta c, c' > = < c, \partial c' >$$

for all $c \in C_{i-1}$ and $c' \in C_i$. Elements of the kernel of ∂ are called cycles, of δ cocycles.

2.3. Remark. If X is a closed differentiable manifold, and if the coefficients of chains c are given by integrating differential forms ω on X of the respective degree then $< \delta c, \sigma_{i+1}^\nu >$ is the integral of ω over ∂c, whence by Stokes' formula the integral of $d\omega$ over σ_{i+1}^ν. In other words, if c corresponds by integration to ω, than the coboundary δc corresponds to the exterior differential $d\omega$; and cocycles correspond to closed differential forms.

2.4. Clearly c_i is a cocycle if and only if $< c_i, \partial c'_{i+1} > = 0$ for all c'_{i+1}: The i-cocycle space in C_i is the orthogonal complement of the boundary space ∂C_{i+1} – and vice-versa, since we are dealing with finite-dimensional vector spaces. Similarly the i-cycle space and the coboundary space δC_{i-1} are orthogonal complements. Since $\partial\partial = 0$ the boundary space is a subspace of the i-cycle space, and the coboundary space of the i-cocycle space. If we call $\mathcal{H}_i$ the intersection of the cycle- and the cocycle-space we have a decomposition of C_i into three mutually orthogonal subspaces

$$C_i = \partial C_{i+1} \oplus \delta C_{i-1} \oplus \mathcal{H}_i.$$

Chains in $\mathcal{H}_i$ are called *harmonic* since they constitute the kernel of the combinatorial Laplace operator

$$\Delta = \delta\partial + \partial\delta.$$

Indeed for $c_i \in \mathcal{H}_i$ one has $\Delta c_i = 0$; conversely $\Delta c_i = 0$ implies that both term in Δc_i are zero since they are orthogonal; then $\delta\partial c_i = 0$ implies

$$< \delta\partial c_i, c_i > = 0 = < \partial c_i, \partial c_i >$$

whence $\partial c_i = 0$, etc.

2.5. We define the Betti number $\beta_i(X)$ of the cell complex X by

$$\beta_i(X) = \dim_{\mathbb{R}} \mathcal{H}_i.$$

Now the Euler-Poincaré formula can easily be read off from the above decomposition of the C_i, as follows. Let n_i be the dimension of ∂C_{i+1}, and n_i' of δC_{i-1}. Since δ and ∂ are dual maps we have $n_{i+1}' = n_i$. Now $\alpha_i = \beta_i + n_i + n_i'$, and thus

$$\sum_i (-1)^i \alpha_i = \sum_i (-1)^i (\beta_i + n_i + n_i').$$

But $\sum_i (-1)^i (n_i + n_i') = \sum (-1)^i (n_i + n_{i-1}) = 0$, which yields the required result.

It is sometimes useful to express the dimension of a subspace S of C_i as trace of the matrix which describes, in any basis, the orthogonal projection $\phi : C_i \longrightarrow C_i$ with image S. In the basis consisting of the cells σ_i^ν one has

$$\dim_{\mathbb{R}} S = \sum_\nu < \phi(\sigma_i^\nu), \sigma_i^\nu >= \sum_\nu < \phi(\sigma_i^\nu), \phi(\sigma_i^\nu) >$$

since ϕ is idempotent and selfadjoint ($\phi^2 = \phi$, $\phi^* = \phi$).

3. Properties of Betti numbers.

3.1. The important property of Betti numbers of X is homotopy invariance, whence topological invariance. It is, of course, due to the relation between the $\mathcal{H}_i$ and *homology*: From the direct sum decomposition of C_i it follows that $\mathcal{H}_i$ is isomorphic to the homology group $H_i(X) = \{i - cycles\}/\partial C_{i+1}$, and similarly to the cohomology group $H^i(X)$.

We do not prove here that (co)homology is homotopy invariant (this can be found in many topology books) but make some remarks. Looking back into history it seems that topological invariance of homology of cell complexes was a very difficult problem until about the time of World War Two. After that the concept of homotopy equivalence and (later) the functorial approach made the phenomenon clear and proofs transparent, as sketched by the following hints.

Homotopy equivalence: This is a continuous map $f : X \longrightarrow Y$ between two spaces X and Y such that there is a map $g : Y \longrightarrow X$ with gf homotopic to the identity of X and fg to the identity of Y. Clearly a homeomorphism is a homotopy equivalence, but the latter is much more general. In our context "homotopy invariance", and thus topological invariance, means that a homotopy equivalence $f : X \longrightarrow Y$ between two cell-complexes induces isomorphisms between the homology groups of X and Y and therefore between the $\mathcal{H}_i(X)$ and $\mathcal{H}_i(Y)$, whence $\beta_i(X) = \beta_i(Y)$.

Given a map $f : X \longrightarrow Y$ between two finite cell-complexes one associates to f, by cellular approximation, linear maps $f_i : C_i(X) \longrightarrow C_i(Y)$ commuting with ∂ and thus inducing linear maps of the homology groups $f_i : H_i(X) \longrightarrow H_i(Y)$. All one has to prove in order to establish the homotopy invariance is that two homotopic maps f and g induce the same homomorphism of the homology groups. This is done by replacing the homotopy by an algebraic homotopy $F : C_i(X) \longrightarrow C_{i+1}(Y)$ with $f_i - g_i = F\partial + \partial F$. Then for $c \in C_i$ with $\partial c = 0$ one has $f_i(c) - g_i(c) = \partial F(c)$, i.e. $f_i - g_i = 0$ in homology.

3.2. Dimension 0 and 1.

If $c \in \mathcal{H}_0$ then c, being a cocycle, is $= 0$ on boundaries $\in \partial C_1$. It thus takes the same value on the endpoints of each 1-cell. If X is *connected*, c is constant on all 0-cells, whence $\mathcal{H}_0 = \mathbb{R}$ and $\beta_0 = 1$.

We consider the fundamental groups $\pi_1(X)$ and $\pi_1(Y)$ of the cell-complexes X and Y. Any homorphism $\phi : \pi_1(X) \longrightarrow \pi_1(Y)$ can be realized by a map between the 2-skeleta X^2 and Y^2, and if f and g are two such maps then f is homotopic to a map which coincides with g on the 1-skeleton of X. This implies that f and g induce the same homomorphism of the H_1, and of the $\mathcal{H}_1$. If the fundamental groups are isomorphic one concludes as above that $\mathcal{H}_1(X)$ is isomorphic to $\mathcal{H}_1(Y)$. In other words, $\beta_1(X)$ *depends on the fundamental group of X only*. [From the relation between loops and 1-cycles one deduces, more precisly, that β_1 is the $\mathbb{Q}$-rank of the abelianized fundamental group.]

3.3. Spheres.

S^1. The fundamental group is $= \mathbb{Z}$. Thus $\beta_1 = 1$ and $\chi(S^1) = 0$.

S^2. As seen above, $\chi(S^2) = 2$.

S^m. There is a cell decomposition with one 1-cell and one m-cell. Thus $\chi(S^m) = 1 + (-1)^m = 0$ if m is odd and $= 2$ if m is even.

3.4. Poincaré duality.

If X is a closed manifold of dimension n then the Betti numbers fulfill the duality relation

$$\beta_i(X) = \beta_{n-i}(X), \quad i = 0, 1, ..., n.$$

Here are some hints about an (old-fashioned) proof: Given a simplicial trianguletion of the manifold X there is a "dual" cell-decomposition of X where the cells are the "stars" of the simplices in the barycentric subdivision of the simplicial complex. To each i-simplex corresponds an $(n - i)$-star. For the numbers of $(n - i)$-stars α'_i one has $\alpha_i = \alpha'_{n-i}$; the chain groups in dimensions i and $n - i$ respectively are isomorphic, and under that isomorphism the two chain complexes are isomorphic, except for the numbering ($n - i$ instead of i). Thus homology in complementary dimensions is the same, and so are the Betti numbers.

If the dimension n is odd then the duality implies that $\chi(X) = 0$.

4. Finite coverings

4.1 This section is just meant as a elementary preparation for the next one where we introduce ℓ_2-Betti numbers. These are different from the ordinary Betti numbers but have very similar properties,

Let Y be a finite covering complex of X, with covering transformation group G. This means that G operates freely on Y by permutation of the cells and that X is the orbit complex. We choose, for each cell σ_i^ν of X a cell s_i^ν in the orbit above σ_i^ν. Then all cells of Y can be written in a unique way as xs_i^ν, $x \in G$, where ν goes from 1 to $\alpha_i(X)$. Clearly

$$\alpha_i(Y) = |G|\alpha_i(X)$$

and

$$\chi(Y) = |G|\chi(X).$$

As before the cells xs_i^ν of Y can be used as an orthonormal besis for the scalar product in $C_i(Y)$; the group G acts on $C_i(Y)$ by linear isometries.

For any subspace S of $C_i(Y)$ we express its dimension as trace of the orthogonal projection $\phi : C_i(Y) \longrightarrow C_i(Y)$ with image S

$$\dim_\mathbb{R} S = \sum_x \sum_\nu < \phi(xs_i^\nu), xs_i^\nu > .$$

Note that in $C_i(Y)$ the scalar product, the boundary ∂, etc, are compatible with the operation of G. Thus $\mathcal{H}_i(Y)$ is a G-invariant subspace. and for $S = \mathcal{H}_i$ all terms in the $\sum_x$ above are equal to the term for $x = 1$. For the Betti numbers of Y we get

$$\beta_i(Y) = \dim_\mathbb{R}\mathcal{H}_i(Y) = |G| \sum_\nu < \phi(s_i^\nu), s_i^\nu > .$$

Omitting the factor $|G|$ we write $\dim_G \mathcal{H}_i$ for $\sum_\nu < \phi(s_i^\nu), s_i^\nu >$ and put

$$\overline{\beta}_i(X) = dim_G \mathcal{H}_i(Y).$$

This is clearly not the usual Betti number of X; indeed

$$\overline{\beta}_i(X) = \frac{1}{|G|}\beta_i(Y),$$

and this will in general not be an integer. Nevertheless the Euler-Poincaré characteristic written with the $\overline{\beta}_i(X)$ is equal to the ordinary Euler characteristic of X: namely

$$\sum_i (-1)^i\alpha_i(X) = \frac{1}{|G|}\sum_i (-1)^i\alpha_i(Y) = \sum_i (-1)^i\frac{1}{|G|}\beta_i(Y),$$

i.e.

$$\sum_i (-1)^i\alpha_i(X) = \sum_i (-1)^i\overline{\beta}_i(X).$$

4.2. The last formula in itself is not very interesting since it just expresses the fact that the covering Y of X is $|G|$-sheeted. It can, however, also be obtained in a second way without using the fact that $\alpha_i(X) = \frac{1}{|G|}\alpha_i(Y)$, as follows. One has to observe that $\dim_G = \sum_\nu < \phi(s_i^\nu), s_i^\nu >$ makes sense for any G-invariant subspace of $C_i(Y)$, where ϕ is the respective orthogonal projection; and that it has all the usual properties of a dimension. In particular $\dim_G C_i(Y) = \alpha_i(X)$. If we apply $\dim_G$ to the decomposition of $C_i(Y)$ into three mutually orthogonal invariant subspaces $\mathcal{H}_i(Y) \oplus \partial C_{i+1} \oplus \delta C_{i-1}$ then exactly the same arguments as in **2.5** yield the "new" Euler-Poincaré formula.

5. ℓ_2-Betti numbers. Atiyah formula

5.1. The foregoing arguments belong all to the algebraization of the elementary homology concepts. But now we make, historically, a jump of some 50 years and consider the case where the covering Y of X is infinite, $|G| = \infty$. Then the situation is of course different. To fix ideas we take for Y the universal covering of X, and $G = \pi_1(X)$ operates as before freely on Y by cell-permutation (any regular infinite covering would also do). Now $\beta_i(Y)$ cannot be used since it may be infinite. But the second procedure above for defining new Betti numbers $\overline{\beta}_i(X)$ can be applied.

To this end one uses Hilbert space methods. Let now $C_i(Y)$ denote the Hilbert space of all square-summable chains (or ℓ_2-chains) in Y. The xs_i^ν $x \in G, \nu = 1, 2, ..., \alpha_i(X)$ (notations as in **4.1**) are used as an orthonormal basis for the scalar product. G acts on $C_i(Y)$ by isometries. They commute with the boundary ∂ and coboundary δ, which are bounded linear operators; kernels are closed subspaces, but images need not be closed. In the decomposition of $C_i(Y)$ into three mutually orthogonal subspaces we have to replace ∂C_{i+1} and δC_{i-1} by their closures in $C_i(Y)$:

$$C_i(Y) = \mathcal{H}_i(Y) \oplus \overline{\partial C_{i+1}} \oplus \overline{\delta C_{i-1}}.$$

These closed subspaces are G-invariant and thus have in $C_i(Y)$ closed G-invariant orthogonal complements. Their dimension $\dim_G$, called the *von Neumann dimension* can be defined exactly as in **4.1**: If S is a G-invariant closed subspace and ϕ the corresponding orthogonal projection then

$$\dim_G S = \sum_\nu < \phi(s_i^\nu), s_i^\nu > = \sum_\nu < \phi(s_i^\nu), \phi(s_i^\nu) > = \sum_\nu ||\phi(s_i^\nu)||^2$$

since ϕ is idempotent and selfadjoint. Clearly $\dim_G$ is ≥ 0. It is $= 0$ only if all $\phi(s_i^\nu)$ are zero; since ϕ commutes with all $x \in G$ it is thus zero on all cells xs_i^ν of Y and on all ordinary chains, and these being dense in the Hilbert space $C_i(Y)$ it follows that $\phi = 0$, whence $S = 0$:

If S is a G-invariant subspace of $C_i(Y)$ then $\dim_G S \geq 0$, and $= 0$ only if $S = 0$.

The ℓ_2-Betti numbers of X are now defined by

$$\overline{\beta}_i(X) = \dim_G \mathcal{H}_i(Y).$$

They are real numbers ≥ 0. The same arguments as in **2.5** yield

$$\sum_i (-1)^i \alpha_i(X) = \sum_i (-1)^i \overline{\beta}_i(X).$$

This is the **Atiyah-ℓ_2-Euler-Poincaré** formula. It was established in 1976, actually for closed manifolds X only, and by a quite different method: Using the index theorem Atiyah proves that the ℓ_2-Euler-Poincaré characteristic is equal to the ordinary Euler-Poincaré characteristic. The above method using cell-decompositions has, apart from being more general, several advantages: Exactly as for ordinary Betti numbers one can prove homotopy invariance (see **3.1**) of the $\overline{\beta}_i(X)$, Poincaré duality and similar properties.

In dimension 0 one has, always assuming G infinite, $\overline{\beta}_0(X) = 0$. Indeed $c \in \mathcal{H}_0(Y)$ being a constant can only be ℓ_2 if it is $= 0$. In dimension 1 one shows as in **3.3** that $\overline{\beta}_1(X)$ depends on the fundamental group G of X only and can be written $\overline{\beta}_1(G)$. It is known to be zero for big classes of groups (e.g. amenable groups).

5.2. General remarks (for details see [E1]). For the procedure described above several things have to be checked. We first recall some of the usual terminology.

If G is a countable group $\ell_2 G$ denotes the Hilbert space of square-summable functions on G, with the group elements as orthonormal basis. G operates on $\ell_2 G$ isometrically from left and right. A Hilbert space S with a (left) isometric G-action is called a Hilbert G-module if it can be imbedded isometrically and G-equivariantly into $\ell_2 G^m$ for some m. Then $\dim_G S$, the von Neumann dimension of S is defined as above through the orthogonal projection. It has the usual dimension properties and it is independent of the imbedding. If two Hilbert G-modules are G-isomorphic by a bounded linear operator then there exists a G-isometry between them.

In our situation $C_i(Y)$ can be written as $\ell_2 G s_i^1 \oplus ... \oplus \ell_2 G s_i^{\alpha_i}$. Thus $\mathcal{H}_i(Y)$ is a Hilbert G-module and its $\dim_G$ is independent of the choice of the cells s_i^ν in the orbits. One easily checks that not only ∂ and δ but also all maps occurring in the proof of homotopy invariance are bounded linear G-operators.

5.3. To illustrate the role of the new Betti numbers we consider here only two examples; further ones will be given in a more special context below (aspherical spaces).

1) Let G be a finitely presented infinite group, and consider a presentation with g gnerators and r defining relations. One easily constructs a cell complex X with 1 vertex, g edges corresponding to the generators and r 2-cells corresponding to the relations. Then

$$\chi(X) = 1 - g + r = \overline{\beta}_0(X) - \overline{\beta}_1(X) + \overline{\beta}_2(X) = -2\overline{\beta}_1(G) + \overline{\beta}_2(X).$$

If $\overline{\beta}_1(G) = 0$ the $\chi(X) \geq 0$ whence $g - r \leq 1$. One calls $g - r$ the deficiency of the presentation; the maximum of the deficiencies for all presentations of G (it exists) is the deficiency of the group.

If G is finitely presented and infinite, and if $\overline{\beta}_1(G) = 0$ then the deficiency of G is ≤ 1.

2) Let X be a connected closed 4-dimensional manifold with infinite fundamental group G. Using Poincaré duality we have

$$\chi(X) = 2 - 2\beta_1(X) + \beta_2(X) = -2\overline{\beta}_1(X) + \overline{\beta}_2(X).$$

If $\overline{\beta}_1(G) = 0$ then $\chi(X) \geq 0$. This has various applicationa, see [E2].

6. Aspherical spaces.

6.1. A space X is called aspherical if its universal covering Y is *contractible*. Here space always means connected finite cell-complex. According to Hurewicz two aspherical spaces with isomorphic fundamental groups are homotopy equivalent. Their homology, Betti numbers, etc. thus depend on the fundamental group only. Moreover for any group G (in our case finitely presented) there exists an aspherical space with fundamental group G, Homology, Betti numbers, etc. are invariants of the group G. Soon after Hurewicz efforts have been made to deduce these invariants directly from the group G, and this has led to the development of homological algebra in the modern sense. A group G, even if it is finitely presented, does in general not admit an aspherical space as above which is a finite cell-complex. If so then the concepts developed here apply to the group G.

We note that if such a group is amenable then all its ℓ_2-Betti numbers are zero (Cheeger-Gromov).

6.2. The following examples show that there are interesting groups fulfilling the above assumption; the examples also illustrate the role of the ℓ_2-Betti numbers.

1) The n-torus $(S^1)^n$ is aspherical since the universal covering is $\mathbb{R}^n$. The fundamental group is $\mathbb{Z}^n$; it is amenable and all its ℓ_2-Betti numbers are zero (there are many other ways to prove this).

2) The wedge (bouquet) of k circles is aspherical, since the universal covering is a tree. The fundamental group G is free on k generators. The Euler characteristic is $= 1 - k$, and on the other hand $= -\beta_1(G)$, whence $\beta_1 = k - 1$. Except for $k = 1$ this is > 0 which shows that the free group on $k > 1$ generators is not amenable.

3) A closed (orientable) surface of genus $g > 1$ is aspherical, since the universal covering is the hyperbolic 2-plane, and the fundamental group G is a well-known group of hyperbolic motions. The ordinary Betti numbers are $\beta_0 = \beta_2 = 1$ and $\beta_1 = 2g$, thus $\chi = 2 - 2g$. On the other hand $\chi = -\overline{\beta}_1$ whence $\overline{\beta}_1 = 2g - 2$. It is > 0 which again shows that the group is not amenable.

4) Let X be a closed Riemannian n-manifold with negative (sectional) curvature. It is well-known that its unversal covering Y is $\mathbb{R}^n$ with the induced metric, whence contractible. The fundamental group of X is a special case of a "hyperbolic" group in the sense of Gromov.

6.3. There are famous conjectures concerning aspherical manifolds.

(I) Let X be a closed aspherical manifold of dimension n.

If n is odd, $= 2k + 1$ then all $\overline{\beta}_i(X)$ are zero.

If n is even$= 2k$ then all $\overline{\beta}_i(X)$ are zero with possible exception $\overline{\beta}_k(X) \geq 0$, whence $\chi(X)(-1)^k \geq 0$.

If the fundamental group of X is amenable then this is true.

(II) Let X be a closed Riemannian manifold of even dimension $n = 2k$, with negative curvature (whence aspherical). Then $\overline{\beta}_k(X) > 0$. From this and (I) it would follow that $\chi(X)(-1)^k > 0$ (Hopf conjecture).

(II) has been proved by Gromov 1991 for hyperbolic Kähler manifolds.

7. Kähler manifolds, different types of Euler-Poincaré characteristics.

7.1. We conclude our survey with short remarks on some "Euler-Poincaré characteristics" of a different type, in the case of complex, in particular Kähler manifolds; the odinary Betti numbers, the dimensions of the spaces of harmonic differential forms, are replaced by other suitable dimensions.

Here X will denote a closed complex-analytic manifold of complex dimension m (real dimension $2m$) with local coordinates $z_1, ..., z_m, \overline{z}_1, ..., \overline{z}_m$. Differential forms in X, here always with complex coefficients, can be expressed by the differentials dz_k and $d\overline{z}_l$ instead of the real differentials. The *holomorphic characteristic* $\gamma(X)$ of X is the analogue of the Euler-Poincaré characteristic, with the Betti numbers β_i replaced by the dimensions of the cohomology of X with coefficients in the sheaf of (germs of local) holomorphic functions on X. We do not explain these notions but restrict attention to Kähler manifolds, where $\gamma(X)$ has a very explicit meaning: Let g_i be the dimension over $\mathbb{C}$ of the space of holomorphic differential forms of degree k,, i.e.of forms in the dz_j alone with holomorphic coeficients. Then

$$\gamma(X) = \sum_{i=0}^{m} (-1)^i g_i.$$

7.2. We recall that a Kähler manifold is a complex manifold X provided with a Hermitean metric $g_{k,l}$ such that the associated 2-form $\omega = \sum g_{k,l} dz_k d\overline{z}_l$ is closed. This condition has very strong implications concerning the harmonic differential forms on X, as follows (see [H], for example); here we have to deal with complex vector spaces and their complex dimension.

The holomorphic differentials forms on X are a special case of harmonic differential forms "of type (p, q)", i.e. containing p differentials dz_k and q differentials $d\overline{z}_l$. Harmonic forms of type $(p, 0)$ are easily seen to be holomorphic, and in a Kähler manifold the converse is true. Moreover $\mathcal{H}_i(X)$ decomposes into a direct sum of spaces of harmonic forms of type p, q with $p + q = i$. denoted by $\mathcal{H}_{p,q}$,

$$\mathcal{H}_i = \oplus_{p+q=i} \mathcal{H}_{p,q}.$$

One writes $h_{p,q} = \dim_{\mathbb{C}} \mathcal{H}_{p,q}$; clearly $\sum_{p+q=i} h_{p,q} = \dim_{\mathbb{C}} \mathcal{H}_i = \beta_i(X)$, and $h_{p,0} = g_p$. The $h_{p,q}$ fulfill various symmetry and other relations, e.g. $h_{p,q} = h_{q,p}$ and $h_{m-p,m-q} = h_{p,q}$; and $h_{p,p} \geq 1$ since ω and its powers, $0 \leq p \leq m$ are harmonic (cf. for example, [H, p.123ff]).

7.3. We now look at the following characteristics involving the $h_{p,q}$. The holomorphic characteristic of X can be written as

$$\gamma(X) = \sum_q (-1)^q h_{0,q} = \sum_q (-1)^q h_{q,0} = \sum_q (-1)^q g_q,$$

The Euler characteristic of X is

$$\chi(X) = \sum_{p,q} (-1)^{p+q} h_{p,q} = \sum_i (-1)^i \beta_i(X).$$

A Theorem of Hodge tells that a similar characteristic yields the *signature* $\sigma(X)$ of the Kähler manifold,

$$\sigma(X) = \sum_{p,q} (-1)^q h_{p,q}.$$

We recall the definition of $\sigma(X)$: The ordinary $\mathcal{H}_m$ with real coefficients for even m (actually for any closed manifold of dimension $4k$) is a quadratic nondegenerate space, thanks to the product of harmonic m-forms evaluated on X; the difference of the dimension β_m^+ of the positive and β_m^- of the negative definite subspace is called the signature $\sigma(X)$; if m is odd, σ is defined to be 0. The proof of the Hodge Theorem is in the framework of the Kähler metric formulae.

We now concentrate on the special case $m = 2$. In terms of ordinary Betti numbers the combination $\chi + \sigma$ is, for any 4-manifold,

$$\chi(X) + \sigma(X) = 2 - 2\beta_1 + \beta_2 + \beta_2^+ - \beta_2^- = 2 - 2\beta_1 + 2\beta_2^+.$$

In 5.1 we have expressed $\chi(X)$ by ℓ_2-Betti numbers, for an infinite fundamental group G. Thanks to Atiyah's ℓ_2-signature theorem a similar expression can be obtained for $\chi + \sigma$

$$\chi(X) + \sigma(X) = -2\overline{\beta_1}(G) + 2\overline{\beta_2}^+(X).$$

7.4. In the Kähler surface case we consider the combination

$$\chi(X) + \sigma(X) = 2 \sum_{p=0,2} \sum_{q=0,1,2} (-1)^q h_{p,q} = 4 \sum_{q=0,1,2} (-1)^q h_{0,q},$$

i.e.

$$\chi(X) + \sigma(X) = 4\gamma(X).$$

(The same formula holds for any complex surface, but the proof requires the use of Chern numbers and the simplest form of the Riemann-Roch theorem.)

If we assume that the (infinite) fundamental group G fulfills $\overline{\beta_1}(G) = 0$ it follows that $\gamma(X) \geq 0$.

According to the Enriques-Kodaira classification of complex (Kähler) surfaces this implies that X is not a ruled surface over a curve of genus ≥ 2. In other words it is not a fiber bundle over a hyperbolic Roiemann surface with $\mathbb{C}P^1$ (2-sphere) fibers. On the other hand it is clear that for such a fibre bundle the fundamental group G is that of the curve, i.e. of a hyperbolic Riemann surface and thus, $\overline{\beta}_1(X) \neq 0$ (example 2) in **6.2**. An interesting theorem of Gromov (1995) yields a converse: If for the fundamental group G of a Kähler surface $\overline{\beta}_1(G) \neq 0$ then G is, up to finite extensions, isomorphic to the fundamental group of a closed surface of genus ≥ 2. Thus, except for that case one has $\overline{\beta}_1(G) = 0$ for the fundamental group G of any Kähler surface.

References

We restrict the references to very few ones indicating complements which could not be treated in the text. The work of Atiyah and more recent work of Gromov et al. can easily be found in the reviewing journals.

[E1] Beno Eckmann, Introduction to ℓ_2-homology (Notes by Guido Mislin), to appear in Israel Journal of Math.

[E2] Beno Eckmann, 4-manifolds, group invariants, and ℓ_2-Betti numbers, Enseign. Math. (2) 43, 271–279 (1997)

[H] Fritz Hirzebruch, Topological Methods in Algebraic Geometry, Springer Verlag Heidelberg (1995)

[S] Hans Samelson, In defense of Euler, Enseign. Math. (2) 42, 377–382 (1996)

Topology, algebra, analysis – relations and missing links

Notices AMS 46 (1999), 520–527

This is largely, but not entirely, a historical survey. It puts various matters together that are usually considered in separate contexts. Moreover, it leads to open, probably quite difficult problems and has analogues in contemporary mathematics. There are three parts.

The **first part** is about a certain range of classical results in algebraic topology concerning continuous real functions and maps, vector fields, etc., that can be stated in a very simple way: just replace "continuous" by "linear". They thus seem to be reduced to problems of algebra, essentially linear algebra, where the solution is relatively easy. The proofs of all these statements, however, do not use such a reduction principle. They are beautiful and in general quite difficult, using elaborate ideas and techniques of topology such as cohomology operations, spectral sequences, K-theory, and so on.

In these cases the absence of a *reduction principle* from continuity to linearity is thus a missing link between two areas. Of course, such a link is not necessary since the results are proved. Still, it might be interesting to have, at least in special cases, a direct reduction to linearity which could throw new light onto old and new mathematics.

We begin, in that first part, with a very elementary example. It contains in a nutshell the problem to be discussed later for analogous but more complex phenomena.

Beno Eckmann is professor emeritus of mathematics and retired founding director of the Institute for Mathematical Research at Eidgenössische Technische Hochschule (ETH) in Zürich.

The **second part** is about the homotopy of the unitary groups. It leads to a different linearization phenomenon stating that maps of a sphere into a unitary group are homotopic in the infinite unitary group to a linear map (in the situation of the first part it is, in general, not true that any continuous solution is *homotopic* to a linear one). Again, no direct proof is known—by differential geometry, by approximation, or by optimization. This linearization result is due to the close relation between the classical Hurwitz-Radon matrices and Bott periodicity. Since the latter is the source of topological K-theory and general homology theories, a direct approach would establish a link between an important old matrix problem and refined methods of algebraic topology.

The **third part** is a short outlook concerning topics of contemporary interest. It is about recent trends in homology theories for spaces with infinite fundamental group G using Hilbert space methods (Hilbert-G-modules, ℓ_2-homology, and ℓ_2-Betti numbers). The presentation, necessarily a little technical, turns around the von Neumann algebra of G considered just as an algebra, forgetting the analysis behind it. We first recall the (weak) Bass conjecture (1976), which has become a theorem for several classes of groups but is still open in general, and give it a simple von Neumann algebra formulation. Application to algebraic topology and group homology again lead, in a very special example, to linearity.

We probably all agree that eventually reducing a difficult problem to a "nice" situation is at the heart of mathematics. What I present here is a modest attempt to list topics where a direct link to linearity or to easy algebra probably exists but is not known. If found, it might, even in different contexts, have some general significance. But in most other areas, nice situations certainly are of a quite different nature; there is no need to list examples.

Vector Fields and Vector Functions

Two preliminary remarks seem appropriate.

1) The results in the first part are all about certain positive integers (dimensions, number of vector fields, etc.). They state that, apart from those values of these integers where (multi-)linear solutions exist, there are no continuous solutions. This negative statement can be turned into a positive one, mainly about the existence of zeros; see Proposition 3. Moreover, in some of those cases where a linear solution exists, there are also other continuous solutions of a different nature (not homotopic to linear ones).

2) We, of course, do not suggest that the respective topological results should have been proved from the outset by reduction to linearity. On the contrary, these problems were a source of stimulation for developing interesting tools that still are very important, e.g., in modern homotopy theory. The linearization phenomenon emerged only a posteriori after the topological theorems had been established.

Tangent Vector Field on a Sphere

A *tangent unit vector field* v on $S^{n-1} \subset \mathbb{R}^n$ (given in coordinates $\xi_1, \ldots, \xi_n$ of $\mathbb{R}^n$ by $\sum \xi_i^2 = 1$) is a function that attaches to $x \in S^{n-1}$ a vector $v(x)$ satisfying

$$(1) \qquad \langle v(x), x \rangle = 0,$$

$$(2) \qquad |v(x)|^2 = \langle v(x), v(x) \rangle = 1$$

for all $x \in S^{n-1}$. Here $\langle \, , \, \rangle$ is the standard scalar product in $\mathbb{R}^n$.

Linearity implies that (2) is equivalent to

$$(2') \qquad |v(x)|^2 = \langle x, x \rangle \quad \text{for all } x \in \mathbb{R}^n.$$

If n is even, $v(x) = (-\xi_2, \xi_1, -\xi_4, \xi_3, \ldots)$ is such a field, *linear* in x. What about linear fields if n is odd? Let

$$v_i = \sum a_{ik} \xi_k$$

be the components of $v(x)$. Then (1) means $\sum a_{ik} \xi_i \xi_k = 0$ for all x; i.e., the matrix a_{ik} is skew-symmetric. If n is odd, its determinant is 0, and thus there is an $x \in S^{n-1}$ with $v(x) = 0$, in contradiction to (2).

Proposition 1. A linear tangent unit vector field on S^{n-1} exists if and only if $n - 1$ is odd.

We now ask the same question for continuous vector fields. These are, of course, more interesting from the viewpoint of geometry and analysis. Given such a field on S^{n-1}, we consider the great circle determined by x and $v(x)$ and move the point x, in the $v(x)$-direction, to its antipode $-x$, for all $x \in S^{n-1}$. This is a homotopy (a continuous deformation) between the identity map and the antipodal map of S^{n-1}.

At this point the homological concept of *degree* comes into play. Its value is 1 for the identity and $(-1)^n$ for the antipodal map. Homotopic maps have the same degree, so we get $1 = (-1)^n$; i.e., n is even.

Proposition 2. A continuous tangent unit vector field on S^{n-1} exists if and only if $n - 1$ is odd.

The crucial point is now to express these facts in a different way. We write P_n for the problem: Does there exist on S^{n-1} a tangent unit vector field?

Theorem L ("Linearization"). *If P_n has a continuous solution, then it also has a linear solution.*

Theorem L is proved "indirectly" by using a method from algebraic topology. There is no harm in doing so. By a "direct" proof we would mean a procedure replacing a continuous field by a linear one (for example, through a variational principle where in the space of all continuous fields there would be an extremal expected to be linear). Such a direct proof would reduce the topological problem to very elementary linear algebra.

In the more complicated situations to be described below, a direct reduction to a transparent algebraic argument, though not necessary, might be even more interesting.

Remarks. 1) The negative statement "There is, for even $n - 1$, no continuous tangent unit vector field on S^{n-1}" can be turned into a nontrivial existence statement for zeros, as follows.

Proposition 3. Let $f_i(\xi_1, \ldots, \xi_n)$, $i = 1, \ldots, n$, be continuous functions satisfying

$$\sum \xi_i f_i(\xi_1, \ldots, \xi_n) = 0$$

for all x with $|x| = 1$. If n is odd, then the f_i have a common zero.

Otherwise the f_i could be normalized so as to be the components of a unit tangent vector field on S^{n-1}. If the f_i are polynomials, one has an algebraic statement for which no algebraic proof seems to be known. For the complex analogue, however, there is an algebraic proof by van der Waerden (1954).

2) The above proof of Proposition 2 is by elementary algebraic topology. For differentiable vector fields there are other classical proofs, using analysis or geometry; they are all based on some version of the concept of degree. A very different analytic proof, however, is due to Milnor (1978). None of these proofs is by reduction to linearity.

Vector Functions of Two (or More) Variables, Multiplications in $\mathbb{R}^n$

The well-known vector cross-product $x \times y$ in $\mathbb{R}^3$ is a function of two vectors that is bilinear and fulfills

(3) $\langle x \times y, x \rangle = \langle x \times y, y \rangle = 0$,

(4) $|x \times y|^2 = |x|^2 |y|^2 - \langle x, y \rangle^2$.

For which n does such a bilinear vector product exist in $\mathbb{R}^n$?

We assume that it exists in $\mathbb{R}^n$ and imbed $\mathbb{R}^n$ in $\mathbb{R}^{n+1} = \mathbb{R} \oplus \mathbb{R}^n$. We write $X \in \mathbb{R}^{n+1}$ as $X = \xi + x$, $\xi \in \mathbb{R}$, $x \in \mathbb{R}^n$, and similarly $Y = \eta + y \in \mathbb{R}^{n+1}$, and put

(5) $X \cdot Y = \xi\eta - \langle x, y \rangle + \xi y + \eta x + x \times y$.

Then $1 + 0 \in \mathbb{R}^{n+1}$ is a two-sided identity for that product, and an easy computation using (3) and (4), namely,

$$|X \cdot Y|^2 = \xi^2 \eta^2 + \langle x, y \rangle^2 - 2\xi\eta\langle x, y \rangle + \xi^2 |y|^2$$
$$+ \eta^2 |x|^2 + 2\xi\eta\langle x, y \rangle + |x|^2 |y|^2 - \langle x, y \rangle^2$$
$$= \xi^2 \eta^2 + \xi^2 |y|^2 + \eta^2 |x|^2 + |x|^2 |y|^2,$$

yields

(6) $|X \cdot Y|^2 = |X|^2 |Y|^2$.

The product (5) turns $\mathbb{R}^{n+1}$ into an "algebra"; the commutative and associative laws are not required. It fulfills, however, the *norm product rule* (6).

We consider for a moment such algebras in $\mathbb{R}^n$ with norm product rule. The product $X \cdot Y$ can easily be modified so as to contain a two-sided identity whose existence we will assume in the following and denote 1. A bilinear product can be given by the multiplication table of a basis of $\mathbb{R}^n$; it is convenient to have 1 as a basis element.

The classical examples for $n = 1, 2, 4$, and 8 are:

$\mathbb{R}^1 = \mathbb{R}$

$\mathbb{R}^2 = \mathbb{C}$

$\mathbb{R}^4$ = quaternion algebra $\mathbb{H}$ (associative but not commutative)

$\mathbb{R}^8$ = "Cayley numbers" or Octonion algebra (not associative, not commutative)

We do not give the well-known multiplication tables for $\mathbb{C}$ (basis 1, i), $\mathbb{H}$ (basis 1, i, j, k), and the Octonions. We recall, however, that all these algebras fulfill the norm product rule (which implies that there are no zero-divisors).

If we write ξ_j for the components of X, η_j of Y, ζ_j of $X \cdot Y$, the norm product rule becomes

(7) $(\xi_1^2 + \cdots + \xi_n^2)(\eta_1^2 + \cdots + \eta_n^2) = \zeta_1^2 + \cdots + \zeta_n^2$.

Because of the "composition of quadratic forms" given by (7), the algebras with norm product rule are also called *composition algebras*. In 1898 Hurwitz proved that such a "composition of quadratic forms" with bilinear functions ζ_j of the ξ_j and η_j, with real or complex coefficients, can exist for $n = 1, 2, 4, 8$ only.

Proposition 4. A bilinear multiplication in $\mathbb{R}^n$ with two-sided identity and with norm product rule exists if and only if $n = 1, 2, 4, 8$.

In the same spirit as before, we consider the corresponding problem for continuous multiplications. One would expect that continuity gives much more flexibility than bilinearity. However, Adams proved in 1960 that:

Theorem A. A continuous multiplication with two-sided identity and norm product rule exists only for $n = 1, 2, 4, 8$.

Thus, if we now write P_n for the continuous multiplication problem in $\mathbb{R}^n$ (with the above properties), one again has Theorem L, except that *linear* is to be replaced by *bilinear*. And a "direct" proof would reduce the proof of Adams's famous Theorem A to the very old Hurwitz argument of linear algebra.

The original proof of Adams's theorem was a real tour de force, using the whole range of methods of algebraic topology known at that time. A very simple proof became available later thanks to the development of topological K-theory and the Atiyah-Hirzebruch integrality theorems; the proof is simple, but the prerequisites are certainly not.

Here too an algebraic corollary can be mentioned for which no algebraic proof is known.

Proposition 5. $\mathbb{R}^n$ is a bilinear division algebra if and only if $n = 1, 2, 4$, or 8.

Division algebra means a product without zero-divisors (associativity and commutativity are not required). If such a product is given, it can be renormalized so as to fulfill the norm product rule, but one loses bilinearity. Theorem A then says that $n = 1, 2, 4$, or 8.

If we return to vector products of two vectors in $\mathbb{R}^n$, the earlier arguments combined with the Hurwitz Theorem (Proposition 4) yield

Proposition 6. A nontrivial bilinear vector product fulfilling (3) and (4) exists in $\mathbb{R}^3$ and in $\mathbb{R}^7$ and in no other $\mathbb{R}^n$.

Indeed, it follows that $n + 1$ must be 1, 2, 4, or 8. We have to show only that such a vector product actually exists in $\mathbb{R}^7$. We first note that for $n = 3$ the product (5) defines the usual quaternion multiplication in $\mathbb{R}^4$, where in $X = \xi + x$ the $\mathbb{R}$-multiple ξ of 1 is the "real part", x the "imaginary part". Conversely, starting from the quaternion product, one considers imaginary quaternions $x, y \in \mathbb{R}^3$ and puts

$$x \times y = x \cdot y + \langle x, y \rangle,$$

which is imaginary. Then

$$(x \cdot y) \cdot y = x \cdot (y \cdot y) = -|y|^2 x$$
$$(\text{since } y \cdot y = -|y|^2 \text{ for } y \in \mathbb{R}^3)$$
$$= (x \times y) \cdot y - \langle x, y \rangle y$$
$$= -\langle x \times y, y \rangle + \text{imaginary terms};$$

whence $\langle x \times y, y \rangle = 0$, and similarly $\langle x \times y, x \rangle = 0$. By the norm product rule valid for the quaternions one has

$$|x \cdot y|^2 = |x|^2 \cdot |y|^2 = |x \times y - \langle x, y \rangle|^2$$
$$= |x \times y|^2 + \langle x, y \rangle^2 ;$$

i.e., (4) holds.

Exactly the same procedure works for the Octonions in $\mathbb{R}^8 = \mathbb{R} \oplus \mathbb{R}^7$. Although the product is *not* associative, the "alternative" law $(X \cdot Y) \cdot Y = X \cdot (Y \cdot Y)$ holds, and only this has been used (actually the alternative law holds in any composition algebra). Thus $x \times y = x \cdot y + \langle x, y \rangle$, $x, y \in \mathbb{R}^7$, is a bilinear vector product.

For a *continuous* vector product in $\mathbb{R}^n$, the formula (5) defines, exactly as in the bilinear case, a continuous product with two-sided identity and norm product rule in $\mathbb{R}^{n+1}$. Again using Adams's Theorem A, we get

Proposition 7. A continuous vector product in $\mathbb{R}^n$ fulfilling (3) and (4) exists if and only if $n = 3$ or 7.

Writing P_n for the existence problem of a vector product of two vectors in $\mathbb{R}^n$, one has Theorem L (with *bilinear* instead of *linear*).

An interesting corollary of Proposition 7 concerns almost-complex structures on S^n. Such a structure is given by a continuous field $J(x)$ of linear transformations of the tangent space at $x \in S^n$ with $J(x)^2 = (\text{minus})$identity. (On a complex-analytic manifold, multiplication of complex vector components by $\sqrt{-1}$ is such a field. But we do not assume that complex-analytic coordinates are given.)

Given the field J on S^n, we consider $x \in S^n$, a unit tangent vector $y(x)$ (i.e., two vectors $x, y \in \mathbb{R}^{n+1}$ with $|x| = |y| = 1$ and $\langle x, y \rangle = 0$), and the oriented tangent 2-plane determined by y and $J(x)y$. We choose $x \times y$ to be the unit vector orthogonal to y in that plane and corresponding to the orientation.

We then have a vector product $x \times y$ defined for $|x| = |y| = 1$ and $\langle x, y \rangle = 0$ only, but it can easily be extended to all vectors $x, y \in \mathbb{R}^{n+1}$ so as to fulfill (3) and (4) and be continuous. Therefore $n + 1$ must be 3 or 7; whence $n = 2$ or 6.

Proposition 8. S^n admits an almost-complex structure only for $n = 2$ and 6.

On S^2 such a structure exists, of course, since S^2 can be turned into the Riemann sphere. On S^6 a (linear) almost-complex structure can be derived from the Octonions in $\mathbb{R}^8$; it has been known since 1951 that it cannot come from a complex-analytic structure on S^6.

A *vector product of r vectors in* $\mathbb{R}^n$, $r < n$, is a (multilinear) vector function $v(x_1, \ldots, x_r) \in \mathbb{R}^n$ that fulfills

(1)$_r$ $\langle v(x_1, .., x_r), x_j \rangle = 0$, $j = 1, \ldots, r$;

(2)$_r$ $|v(x_1, \ldots x_r)|^2 = \text{determinant of the } \langle x_j, x_k \rangle$.

Condition (2)$_r$ above implies that $v \neq 0$ if and only if the r vectors x_j are linearly independent.

For $r = 2$ this is the vector product above. The case $r = 1$, vector field on a sphere, has been treated in the very beginning; a solution exists only if n is even.

For which (r, n) does there exist such a vector product? We assume $r \geq 2$ and fix an arbitrary unit vector x_r. Restricting the other variables to the $\mathbb{R}^{n-1}$ orthogonal to x_r clearly yields a vector product of $r - 1$ vectors in $\mathbb{R}^{n-1}$. Continuing with the reduction, we get a vector product of 1 vector in $\mathbb{R}^{n-r+1}$, which implies that $n - r$ must be odd. For $r \geq 2$, reducing to 2 vectors yields $n - r + 2 = 3$ or 7; i.e., $r = n - 1$ or $r = n - 5$.

For arbitrary $n > 1$ and $r = n - 1$ there is a well-known multilinear solution: Take for v the vector orthogonal to the hyperplane spanned by the vectors x_j, $j = 1, \ldots, n - 1$ (if they are linearly independent), with suitable orientation and suitably normalized. In terms of the $n \times (n - 1)$-matrix of the components of the x_j, the components of v are given by the $(n - 1) \times (n - 1)$-minors with the usual signs.

For $r = n - 5$ we know that $(2, 7)$ has a bilinear solution. What about $(3, 8)$? Here again, a multilinear vector product can be given explicitly in terms of the Octonions. A more elaborate argument, using Octonions again, shows that $(4, 9)$ does not have any multilinear solution. On the other hand, a topological method (cross-section of a Stiefel-manifold fibering, Steenrod squares; see survey [E1]) proves that even the continuous $(4, 9)$-problem does not have a solution.

In summary: *A continuous vector product of r vectors in $\mathbb{R}^n$ exists only in the cases $(1, n)$ with n even, $(n - 1, n)$, $(2, 7)$, and $(3, 8)$, and in these cases there is a (multi-)linear solution.*

Writing $P_{(r,n)}$ for the problem, Is there a vector product of r vectors in $\mathbb{R}^n$? we have:

Theorem L. If $P_{(r,n)}$ admits a continuous solution, then for that pair (r, n) it also has a (multi-)linear solution.

Vector Functions of One Variable, Maximal Number of Orthonormal Solutions

We return to one vector function of one variable in $\mathbb{R}^n$ (tangent vector field problem on S^{n-1}) and consider s orthonormal solutions. In other words, we have $s+1$ vector functions $v_j(x)$, $j = 0, 1, \ldots, s$ in $\mathbb{R}^n$, defined for $|x| = 1$ with $v_0(x) = x$, and

$$\langle v_j(x), v_k(x) \rangle = \delta_{jk}, \quad j, k = 0, 1, \ldots, s.$$

We assume that the vectors v_j are linear functions of x and write $v_j(x) = A_j x$ where A_j is a real $n \times n$-matrix, $A_0 = E$ (unit matrix). Then

$$\langle v_j(x), v_k(x) \rangle = \langle A_j x, A_k x \rangle = \delta_{jk} \langle x, x \rangle$$

for all $x \in \mathbb{R}^n$. This implies that all A_j are orthogonal matrices, that $A_j^T A_j = E$ where A_j^T denotes the transposed matrix, and that for $j \neq k$

$$A_j^T A_k + A_k^T A_j = 0.$$

For $k = 0$ this yields $A_j^T + A_j = 0$, $j = 1, \ldots, s$; whence

(8) $\quad A_j^2 = -E, \quad A_j A_k + A_k A_j = 0$

$$\text{for } j \neq k, \ j, k = 1, \ldots s.$$

In addition the matrices have to be orthogonal or, equivalently, skew-symmetric.

Such matrices, with real or complex entries, are called *Hurwitz-Radon matrices*. They were independently[1] examined around 1920 by Hurwitz and Radon; they determined for given n the maximum possible number s. If $n = \text{odd}.16^\alpha 2^\beta$, $\beta = 0, 1, 2, 3$, then

$$s_{\max} = 8\alpha + 2^\beta - 1.$$

The quantity $\rho(n) = 8\alpha + 2^\beta$ is called the Radon number of n, and the relations (8) are called the Hurwitz matrix equations.

Proposition 9. The maximum number of orthonormal tangent vector fields on S^{n-1} depending linearly on $y \in S^{n-1}$ is $\rho(n) - 1$.

Again, the continuous analogue was established by Adams (1962) in his famous

Theorem B. The maximum number of continuous orthonormal tangent vector fields on S^{n-1} is $\rho(n) - 1$.

The proof is yet more difficult and technical than the original proof of Theorem A. So far no simpler argument of algebraic topology has been found. If, by analogy to the foregoing, we write $P_{n,s}$ for the problem, Is there a system of s orthonormal tan-

---

[1]*Hurwitz died in 1919. His paper appeared in 1923. Radon's work was submitted in 1922 and also was published in 1923.*

gent vector fields (s-frames) on S^{n-1}? then we again have Theorem L, this time with P_n replaced by $P_{n,s}$.

In the continuous case the same holds for linearly independent fields instead of orthonormal ones, since orthonormalization does not affect continuity.

Hurwitz-Radon Matrices and Homotopy Groups

We begin with some remarks concerning the Hurwitz-Radon matrix problem (8).

The proofs by Hurwitz and Radon were by matrix computations. A different proof, of a more conceptual nature, was given in 1942 by the author using classical representation theory applied to a certain finite group G_s (generated by symbols $A_1, \ldots, A_s, \varepsilon$ with relations dictated by (8), i.e., $\varepsilon^2 = 1, A_j^2 = \varepsilon, A_j A_k = \varepsilon A_k A_j, j \neq k$). Instead of looking for maximal s given n, one asks for minimal n given s. Minimal n is provided by irreducible orthogonal representations with $\varepsilon \mapsto -E$. The advantage of this method is that it gives explicitly *all* solutions and shows very simply that there exist solutions with matrix entries 0, $+1$, and -1 only; see [E2]. We note here, for use in the next section, that the same matrix problem can, of course, be formulated for *unitary* representations and this is simpler than the orthogonal problem. In that case the minimal n is $2^{\frac{s}{2}}$ for even s and $2^{\frac{s-1}{2}}$ for odd s. All solutions are direct sums of the minimal ones. In a solution for $s+1$, omitting the last matrix A_{s+1} of course yields a solution for s. For even s, a minimal solution is obtained in this way from a minimal solution for $s+1$, since $2^{\frac{s}{2}} = 2^{\frac{(s+1)-1}{2}}$. In other words, the solutions for even s are not essential; they all come from $s+1$.

Let $A_1, \ldots, A_s$ be a set of orthogonal Hurwitz-Radon matrices, i.e., a solution of (8); let $A_0 = E$; and let $\alpha_0, \alpha_1, \ldots, \alpha_s$ be real numbers with $\Sigma \alpha_j^2 = 1$. From (8) it follows easily that the $n \times n$-matrix

(9) $\qquad\qquad f(a) = \Sigma \alpha_j A_j$

is orthogonal, and this is equivalent to (8). We write $a = (\alpha_0, \alpha_1, \ldots, \alpha_s) \in S^s$ in $\mathbb{R}^{s+1}$ and consider f as a (linear) map of S^s into the orthogonal group $O(n)$. Combining with the natural imbedding of $O(n)$ into the infinite orthogonal group O (the limit of the usual inclusions $O(n) \to O(n+1)$), we get a map $F : S^s \to O$. Conversely, any linear map $F : S^s \to O$ of the form (9) (the image necessarily lies in some $O(n)$), with $A_0 = E$, is given by orthogonal Hurwitz-Radon $n \times n$-matrices. The homotopy class of F is an element of the homotopy group $\pi_s O$.

[Remark: The matrix (9) being orthogonal shows that a solution of the Hurwitz equations (8) is equivalent to the composition of quadratic forms, generalizing (7),

$$(\alpha_0^2 + \cdots + \alpha_s^2)(\xi_1^2 +, \cdots, \xi_n^2) = (\zeta_1^2 + \cdots + \zeta_n^2),$$

where the ζ_i are bilinear in the α_j and ξ_i.]

One can proceed in exactly the same way with a *unitary* solution of (8). Then f and F (we use the same letters) yield a homotopy class in $\pi_s(U)$ where U is the infinite unitary group, the limit of the inclusions $U(n) \to U(n+1)$. Both cases are closely related to *Bott periodicity* (1956). Here we restrict ourselves, for simplicity, to the unitary case; the orthogonal case can be dealt with in the same way, though the details are a little more complicated.

As for the homotopy group $\pi_s U(n)$, it has been known since around 1940 that

$$\pi_s U(n) \cong \pi_s U\left(\frac{s+1}{2}\right)$$

$$\text{if } s \text{ is odd and } n \geq \frac{s+1}{2},$$

$$\pi_s U(n) \cong \pi_s U\left(\frac{s+2}{2}\right)$$

$$\text{if } s \text{ is even and } n \geq \frac{s+2}{2}.$$

The isomorphisms are given by the imbedding $U(n) \to U(n+1)$. These "stable" groups are the homotopy groups $\pi_s U$.

The Bott periodicity theorem determined the groups $\pi_s U$ completely:

$$\pi_s U = \mathbb{Z} \quad \text{if } s \text{ is odd,}$$
$$= 0 \quad \text{if } s \text{ is even.}$$

If s is even, any solution of (8) yields, as it must, a nullhomotopic map $f : S^s \to U(n)$, since it comes from a solution for $s + 1$, so that f can be extended to a linear map $S^{s+1} \to U(n)$ and is therefore null-homotopic (even in a *linear* way).

For odd s, however, one has the interesting result:

A minimal solution of the Hurwitz-Radon problem yields, for odd s, through $S^s \to U(n) \to U$ given by f above, a generator of $\pi_s U$.

Note that here $n = 2^{\frac{s-1}{2}}$, while in the usual approach the generator, in the stable group $\pi_s U(\frac{s+1}{2})$, lies in dimension $n = \frac{s+1}{2}$. For the lowest cases $s = 1$ and 3, these n are equal, and the generators are easily recognized to be identical. This is not so for $s > 3$, so that a proof is needed. It makes use (see [E2]) of Bott's theorem in its full topological statement. On the algebraic side it is based on a simultaneous analysis of the solutions of the matrix problem (8) for all values of s.

What about the multiples of the generator? The group operation (addition) for two elements of

$\pi_s U$ can be described in a simple way: One just places the two maps $S^s \to U(n)$ over the diagonal in $U(2n)$. One therefore can obtain all elements of $\pi_s U$, i.e., of $\pi_s U(n)$ for sufficiently high n, through unitary Hurwitz-Radon matrices. This can be expressed again as a linearization result of a different nature:

Theorem L'. Any continuous map $S^s \to U(n)$, $n \geq \frac{s+1}{2}$, or $\geq \frac{s+2}{2}$ respectively is homotopic in U to a linear map.

A direct proof of that theorem would reduce Bott periodicity to the purely algebraic discussion of Hurwitz-Radon matrices.

What Bott proved was actually more than the periodicity of homotopy groups: One considers ΩU, the space of loops in U beginning and ending in $1 \in U$. The periodicity of homotopy groups $\pi_{s+2}(U) = \pi_s(U)$ for all $s \geq 0$ is essentially the same as a homotopy equivalence between $\Omega\Omega U$ and U and thus a periodicity with period 2 for all iterated loop-spaces of U. For the groups of homotopy classes of maps of arbitrary spaces (cell complexes) X into U and into the iterated loop-spaces of U, one therefore has the same periodicity. The (abelian) groups thus obtained constitute a cohomology functor called topological K-theory. This was the first example of an "extraordinary" cohomology theory, and it seems interesting that it is closely related to the unitary Hurwitz-Radon matrices.

Everything can also be said, mutatis mutandis, about the orthogonal (or the symplectic) Hurwitz-Radon matrices and the infinite orthogonal (or symplectic) group and the corresponding K-theory; here the periodicity has period 8.

Von Neumann Algebra of a Group

About the Bass Conjecture

Here we consider the complex group algebra $\mathbb{C}G$ of a discrete group G. Recall that it is the complex vector space having the group elements as basis, with product given by the group multiplication of the basis elements. The identity element $1 \in G$ is the identity for the algebra product. An idempotent $a \in \mathbb{C}G$ is an element fulfilling $a^2 = a$. The *idempotent conjecture* says that the only idempotents in $\mathbb{C}G$ are 0 and 1, as in $\mathbb{C}$ or any division ring, provided the group G is torsion-free, i.e., has no elements of finite order $\neq 1$ (elements of finite order easily yield nontrivial idempotents of $\mathbb{C}G$). In the following we always assume G to be torsion-free.

A strong tool to deal with this problem is the "canonical" trace $\kappa(a)$, also called the Kaplansky trace of $a \in \mathbb{C}G$; it is the coefficient of $1 \in G$ of a. A little more generally, let $A = (a_{ij})$ be an idempotent $(n \times n)$-matrix with entries in $\mathbb{C}G$ and $\kappa(A) = \Sigma\kappa(a_{ii})$. The image of A in $\mathbb{C}G^n$ is a finitely generated projective $\mathbb{C}G$-module P, and we write also $\kappa(P)$ for $\kappa(A)$, since κ is independent of the imbedding of P in some $\mathbb{C}G^n$.

Kaplansky Theorem. $\kappa(A)$ is a nonnegative real number, equal to 0 only if $A = 0$.

Actually $\kappa(A)$ is known to be rational, but here we will not make use of this.

We recall that the von Neumann algebra $N(G)$ of G can be defined as the algebra of all G-equivariant bounded linear operators on the Hilbert space $\ell_2 G$ of square-summable complex functions on G. A simple proof of the above theorem is obtained by imbedding $\mathbb{C}G$ into $N(G)$; then $\kappa(A)$ can be identified with the von Neumann trace of A. Although the idempotent map defined by A need not be selfadjoint, it is equivalent to a selfadjoint one (orthogonal projection). In other words, $\kappa(A) = \kappa(P)$ is the von Neumann dimension of the Hilbert-G-module $\ell_2 G \otimes_{cG} P$.

Another notion of trace is given by the augmentation of Σa_{ii}. It is an integer, namely, the dimension of the $\mathbb{C}$-vector space $\mathbb{C} \otimes_{cG} P$, which we write in short $d(P)$. The *Bass conjecture* says that these two traces are equal:

$$(10) \qquad \kappa(P) = d(P).$$

This is the weak form of the conjecture, implied by the strong one, which we do not formulate here; see [B]. It has been proved for several big classes of groups G, such as linear groups, solvable groups (of finite homological dimension), hyperbolic groups, 3-manifold groups, groups of cohomological dimension 2 (over $\mathbb{Q}$). The proof of the simple equation (10) is very indirect and different for the various classes, using arithmetic methods, cyclic homology of groups, homological dimension, etc. (Bass 1976, Eckmann 1986).

Note that for an idempotent $a \in \mathbb{C}G$ the projective P is the left ideal $\mathbb{C}Ga \subset \mathbb{C}G$ and $d(P)$ is necessarily equal to 0 or 1. Thus $\kappa(a) = 0$ or $\kappa(1 - a) = 0$, which proves the idempotent conjecture for the respective groups.

Projective Modules over $N(G)$

Beyond the equality (10), i.e., the weak Bass conjecture, one can say more: Through the imbedding of $\mathbb{C}G$ in $N(G)$ the projective module P becomes a finitely generated projective $N(G)$-module $N(G) \otimes_{cG} P$ that turns out to be a *free* $N(G)$-module of rank equal to $d(P) = \dim_{\mathbb{C}}\mathbb{C} \otimes_{cG} P$. This can be expressed as an isomorphism of $N(G)$-modules

$$N(G) \otimes_{cG} P = N(G) \otimes_{\mathbb{C}} (\mathbb{C} \otimes_{cG} P),$$

or more intuitively as an associativity formula

$$(11) \qquad (N(G) \otimes_{\mathbb{C}} \mathbb{C}) \otimes_{cG} P = N(G) \otimes_{\mathbb{C}} (\mathbb{C} \otimes_{cG} P),$$

where everything is purely algebraic. In particular, $N(G)$ is considered just as an algebra.

For the proof, however, one uses the fact that the $N(G)$ is a finite von Neumann algebra and thus admits a center-valued trace: For the projection onto $N(G) \otimes_{cG} P$ its value is precisely $\kappa(P)$.identity (here a deep result on finite conjugacy classes in G following from [B], Theorem 8.1, is used). If the weak Bass conjecture holds, it is equal to $d(P)$.identity, which is the center-valued trace of the above free module. And projective modules having the same center-valued trace are isomorphic.

[Actually, more is true. The algebraic category of finitely generated projective $N(G)$-modules is equivalent to the category of finitely generated Hilbert-G-modules (and G-equivariant bounded linear operators as morphisms). And if two $N(G)$-projectives have the same center-valued trace, then the corresponding Hilbert-G-modules are isometrically G-isomorphic. The von Neumann dimension can be carried over to finitely generated projective $N(G)$-modules; a special feature is that submodules of finitely generated projective $N(G)$-modules are again projective.]

Conversely, the center-valued traces show that (11) implies the (weak) Bass conjecture for G, $\kappa(P) = d(P)$.

Theorem L''. The associativity formula (11) is equivalent to the weak Bass conjecture $\kappa(P) = d(P)$ for all finitely generated projective $\mathbb{C}G$-modules.

For many classes of groups, $\kappa(P) = d(P)$; whence (11) is a theorem, and certainly not an easy one, with different proofs according to the respective class. Dare one ask here for a more direct approach to Theorem L''? For which groups? In what generality?

Return to Topology: Poincaré-2-Complex

First, some very short technical remarks about Hilbert space methods in algebraic topology.

In recent years homotopy invariants of a space X (cell-complexes of finite type with infinite fundamental group G) have been introduced and applied with the help of Hilbert-G-modules: ℓ_2-homology modules (reduced, i.e., cycles modulo the closure of the boundary space) and ℓ_2-Betti numbers (their von Neumann dimension). These concepts actually go back to Atiyah (1976), but were fully developed much later.

In view of the category equivalence above, one gets a purely algebraic approach to all this: G operates as a covering transformation group on the universal covering $\tilde{X}$ of X; the chain groups of $\tilde{X}$ are free $\mathbb{Z}G$-modules, and tensoring them over $\mathbb{C}G$ with the algebra $N(G)$, one obtains a complex of finitely generated free $N(G)$-modules. Its homology groups are not projective in general, but finitely presented $N(G)$-modules. Their "projective part", corresponding to reduced ℓ_2-homology, yields the ℓ_2-Betti numbers, and they also yield the Novikov-Shubin invariants (Farber, Lück, 1995).

If G fulfills the Bass conjecture, then the same procedure can be applied to a *finitely dominated* space X, since the chain groups of $\tilde{X}$ are finitely generated projective $\mathbb{Z}G$-modules, and the tensor product over $\mathbb{C}G$ yields free $N(G)$-modules. This is interesting, for example, for Poincaré complexes.

A connected space is called a Poincaré-n-complex if it fulfills the classical Poincaré duality relations well known for closed n-manifolds—an approximation to the latter. We restrict attention to a very special application. It concerns the theorem (the author et al.; cf. the survey [E3]):

A Poincaré-2-complex X with infinite fundamental group G is homotopy equivalent to a closed surface of genus ≥ 1.

No finiteness assumptions are required; X is finitely dominated. An important ingredient in the proof (we mention here only the orientable case) is to show that the first ordinary Betti number $\beta_1(X)$ is ≥ 2. The methods above greatly simplify the argument, as follows.

The ℓ_2-Betti numbers $b_i(X)$ compute the Euler characteristic of the space and fulfill Poincaré duality in the manifold- or Poincaré complex-case, exactly as the ordinary Betti numbers do. The Betti number b_0 in case of an infinite group G is easily seen to be $= 0$. Thus in our situation

$$\chi(X) = \beta_0 - \beta_1 + \beta_2 = 2 - \beta_1 = -b_1;$$

whence indeed $\beta_1 \geq 2$.

The Poincaré complex above is aspherical; i.e., all homotopy groups in dimensions ≥ 2 are 0. It is thus a classifying space for G, and the homology of G is the same as the homology of X. (The cohomological dimension being 2, the Bass conjecture is fulfilled; this we have already used above implicitly.) Passing to the universal cover of the surface, one can express the result in terms of the group G, yet another linearity statement:

Theorem L'''. A group whose homology fulfills Poincaré duality of dimension 2 is isomorphic to a plane motion group operating freely with compact fundamental domain on the Euclidean or hyperbolic plane.

If a group G is the fundamental group of a closed aspherical n-manifold, then its homology fulfills, of course, Poincaré duality of dimension n. Is the converse true? For dimensions $n \geq 3$ this problem is still unsolved, except for partial results in dimension 3.

References

For the first part we refer to the survey:

[E1] B. ECKMANN, Continuous solutions of linear equations, *Expo. Math.* 9 (1991), 351–365, where the relevant references can be found.

Similarly for the second part:

[E2] ____, Hurwitz–Radon matrices revisited, *The Hilton Symposium* 1993 (Montreal, PQ) CRM Proceedings and Lecture Notes, vol. 6, Amer. Math. Soc., Providence, RI, 1994, pp. 23–35.

For the third part we refer to recent papers by Wolfgang Lück, Michael Farber, and the author, and to:

[B] H. BASS, Euler characteristics and characters of discrete groups, *Invent. Math.* 35 (1976), 155–196.

[E3] B. ECKMANN, Poincaré duality groups of dimension two are surface groups, *Combinatorial Group Theory and Topology*, Ann. of Math. Stud., Princeton Univ. Press, Princeton, NJ, 1986, pp. 35–51.

Introduction to ℓ_2-methods in topology: Reduced ℓ_2-homology, harmonic chains, ℓ_2-Betti numbers

Israel Journal of Mathematics 117 (2000), 183–219

Preface

These are notes from a mini-course at the ETH Zurich addressed to faculty and advanced students. Its purpose was to provide a first acquaintance of the Hilbert space methods in algebraic topology which were initated by Atiyah in 1976 and have become a quite general and important tool during more recent years. Prerequisites are basic algebraic topology of cell-complexes and basic concepts of Hilbert spaces. The definitions (Hilbert-G-module, von Neumann dimension, reduced (co)homology, ℓ_2-Betti numbers of finite complexes) are given, as well as complete proofs of main properties such as homotopy invariance, Poincaré duality, etc. Applications which cannot, or not easily, be done without ℓ_2-Betti numbers concern (partial) Euler characteristic, finitely presented groups, and 4-manifolds; the Cheeger–Gromov lemma on amenable groups is stated and proved. The integrality conjecture known as "Atiyah conjecture" is formulated in a most general way and discussed.

A word about our systematic use of the group of *harmonic* chains, isomorphic to both homology and cohomology groups. To prepare the ground this is illustrated, in a preliminary chapter, by the elementary case of (co-)homology with real coefficients of a finite cell-complex X. The chain groups $C_i(X)$ are finite dimensional vector spaces with a natural scalar product where the cells form an orthonormal basis. Boundary d and coboundary δ are adjoint maps; C_i decomposes into three mutually orthogonal subspaces: dC_{i+1}, δC_{i-1}, and the kernel

---

* Notes by Guido Mislin, based on lectures by Beno Eckmann, autumn 1997, at the Mathematical Research Institute, ETH Zurich.

$\mathcal{H}_i$ of the combinatorial Laplacian $\Delta = d\delta + \delta d$ (or equivalently the intersection of the i-cycle and the i-cocycle space), as described in the intuitive picture

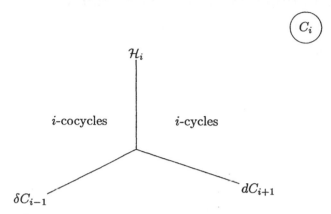

Interesting use of the "harmonic" chains $\in \mathcal{H}_i$ representing (co-)homology classes can already be made in that elementary situation.

The ℓ_2-methods appear if a regular covering Y of X is considered, in general an infinite cell-complex, with the covering transformation group G operating freely. The same decomposition, as above, of the Hilbert space of ℓ_2-chains is obtained with the only difference that one has to replace the i-boundary space and the i-coboundary space by their closures. Thus $\mathcal{H}_i$ is isomorphic to *reduced* homology: cycles modulo the closure of the boundaries, and also to reduced cohomology. All these Hilbert spaces admit isometric G-action; they are Hilbert-G-modules, and their von Neumann dimension relative to G, a real non-negative number, plays the role of the vector space dimensions in the finite complex case. In particular, the von Neumann dimension of $\mathcal{H}_i$ is the i-th ℓ_2-Betti number β_i of Y relative to G. If $Y = \tilde{X}$, the universal covering, G the fundamental group of X, it is just called $\beta_i(X)$. For $G = 1$, $Y = X$, one is in the elementary case above. If G is finite, $\beta_i(X)$ is the ordinary Betti number of Y divided by $|G|$. For infinite G the values of the ℓ_2-Betti numbers are more complicated but they nevertheless compute the Euler characteristic exactly as the ordinary Betti numbers do.

Many thanks to Guido Mislin for writing up these notes, very carefully and with many improvements; and to Emmanuel Dror Farjoun for asking us to publish them in this journal despite their introductory character.

Beno Eckmann

1. Finite CW-complexes and $\mathbb{R}$-homology

1.1.   Let X be a finite CW-complex with cellular chain complex $(K_*(X), d)$. We write $C_*(X) = \mathbb{R} \otimes K_*(X)$ for the associated chain complex over $\mathbb{R}$. If α_i denotes the number of i-cells of X, then

$$\dim_{\mathbb{R}} C_i(X) = \alpha_i,$$

and $C_i(X)$ has a natural basis $\sigma_1, \ldots, \sigma_{\alpha_i}$ consisting of the i-cells of X. We will consider $C_i(X)$ as a (real) Hilbert space with orthonormal basis $\sigma_1, \ldots, \sigma_{\alpha_i}$ and associated inner product

$$\langle \ , \ \rangle \colon C_i \otimes_{\mathbb{R}} C_i \longrightarrow \mathbb{R}.$$

The boundary operator

$$d_i \colon C_i(X) \longrightarrow C_{i-1}(X)$$

then has an **adjoint**

$$d_i^* = \delta_{i-1} \colon C_{i-1}(X) \longrightarrow C_i(X),$$

given by $\langle \delta_{i-1} x, y \rangle = \langle x, d_i y \rangle$. Thus

$$\ker \delta_i = (\operatorname{im} d_{i+1})^{\perp} \subset C_i(X), \quad \text{and}$$
$$\ker d_i = (\operatorname{im} \delta_{i-1})^{\perp} \subset C_i(X).$$

Putting $Z_i = \ker d_i$, $B_i = \operatorname{im} d_{i+1}$, $Z^i = \ker \delta_i$, $B^i = \operatorname{im} \delta_{i-1}$ and $C_i = C_i(X)$, one finds orthogonal decompositions

$$C_i = B^i \perp Z_i = B_i \perp Z^i.$$

Since $\langle \delta_{i-1} x, d_{i+1} y \rangle = 0$ for all $x, y \in C_i$, one has $B^i \perp B_i$ and therefore

$$C_i(X) = B_i \perp B^i \perp (Z_i \cap Z^i),$$

the **Hodge–de Rham decomposition** of $C_i(X)$. The groups

$$\mathcal{H}_i(X) := Z_i(X) \cap Z^i(X)$$

are called the **harmonic i-chains** of X. One defines the **Laplacian** by

$$\Delta_i = d_{i+1}\delta_i + \delta_{i-1}d_i \colon C_i \longrightarrow C_i.$$

It has the property that

$$\mathcal{H}_i(X) = \{x \in C_i(X) \mid \Delta_i(x) = 0\}.$$

Indeed, it is plain that $Z_i \cap Z^i \subset \ker \Delta_i$. Conversely, if $\Delta_i x = 0$ then

$$d_{i+1}\delta_i x = -\delta_{i-1}d_i x \in B_i \cap B^i = \{0\},$$

thus $\delta_i x \in B^{i+1} \cap Z_{i+1} = \{0\}$ and $d_i x \in B_{i-1} \cap Z^{i-1} = \{0\}$, implying that $x \in Z_i \cap Z^i$.

1.2. The Euler characteristic of X is, as usual, defined by

$$\chi(X) = \sum_i (-1)^i \alpha_i,$$

and we define the Betti numbers by putting

$$b_i(X) = \dim_{\mathbb{R}} \mathcal{H}_i(X).$$

COROLLARY 1.2.1: *The Euler characteristic of X satisfies*

$$\chi(X) = \sum_i (-1)^i b_i(X).$$

Proof: Since $C_i(X) = B^i \perp B_i \perp \mathcal{H}_i(X)$ and $Z^i = B^i \perp \mathcal{H}_i(X)$, we see that δ_i maps $(Z^i)^\perp \subset C_i(X)$ onto B^{i+1}, inducing an isomorphism

$$(1) \qquad\qquad\qquad B_i \overset{\cong}{\longrightarrow} B^{i+1}$$

so that $\dim_{\mathbb{R}} B_i = \dim_{\mathbb{R}} B^{i+1}$ for all i. Since $\alpha_i = \dim_{\mathbb{R}} C_i(X)$, it follows then that

$$\chi(X) = \sum_i (-1)^i \dim_{\mathbb{R}} C_i(X) = \sum_i (-1)^i \dim_{\mathbb{R}} \mathcal{H}_i(X) + r,$$

where $r = \sum_i (-1)^i (\dim_{\mathbb{R}} B_i + \dim_{\mathbb{R}} B^i) = 0$. ∎

Similarly, we obtain the following inequalities.

COROLLARY 1.2.1 (Morse Inequalities): *Let X be a finite CW-complex with α_i i-cells and Betti numbers $b_i, i \in \mathbb{N}$. Then, for every $k \geq 0$,*

$$\alpha_k - \alpha_{k-1} + \alpha_{k-2} - \cdots (-1)^k \alpha_0 \geq b_k - b_{k-1} + b_{k-2} \cdots (-1)^k b_0.$$

Proof: Indeed, using (1), $C_i = B_i \perp B^i \perp \mathcal{H}_i \cong B_i \oplus B_{i-1} \oplus \mathcal{H}_i$ and therefore

$$\sum_{i=0}^k (-1)^{k-i} \alpha_i - \sum_{i=0}^k (-1)^{k-i} b_i = \dim_{\mathbb{R}} B_k \geq 0. \quad\blacksquare$$

1.3.   The cellular $\mathbb{R}$-homology groups are defined by

$$H_i(X;\mathbb{R}) = Z_i(X)/B_i(X).$$

Because $Z_i(X) = B_i(X) \perp \mathcal{H}_i(X)$, the orthogonal projection $Z_i(X) \to \mathcal{H}_i(X)$ induces an isomorphism

(2) $$H_i(X;\mathbb{R}) \overset{\cong}{\longrightarrow} \mathcal{H}_i(X).$$

In particular, our Betti numbers agree with the usual Betti numbers, $b_i(X) = \dim_{\mathbb{R}} H_i(X;\mathbb{R})$, and they are therefore homotopy invariants.

The real cellular cochain complex $C^* = C^*(X) = \operatorname{Hom}_{\mathbb{R}}(C_*(X),\mathbb{R})$ has differential defined by $\delta^{i-1} = \operatorname{Hom}_{\mathbb{R}}(d_i,\mathbb{R})\colon C^{i-1} \to C^i$. Using the inner product of C_*, one obtains natural isomorphism

$$\Lambda_i\colon C_iX \longrightarrow C^iX, \quad \sigma \mapsto \langle \sigma, \ \rangle,$$

σ an i-cell of X. Since

$$((\Lambda_{i+1}\delta_i)(\sigma))(c) = \langle \delta_i\sigma, c \rangle = \langle \sigma, d_{i+1}c \rangle$$
$$= (\Lambda_i(\sigma))(d_{i+1}c) = ((\delta^i\Lambda_i)(\sigma))(c),$$

$\Lambda_*\colon (C_*,\delta) \to (C^*,\delta)$ defines an isomorphism of cochain complexes, mapping $Z^i(X)$ isomorphically onto $\ker\delta^i$, and $B^i(X)$ onto $\operatorname{im}\delta^{i-1}$. We may therefore refer, under that isomorphism, to the elements of $Z^i(X) \subset C_i(X)$ as **cocycles**, and $B^i(X) \subset C_i(X)$ as **coboundaries**. The harmonic chains are then those chains, which are simultaneously cycles and cocycles.

1.4.   A cellular map $f\colon X \to Y$ induces $f_i\colon C_i(X) \to C_i(Y)$ mapping cycles to cycles, but in general, cocycles are not mapped to cocycles (of course, the adjoint f_i^* maps cocycles to cocycles). If we wish to view $\mathcal{H}_i$ as a (co)functor, we may proceed as follows. Using the identification (2): $H_i(X;\mathbb{R}) \to \mathcal{H}_i(X)$ induced by the orthogonal projection $Z_i(X) \to \mathcal{H}_i(X)$, we obtain a **functor** $\mathcal{H}_i$ on the category of finite CW-complexes and cellular maps; the so induced maps

$$f_!\colon \mathcal{H}_i(X) \to \mathcal{H}_i(Y)$$

given by

$$f_!\colon \mathcal{H}_i(X) \overset{\cong}{\longrightarrow} H_i(X;\mathbb{R}) \overset{H_i(f)}{\longrightarrow} H_i(Y;\mathbb{R}) \overset{\cong}{\longleftarrow} \mathcal{H}_i(Y)$$

depend obviously on the homotopy class of f only, showing that $\mathcal{H}_i$ is a functor on the category of finite CW-complexes and homotopy classes of (not necessarily cellular) maps. In a similar way, using the orthogonal projection $Z^i(X) \to \mathcal{H}_i(X)$, we obtain isomorphisms

(3) $$H^i(X;\mathbb{R}) \overset{\cong}{\longrightarrow} \mathcal{H}_i(X).$$

yielding **cofunctors** (i.e., contravariant functors) $\mathcal{H}_i$ with induced maps

$$f^!\colon \mathcal{H}_i(Y) \to \mathcal{H}_i(X),$$

and one checks easily that $f_!$ and $f^!$ are adjoints of each other.

1.5.   To illustrate the use of harmonic chains, we consider the following example. Let $pr\colon \bar{X} \to X$ be the projection of a finite, regular, connected covering space, with X a finite CW-complex and $\bar{X}$ carrying the cell structure induced from the cell structure of the base space X. If $\bar{\sigma}$ denotes an i-cell over σ and $\bar{\tau}$ one over τ, the projection $pr_i\colon C_i\bar{X} \to C_iX$ satisfies

$$\langle pr_i\bar{\sigma}, \tau \rangle = \delta_{\sigma,\tau} = \langle \bar{\sigma}, \sum_{g \in G} g\bar{\tau} \rangle,$$

where G denotes the covering transformation group. It follows that the adjoint $(pr_i)^*\colon C_i(X) \to C_i(\bar{X})$ is given by $c \mapsto \sum_{g \in G} g\bar{c}$, where $pr_i\bar{c} = c$. We will use the notation $\sum g\bar{c} = N\bar{c}$, $N \in \mathbb{Z}[G]$ the norm element, thus

$$pr_i^* \circ pr_i = N, \quad pr_i \circ pr_i^* = |G|.$$

Therefore, the adjoint $pr_i^*\colon C_i(X) \to C_i(\bar{X})$ is injective, and

$$pr_i^*(pr_i \circ \delta_{i-1}) = N\delta_{i-1} = \delta_{i-1}N = pr_i^*(\delta_{i-1} \circ pr_i)$$

so that, in this case, pr_i commutes with δ and d, inducing

$$\mathcal{H}_i(pr_i) = pr_!\colon \mathcal{H}_i(\bar{X}) \to \mathcal{H}_i(X).$$

The adjoint $(pr_!)^* = pr_i^!$ is then induced by pr_i^* and satisfies $pr_! \circ pr^! = |G|$ as well as $pr^! \circ pr_! = N$. We thus obtain isomorphisms

$$pr_!\colon \mathcal{H}_i(\bar{X})^G \xrightarrow{\cong} \mathcal{H}_i(X), \quad \text{and} \quad pr^!\colon \mathcal{H}_i(X) \xrightarrow{\cong} \mathcal{H}_i(\bar{X})^G.$$

If one writes $\mathcal{H}_i(g)$ for the map $g_!\colon \mathcal{H}_i(\bar{X}) \to \mathcal{H}_i(\bar{X})$ induced by the covering transformation $g\colon \bar{X} \to \bar{X}, (g \in G)$, then $\mathcal{H}_i(\bar{X})$ is an $\mathbb{R}[G]$-module and

$$(4) \qquad\qquad b_i(X) = \frac{1}{|G|} \sum_{g \in G} tr\big(\mathcal{H}_i(g)\big),$$

because $\frac{1}{|G|} \sum tr\big(\mathcal{H}_i(g)\big)$ equals the multiplicity of the trivial representation in the G-representation $\mathcal{H}_i(\bar{X})$.

Remark 1.5.1:   The idea of using the natural inner product structure on chain groups goes back to [6], where the notion of harmonic chains for finite simplicial complexes was first introduced and discussed. Various applications, in particular of (4), can be found in [7].

2. Regular coverings of finite CW-complexes; ℓ_2-chains

Let Y be a connected CW-complex and G a group acting freely on Y by permuting the cells. We assume* the action on Y to be cocompact so that $X = Y/G$ is a finite CW-complex. Note that then G must be countable, being a factor group of the finitely generated fundamental group $\pi_1(X)$. We write $\ell_2 G$ for the (real separable) Hilbert space of square summable functions** $f : G \to \mathbb{R}$; sometimes we use the notation $\sum_{x \in G} f(x)x$ for such an f, with $f(x) \in \mathbb{R}$ and $\sum f(x)^2 < \infty$. (The general facts which follow do not depend on the condition of G being countable. If the discrete group G is not countable, $\ell_2 G$ is defined to be the Hilbert space of real valued functions on G with countable support.) The inner product on $\ell_2 G$ is given by

$$\ell_2 G \times \ell_2 G \xrightarrow{\langle \, , \, \rangle} \mathbb{R}, \quad \langle f, g \rangle = \sum_{x \in G} f(x)g(x).$$

Note that the group algebra $\mathbb{R}G$ can then be viewed as a dense subspace of $\ell_2 G$, consisting of all functions $G \to \mathbb{R}$ with finite support. In this way we may consider $G \subset \ell_2 G$ as a subset, and we write $1 \in \ell_2 G$ for the image of $1 \in G$. We like to stress that the inclusion $\mathbb{R}[G] \subset \ell_2 G$ is not an inclusion of rings: the multiplication in $\mathbb{R}[G]$ does in general not extend in a natural way to a multiplication in $\ell_2 G$. The elements $y \in G$ operate then via isometries on $\ell_2 G$, from the left and from the right: for $f \in \ell_2 G$ one has

$$y \cdot \sum_x f(x)x = \sum_x f(y^{-1}x)x, \quad \sum_x f(x)x \cdot y = \sum_x f(xy^{-1})x$$

yielding a $\mathbb{Z}[G]$-bimodule structure on $\ell_2 G$. Note also that the associated action of $\mathbb{R}G$ is an action by bounded operators. Indeed if $\alpha = \sum r(x)x \in \mathbb{R}G$ and $f \in \ell_2 G$, then

$$\|\alpha \cdot f\| \leq \sum |r(x)| \|xf\| = |\alpha| \cdot \|f\|,$$

where $|\alpha| = (\sum |r(x)|)$; similarly for $\|f \cdot \alpha\|$.

2.2.   Because G acts cocompactly on Y, the cellular chain group $K_i Y$ is a finitely generated free $\mathbb{Z}[G]$-module of rank equal to the number of i-cells of $X = Y/G$. We put

$$C_i(Y, G) = \ell_2 G \otimes_G K_i(Y).$$

---

   * Cf. Remark 2.6.3.
   ** For simplicity we work throughout over $\mathbb{R}$. Everything could be done over $\mathbb{C}$ (which is relevant in a more general context, but not here).

If the G-action on Y is clear from the context, we write just $C_i(Y)$ for $C_i(Y, G)$. Note that $C_i(Y)$ is a (left) $\mathbb{R}G$-module. We define a Hilbert space structure on $C_i(Y)$ by exhibiting an orthonormal Hilbert basis. For this we choose from each G-orbit of i-cells a representative $\bar{\tau}_i^\mu, \mu \in \{1, \ldots, \alpha_i\}$, with α_i the number of i-cells of X. Then

$$\{x \otimes \bar{\tau}_i^\mu | x \in G, \mu \in \{1, \ldots, \alpha_i\}\}$$

constitutes an orthonormal Hilbert basis for $C_i(Y)$; obviously, the Hilbert space structure on $C_i(Y)$ does not depend on the choice of the representatives $\bar{\tau}_i^\mu$. (As a matter of fact, $C_i(Y)$ is naturally isomorphic as a Hilbert space to the space of square summable chains

$$C_i^{(2)}(Y) := \{\sum_{\sigma \in J_i} f(\sigma)\sigma | f(\sigma) \in \mathbb{R}, \sum_{\sigma \in J_i} f(\sigma)^2 < \infty\},$$

with orthonormal Hilbert basis $\{\sigma\}_{\sigma \in J_i}$, J_i denoting the set of i-cells of Y.) Note also that for $f \in \ell_2 G$, the elements $f \otimes \bar{\tau}_i^\mu \in C_i(Y)$ satisfy $\|f \otimes \bar{\tau}_i^\mu\| = \|\sum f(x)x \otimes \bar{\tau}_i^\mu\| = \|f\|$ and therefore

$$(\ell_2 G)^{\alpha_i} \to C_i(Y)$$

$$(f_1, \ldots, f_{\alpha_i}) \mapsto \sum_{\mu=1}^{\alpha_i} f_\mu \otimes \bar{\tau}_i^\mu$$

defines an isometric G-equivariant isomorphism of Hilbert spaces (here $(\ell_2 G)^{\alpha_i}$ is considered as a Hilbert-space in the usual way, with

$$\|(f_1, \ldots, f_{\alpha_i})\|^2 = \sum_{\mu=1}^{\alpha_i} \|f_\mu\|^2$$

so that the inclusions $\ell_2 G \to (\ell_2 G)^{\alpha_i}$ are isometric G-equivariant embeddings). The induced boundary maps

$$\ell_2 G \otimes_G d_i : C_i(Y) \to C_{i-1}(Y),$$

which we denote by d_i too, if no confusion can arise, are bounded operators. Indeed, the following more general result is easy to prove.

LEMMA 2.2.1: Let $\varphi: (\mathbb{R}G)^n \to (\mathbb{R}G)^m$ be a morphism of $\mathbb{R}G$-modules. Then the induced operator $\tilde{\varphi} := \ell_2 G \otimes_{\mathbb{R}G} \varphi: (\ell_2 G)^n \to (\ell_2 G)^m$ is bounded.

Proof: Let $[\varphi_{ij}]$ denote the matrix of φ so that

$$\varphi(a_1, \ldots, a_n) = (\sum a_i \varphi_{i1}, \ldots, \sum a_i \varphi_{im}), \quad (a_1, \ldots, a_n) \in (\mathbb{R}G)^n$$

with $\varphi_{ij} \in \mathbb{R}G$. Each φ_{ij} has the form $\sum_x c_{ij}(x)x$, with $c_{ij}(x) \in \mathbb{R}$ and $x \in G$. We write as before

$$|\varphi_{ij}| = \sum_x |c_{ij}(x)|.$$

Then for $f \in \ell_2 G$ one has $\|f \cdot \varphi_{ij}\| \leq |\varphi_{ij}| \cdot \|f\|$, and

$$\|\tilde{\varphi}(f_1, \ldots, f_n)\|^2 = \sum_j \| \sum_i f_i \varphi_{ij} \|^2$$
$$\leq \sum_{i,j} |\varphi_{ij}|^2 \|f_i\|^2$$
$$\leq (\sum_{i,j} |\varphi_{ij}|^2) \|(f_1, \ldots, f_n)\|^2.$$

Thus $\tilde{\varphi}$ is bounded.   ∎

2.3.   In particular, the operators $d_i \colon C_i(Y) \to C_{i-1}(Y)$ and their adjoints $\delta_{i-1} = d_i^*$ are continuous. We put $\ker d_i = Z_i(Y)$, and $\ker \delta_i = Z^i(Y)$; these are thus closed subspaces of $C_i(Y)$ so that

$$\mathcal{H}_i(Y, G) := Z_i(Y) \cap Z^i(Y)$$

is a Hilbert subspace (i.e., a closed linear G-subspace) of $C_i(Y)$. The images $B_i(Y) := \mathrm{im}(d_{i+1} \colon C_{i+1}(Y) \to C_i(Y))$ and $B^i(Y) = \mathrm{im}(\delta_{i-1} \colon C_{i-1}(Y) \to C_i(Y))$ need not be closed; we write $\bar{B}_i$ and $\bar{B}^i$ for their closures, respectively. One finds then, as in 1.2, orthogonal decompositions

$$C_i(Y) = \bar{B}^i \perp Z_i = \bar{B}_i \perp Z^i = \bar{B}^i \perp \bar{B}_i \perp \mathcal{H}_i$$

(ℓ_2-Hodge–de Rham decomposition).   As in the first section, we define the Laplacian by

$$\Delta_i = d_{i+1}\delta_i + \delta_{i-1}d_i \colon C_i(Y) \to C_i(Y)$$

and

$$\mathcal{H}_i(Y, G) = \{c \in C_i(Y) | \Delta_i c = 0\}.$$

The proof is essentially the same as in the situation described in section 1. One refers to $\mathcal{H}_i(Y, G)$ as the space of **harmonic** ℓ_2-chains on Y, and we will often just write $\mathcal{H}_i(Y)$, if the G-action is plain from the context.

By analogy with 1.3 it is now natural to introduce **reduced** ℓ_2-homology groups of Y by

$$^{red}H_i(Y) = Z_i(Y)/\bar{B}_i(Y).$$

Then the projection $Z_i \to {}^{red}H_i(Y)$ induces an isomorphism $\mathcal{H}_i(Y) \cong {}^{red}H_i(Y)$ as Hilbert spaces. Similarly reduced cohomology groups are defined by

$$^{red}H^i(Y) = Z^i(Y)/\bar{B}^i(Y)$$

and one has an isomorphism $\mathcal{H}_i(Y) \cong {}^{red}H^i(Y)$. In case of a finite group G it is clear from section 1 that

$$\mathcal{H}_i(Y) \cong H_i(Y;\mathbb{R}) \cong H^i(Y;\mathbb{R}).$$

2.4.   One might also look at non-reduced (co)-homology groups Z_i/B_i and Z^i/B^i. As for the homology groups,

$$Z_i/B_i \cong H_i(\ell_2G \otimes_G K_*(Y)) =: H_i^G(Y;\ell_2G),$$

the **equivariant** homology of Y with coefficients the G-module ℓ_2G. There is thus a natural surjection

$$H_i^G(Y;\ell_2G) \to {}^{red}H_i(Y).$$

In a similar way, using the cochain complex

$$C^i(Y) := \mathrm{Hom}_G(K_i(Y), \ell_2G),$$

one defines equivariant cohomology groups

$$H_G^i(Y;\ell_2G) := H^i(C^*(Y)) = H^i(\mathrm{Hom}_G(K_*(Y), \ell_2G)).$$

Because $K_i(Y)$ is finitely generated and free as a $\mathbb{Z}[G]$-module, $K_i(Y)$ is isomorphic to its dual $\mathrm{Hom}_G(K_i(Y), \mathbb{Z}[G])$. Similarly, using the Hilbert basis of $C_i(Y)$ corresponding to the i-cells of Y, we can identify the Hilbert spaces $C_i(Y)$ and $C^i(Y) = \mathrm{Hom}_G(K_i(Y), \ell_2G)$. This leads to isomorphisms

$$Z^i/B^i \cong H^i(\mathrm{Hom}_G(K_i(Y), \ell_2G)) =: H_G^i(Y;\ell_2G),$$

and a natural surjection

$$H_G^i(Y;\ell_2G) \to {}^{red}H^i(Y).$$

We also point out that, analogous to the isomorphism $C_i^{(2)}(Y) \cong C_i(Y)$, the cochains $C^i(Y)$ can be identified with the ℓ_2-cochains $C_{(2)}^i(Y)$ (defined as those real cochains $\phi\colon K_i(Y) \to \mathbb{R}$, which are square summable, $\sum |\phi(\sigma)|^2 < \infty$, the sum being taken over all i-cells of Y). Clearly, $C_{(2)}^i(Y)$ is naturally isomorphic to the Hilbert space dual $\mathrm{Hom}_{cont}(C_i^{(2)}(Y), \mathbb{R})$, and one has isomorphisms of Hilbert spaces

$$^{red}H_i(Y) \cong \mathrm{Hom}_{cont}(^{red}H_i(Y), \mathbb{R}) \cong {}^{red}H^i(Y), \quad h \mapsto \langle h, \ \rangle.$$

Finally, we observe that there are obvious maps

$$\mathrm{can}_i : H_i(Y; \mathbb{R}) \longrightarrow {}^{red}H_i(Y)$$

given by considering an ordinary real cycle as an ℓ_2-cycle, and

$$\mathrm{can}^i : {}^{red}H^i(Y) \longrightarrow H^i(Y; \mathbb{R}),$$

which can be described as follows: $\mathrm{can}^i(x) = [\tilde{x}]$, where $\tilde{x}$ denotes the unique harmonic cocycle in $C_{(2)}^i$ representing x, and $[\tilde{x}]$ is its ordinary $\mathbb{R}$-cohomology class.

As is clear from examples (sec 2.7.3), the unreduced groups are indeed different from the reduced ones, and they do not easily yield numerical invariants. The advantage of the reduced groups is that they are *Hilbert G-modules*, as explained in the next section.

2.5.   If M is a Hilbert space, we call $V \subset M$ a **Hilbert subspace**, if V is a closed linear subspace with induced Hilbert space structure.

Definition 2.5.1:   A Hilbert G-module is a left G-module M, which is a Hilbert space on which G acts by isometries such that M is isometrically G-isomorphic to a G-stable Hilbert subspace of $(\ell_2 G)^n$ for some n.

It follows that $C_i(Y), Z_i(Y), Z^i(Y), \bar{B}_i(Y), \bar{B}^i(Y)$ and $\mathcal{H}_i(Y)$ are all Hilbert G-modules. If M is a Hilbert G-module and $V \subset M$ a G-stable linear subspace, then $M/\bar{V}$, $\bar{V}$ the closure of V, has a natural Hilbert G-module structure, with norm given by

$$\|w\| = \min\{\|\tilde{w}\|| \ \pi(\tilde{w}) = w\},$$

where $\pi: M \to M/\bar{V}$ denotes the projection. Note that π induces a G-equivariant isometric isomorphism of Hilbert G-modules

$$V^\perp \xrightarrow{\cong} M/\bar{V}.$$

Definition 2.5.2:   A map $f: M_1 \to M_2$ of Hilbert-G-modules is a
- **weak isomorphism**, if f is an injective, bounded G-equivariant operator, with $\mathrm{im}(f)$ dense in M_2;

- strong isomorphism, if f is an isometric G-equivariant isomorphism of Hilbert spaces.

Using the polar decomposition of bounded operators, one easily deduces the following crucial fact.

LEMMA 2.5.3: *Suppose there exists a weak isomorphism $M_1 \to M_2$ of Hilbert G-modules. Then there exists also a strong isomorphism.*

Proof: Let $f \colon M_1 \to M_2$ be a weak isomorphism. Then $\langle f^* f v, v \rangle$ is > 0 for all $v \in M_1 \smallsetminus \{0\}$ and $f^* f \colon M_1 \to M_1$ is a positive operator with $\mathrm{im}(f^* f)$ dense. It follows that there exists a unique positive self-adjoint operator $g \colon M_1 \to M_1$ with $g^2 = f^* f$, and $\mathrm{im}(g) \supset \mathrm{im}(g^2)$ is dense in M_1. Put $\bar{h} = f \circ g^{-1} \colon \mathrm{im}(g) \to M_2$ (g^{-1} exists since g is injective). Then $\mathrm{im}(\bar{h}) = \mathrm{im}(f) \subset M_2$ is dense, and for $x, y \in \mathrm{im}(g)$

$$
\begin{aligned}
\langle \bar{h}x, \bar{h}y \rangle &= \langle f^* f \circ g^{-1} x, g^{-1} y \rangle \\
&= \langle g^2 \circ g^{-1} x, g^{-1} y \rangle = \langle g \circ g^{-1} x, g^* \circ g^{-1} y \rangle \\
&= \langle x, y \rangle,
\end{aligned}
$$

where we used the fact that $g^* = g$. It follows that $\bar{h}$ is an isometric isomorphism $\mathrm{im}(g) \to \mathrm{im}(f)$. Since $\mathrm{im}(g)$ is dense in M_1 and $\mathrm{im}(f)$ is dense in M_2, $\bar{h}$ extends by continuity to an isometric isomorphism $h \colon M_1 \to M_2$. Since f and f^* are G-equivariant, g is G-equivariant too and so is $\bar{h}$. It follows that h is a strong isomorphism of Hilbert G-modules. ∎

Definition 2.5.4: Two Hilbert G-modules M_1 and M_2 will be called isomorphic, and we will write $M_1 \cong M_2$, if there exists a weak and therefore also a strong isomorphism $M_1 \to M_2$.

COROLLARY 2.5.5: *Let $\varphi \colon M_1 \to M_2$ be a bounded G-equivariant operator of Hilbert G-modules. Then*

$$
(\ker \varphi)^{\perp} \cong M_1 / \ker \varphi \cong \overline{\mathrm{im}\,\varphi}
$$

as Hilbert G-modules.

Proof: We have already seen that the projection $M_1 \to M_1 / \ker \varphi$ defines a strong isomorphism $(\ker \varphi)^{\perp} \to M_1 / \ker \varphi$. The canonical map $M_1 / \ker \varphi \to \overline{\mathrm{im}\,\varphi}$ is a weak isomorphism. Thus $M_1 / \ker \varphi \cong \overline{\mathrm{im}\,\varphi}$. ∎

2.6.   Our next goal is to show that the isomorphism type of the Hilbert G-module $\mathcal{H}_i(Y)$ depends only upon the G-homotopy type of Y. It is plain that the projection $Z_i(Y) \to {}^{red}H_i(Y)$ induces a strong isomorphism of Hilbert G-modules

$$\mathcal{H}_i(Y) \xrightarrow{\cong} {}^{red}H_i(Y).$$

LEMMA 2.6.1: ${}^{red}H_i(\ \ )$ defines a functor from the category of free cocompact G-CW-complexes and G-homotopy classes of maps, to the category of Hilbert G-modules and bounded G-equivariant operators.

In particular, the Hilbert G-modules $\mathcal{H}_i(Y)$ of harmonic ℓ_2-chains give rise to G-homotopy invariants.

COROLLARY 2.6.2: If $f: Y \to Z$ is a G-map between free cocompact G-CW-complexes and f is a homotopy equivalence, then the Hilbert G-modules $\mathcal{H}_i(Y)$ and $\mathcal{H}_i(Z)$ are isomorphic.

Indeed, f induces a weak equivalence ${}^{red}H_i(Y) \to {}^{red}H_i(Z)$ and therefore ${}^{red}H_i(Y) \cong {}^{red}H_i(Z)$, thus $\mathcal{H}_i(Y) \cong \mathcal{H}_i(Z)$ as Hilbert G-modules. (There is no need to assume that the map f in the corollary is a G-homotopy equivalence; as a matter of fact, it is well known that any G-map between free G-CW-complexes, which is a homotopy equivalence, is also a G-homotopy equivalence.)

Proof of the Lemma: Let $f: Y \to Z$ be a G-map of free cocompact G-CW-complexes. Then, by the G-cellular approximation theorem, f is G-homotopic to a cellular G-map $\tilde{f}: Y \to Z$, inducing bounded operators (cf. 2.2.1) and chain maps

$$\tilde{f}_i: C_i(Y) \to C_i(Z).$$

Since $\tilde{f}_i$ is continuous, it maps $\bar{B}_i(Y)$ to $\bar{B}_i(Z)$ and induces therefore ${}^{red}H_i(\tilde{f})$: ${}^{red}H_i(Y) \to {}^{red}H_i(Z)$. If $\tilde{\tilde{f}}: Y \to Z$ is a cellular G-map G-homotopic to $\tilde{f}$, then

$$K_i\tilde{f}, K_i\tilde{\tilde{f}}: K_iY \to K_iZ$$

are chain homotopic morphisms of G-chain complexes.   It follows that $\ell_2G \otimes_{RG} K_*\tilde{f} =: \tilde{f}_*$ and $\ell_2G \otimes_{RG} K_*\tilde{\tilde{f}} =: \tilde{\tilde{f}}_*$ are chain homotopic too. Thus $(\tilde{f}_* - \tilde{\tilde{f}}_*)(Z_i(Y)) \subset B_i(Y) \subset \bar{B}_i(Y)$ and, therefore, ${}^{red}H_i(\tilde{f}_*) = {}^{red}H_i(\tilde{\tilde{f}}_*)$ for all i, showing that ${}^{red}H_i(\tilde{f}_*)$ depends on the G-homotopy class of f only.   ∎

Similarly the reduced ℓ_2-cohomology groups,

$$^{red}H^i(Y) = Z^i(Y)/\bar{B}^i(Y),$$

define cofunctors from free cocompact G-CW-complexes (and G-homotopy classes of G-maps) to the category of Hilbert G-modules (and bounded G-equivariant operators), with $f: Y \to Z$ inducing

$$^{red}H^i(f): {}^{red}H^i(Z) \to {}^{red}H^i(Y)$$

via the adjoint $f_i^*: C_i(Z) \to C_i(Y)$. The reader should be careful to notice that the left G-module structure on $^{red}H^i(Y)$, which we are considering, is the one induced from considering the Hilbert G-module $^{red}H^i(Y)$ as a G-subspace of $C_i(Y)$. We could also let $g \in G$ act via $^{red}H^i(g)$, which would define a **right** G-action on $^{red}H^i(Y)$. Passing to the associated left G-action using the inverse, would result in the left G-action considered first, because G acts by isometries on $^{red}H^i(Y)$ and therefore the adjoint of $^{red}H^i(g)$ equals its inverse. In any case, the left G-spaces $^{red}H_i(Y)$, $\mathcal{H}_i(Y)$, and $^{red}H^i(Y)$ are all isomorphic as Hilbert G-modules.

Remark 2.6.3:   One can extend the definitions of the previous sections in the following obvious way. If Y is a G-CW-complex such that for a given $k \in \mathbb{N}$ the k-skeleton Y^k is a free cocompact G-CW-complex, then we define for $i < k$ the i-th (co)-homology groups (with all variations considered above), to be those of Y^k. In particular, for any finitely presented group G, the groups $\mathcal{H}_0(EG)$ and $\mathcal{H}_1(EG)$ are well defined, by choosing a model for EG with cocompact 2-skeleton.

2.7.   The following examples serve as an illustration. But first we need an elementary fact.

LEMMA 2.7.1: *If G is an infinite group, then for $n \geq 1$ the left G-module $(\ell_2 G)^n$ contains no G-invariant element $\neq 0$.*

Proof: If $\sum_{x \in G} f(x)x \in \ell_2 G$ is G-invariant, then $f(x)$ must be independent of x and, since G is infinite, $f(x)$ must be zero for all x; similarly for the case $n > 1$. ∎

Example 2.7.2:   Let Y be a connected G-CW-complex with cocompact 1-skeleton and G an infinite group. Then $^{red}H_0(Y) = \mathcal{H}_0(Y) = {}^{red}H^0(Y) = 0$. To see this, we consider the right-exact sequence

$$C_1(Y) \xrightarrow{d_1} C_0(Y) \longrightarrow \ell_2 G \otimes_G \mathbb{Z} \longrightarrow 0.$$

It shows that $\ker \delta_0 = (\operatorname{im} d_1)^\perp$ gets mapped injectively into the coinvariants $\ell_2 G \otimes_G \mathbb{Z}$, showing that $\ker \delta_0$ consists of G-invariant elements and is therefore trivial. Whence, $^{red}H^0(Y) = 0$. In particular, for any finitely generated infinite group G one has $\mathcal{H}_0(EG) = 0$ (for G finite, $\mathcal{H}_0(EG) = \mathbb{R}$).

But the unreduced H_0 need not be zero, as the following example shows.

Example 2.7.3: Let $Y = \mathbb{R}$, the universal cover of S^1, where $G = \mathbb{Z}$ acts by covering transformations on $\mathbb{R}$ and Y/G is considered as a CW-complex with 2 cells: $S^1 = e^0 \cup e^1$. The complex of ℓ_2-chains $C_*(Y)$ then takes the form

$$0 \longrightarrow \ell_2(\mathbb{Z}) \xrightarrow{d_1} \ell_2(\mathbb{Z}) \longrightarrow 0.$$

If x denotes a generator of $\mathbb{Z}$ we can write $f \in \ell_2(\mathbb{Z})$ as $f = \sum_{n \in \mathbb{Z}} a_n x^n$, and d_1 is given by

$$d_1(f) = (1 - x) \sum_{n \in \mathbb{Z}} a_n x^n \in \ell_2(\mathbb{Z}).$$

Clearly, d_1 is injective, and $\operatorname{im} d_1$ is dense in $\ell_2(\mathbb{Z})$ by 2.7.2. However, d_1 is not surjective. For instance, $1 \in \ell_2(\mathbb{Z})$ is not in the image of d_1, because $1 = (1 - x) \sum_{i \in \mathbb{Z}} a_i x^i$ would imply that all a_j, $j < 0$, are equal whence 0, and all $a_j, j \geq 0$ are equal whence 0, which is not possible. Therefore $^{red}H_*(Y) = 0$ whereas

$$Z_0(Y)/B_0(Y) = H_0^G(Y; \ell_2 G) \neq 0.$$

Expressed in another way, the example shows that $\mathcal{H}_i(E\mathbb{Z}) = 0$ for all $i \geq 0$, whereas $H_i^{\mathbb{Z}}(E\mathbb{Z}; \ell_2\mathbb{Z}) = H_i(\mathbb{Z}; \ell_2\mathbb{Z}) = 0$ for $i > 0$, and $H_0(\mathbb{Z}; \ell_2\mathbb{Z}) \neq 0$ (and $H^i(\mathbb{Z}; \ell_2\mathbb{Z}) = 0$ for $i \neq 1$, $H^1(\mathbb{Z}; \ell_2\mathbb{Z}) \neq 0$).

Remark 2.7.4: A systematic study of the difference between reduced and unreduced ℓ_2-homology leads to the notion of *torsion Hilbert modules*, with associated *Novikov–Shubin* invariants (cf. [21]). For a systematic treatment of these matters the reader is referred to [13, 19].

3. Von Neumann dimension; ℓ_2-Betti numbers

The goal of this section is to define a real valued function "$\dim_G$" (von Neumann dimension) on Hilbert G-modules satisfying the following basic properties:

- $\dim_G M \geq 0$,
- $\dim_G M = 0 \Longleftrightarrow M = 0$,
- $\dim_G M = \dim_G N$ if $M \cong N$,
- $\dim_G M \oplus N = \dim_G M + \dim_G N$,
- $\dim_G M \leq \dim_G N$ if $M \subset N$,

- $\dim_G \ell_2 G = 1$,
- $\dim_G M = \frac{1}{|G|} \dim_{\mathbb{R}} M$, if G is finite,
- $\dim_G M = \frac{1}{[G:S]} \dim_S M$, if $S < G$ has finite index.

The function $\dim_G$ will be derived from a generalization of the standard Kaplansky trace map

$$\rho: \mathbb{R}G \longrightarrow \mathbb{R}, \quad \sum_{x \in G} r(x)x \longmapsto r(1),$$

with $1 \in G$ the neutral element.

Definition 3.1.1:   The von Neumann algebra $N(G)$ is the algebra of bounded (left) G-equivariant operators $\ell_2 G \to \ell_2 G$.

Recall that $\ell_2 G$ is an $\mathbb{R}G$-bimodule. Since the right action of $\mathbb{R}G$ on $\ell_2 G$ is an action by bounded left G-equivariant operators, we may consider $\mathbb{R}G$ as a subalgebra of the von Neumann algebra $N(G)$. Mapping an operator ϕ to its adjoint ϕ^* defines an involution on $N(G)$ (turning it into a real C^*-algebra). Because the adjoint of the right action of $x \in G$ on $\ell_2 G$ is right multiplication by x^{-1}, passing to the adjoint in $N(G)$ corresponds under the inclusion $\mathbb{R}G \subset N(G)$ to **conjugation** in $\mathbb{R}G$. By conjugation on $\mathbb{R}G$ (or $\ell_2 G$) we mean the map

$$f = \sum f(x)x \longmapsto \bar{f} = \sum f(x)x^{-1}.$$

The Kaplansky trace on $\mathbb{R}G$ can now be extended to a **trace** on $N(G)$ as follows.

Definition 3.1.2:   Let $\varphi \in N(G)$ and $1 \in \mathbb{R}G \subset \ell_2 G$. Then

$$\mathrm{trace}_G(\varphi) = \langle \varphi(1), 1 \rangle \in \mathbb{R}.$$

It follows that if we consider $w = \Sigma r(x)x \in \mathbb{R}G$ as an element of $N(G)$,

$$\mathrm{trace}_G(w) = \langle 1 \cdot \Sigma r(x)x, 1 \rangle = \langle \Sigma r(x)x, 1 \rangle = r(1) = \rho(w)$$

where, as before, $\rho(w)$ stands for the Kaplansky trace of w. We also observe that the trace of $\varphi \in N(G)$ satisfies

$$\mathrm{trace}_G(\varphi) = \langle \varphi(1), 1 \rangle = \langle 1, \varphi^*(1) \rangle = \mathrm{trace}_G(\varphi^*).$$

The following lemma shows that the inclusion $\mathbb{R}G \subset \ell_2 G$ extends to an embedding of G-modules $N(G) \subset \ell_2 G$ ($\ell_2 G$ is not a ring!) under which the $*$-involution of $N(G)$ corresponds to conjugation in $\ell_2 G$; in the course of its proof we will make use of the obvious fact that for $f, g \in \ell_2 G$ conjugation satisfies $\langle f, g \rangle = \langle \bar{f}, \bar{g} \rangle$.

LEMMA 3.1.3: *The $\mathbb{R}$-linear map*

$$\theta: N(G) \to \ell_2(G), \quad \varphi \mapsto \varphi(1),$$

is injective and satisfies $\theta(\varphi^*) = \overline{\varphi(1)}$.

Proof: Let $\varphi \in N(G)$ be such that $\varphi(1) = 0$. Then, for all $x \in G \subset \ell_2 G$ one has $\varphi(x) = x\varphi(1) = 0$, showing that θ is injective. Furthermore, for all $x \in G$

$$\begin{aligned}
\langle \varphi^*(1), x \rangle &= \langle 1, \varphi(x) \rangle = \langle 1, x\varphi(1) \rangle \\
&= \langle x^{-1}, \varphi(1) \rangle = \langle \bar{x}, \varphi(1) \rangle = \langle \varphi(1), \bar{x} \rangle \\
&= \langle \overline{\varphi(1)}, x \rangle,
\end{aligned}$$

and it follows that $\varphi^*(1) = \overline{\varphi(1)}$.   ∎

Our "trace_G" has indeed the basic property one requires of a trace:

COROLLARY 3.1.4: *Let $\varphi, \psi \in N(G)$. Then*

$$\text{trace}_G(\varphi\psi) = \text{trace}_G(\psi\varphi).$$

Proof: We have

$$\begin{aligned}
\text{trace}_G(\varphi\psi) &= \langle \varphi(\psi(1)), 1 \rangle = \langle \psi(1), \varphi^*(1) \rangle \\
&= \langle \psi(1), \overline{\varphi(1)} \rangle = \langle \overline{\psi(1)}, \varphi(1) \rangle \\
&= \langle \psi^*(1), \varphi(1) \rangle = \langle 1, \psi(\varphi(1)) \rangle \\
&= \text{trace}_G(\psi\varphi).   ∎
\end{aligned}$$

3.2.   Let $M_n(N(G))$ denote the algebra of bounded (left) G-equivariant operators $(\ell_2 G)^n \to (\ell_2 G)^n$. An operator $F \in M_n(N(G))$ is determined in the usual way by a matrix $[F_{i,j}]$ of operators $F_{i,j}$ in $N(G)$ satisfying

$$F(a_1, \ldots, a_n) = \left(\sum F_{1,k} a_k, \ldots, \sum F_{n,k} a_k \right) \in (\ell_2 G)^n.$$

Note that the adjoint F^* corresponds to the matrix of operators $[(F^*)_{i,j}] = [F^*_{j,i}]$. We extend the definition of the trace to operators F in $M_n(N(G))$ by putting

$$\text{trace}_G(F) := \sum_{i=1}^n \text{trace}_G(F_{i,i}).$$

Clearly $\text{trace}_G(F) = \text{trace}_G(F^*)$ and, using Corollary 3.1.4, we see that for all $F_1, F_2 \in M_n(N(G))$

$$\text{trace}_G(F_1 \circ F_2) = \text{trace}_G(F_2 \circ F_1).$$

LEMMA 3.2.1: *If $F \in M_n(N(G))$ is self-adjoint and idempotent, then*

$$\text{trace}_G(F) = \sum_{i,j} \|F_{i,j}(1)\|^2.$$

Proof: It follows from the definitions that

$$\text{trace}_G(F) = \sum_j \langle F_{j,j}(1), 1 \rangle = \sum_j \langle (F^2)_{j,j}(1), 1 \rangle$$

$$= \sum_{i,j} \langle F_{j,i} F_{i,j}(1), 1 \rangle = \sum_{i,j} \langle F_{i,j}(1), F_{j,i}^*(1) \rangle$$

$$= \sum_{i,j} \langle F_{i,j}(1), F_{i,j}(1) \rangle = \sum_{i,j} \|F_{i,j}(1)\|^2. \quad \blacksquare$$

The following is a simple but important consequence.

COROLLARY 3.2.2: *Let $F \in M_n(N(G))$ be self-adjoint and idempotent. Then* $\text{trace}_G F$ *is non-negative and* $\text{trace}_G F = 0$ *implies* $F = 0$.

Remark 3.2.3: The corollary holds too if F is only assumed to be an idempotent. This is easily seen by using the orthogonal projection $\pi \in M_n(N(G))$ onto $\text{im } F$, observing that $F \circ \pi = \pi$ and $\pi \circ F = F$ so that $\text{trace}_G \pi = \text{trace}_G F$. If $e \in N(G)$ is an idempotent, then $\text{trace}_G(e) + \text{trace}_G(1 - e) = 1$ so that

$$\text{trace}_G e \le 1, \quad \text{and} \quad \text{trace}_G e = 1 \Rightarrow e = 1.$$

Thus, if $e \in \mathbb{Z}[G]$ is an idempotent then, since in that case $\text{trace}_G e$ must be an integer, one has $e = 0$ or $e = 1$, yielding the following classical result:
- the only idempotents in $\mathbb{Z}[G]$ are 0 and 1; or, equivalently, the integral group ring $\mathbb{Z}[G]$ does not admit a non-trivial decomposition into a direct sum of two left ideals.

The **Kaplansky Conjecture** states that the same conclusion holds for $\mathbb{R}[G]$ too, if one assumes G to be torsion free. This would follow, if one could prove that in the torsion free case $\text{trace}_G e$ is an integer for any idempotent $e \in \mathbb{R}[G]$. It is known (Zalesskii [23], see also [2]) that for an arbitrary group G and idempotent $e \in \mathbb{R}[G]$ the value $\text{trace}_G e$ is a rational number.

3.3.   We are now ready to define the function $\dim_G$. First we consider a special case. Let $V \subset (\ell_2 G)^n$ be a G-invariant Hilbert subspace and let π_V denote the orthogonal projection onto V. Because for all $a \in (\ell_2)^n$ and $x \in G$

$$xa = \pi_V(xa) + (xa - \pi_V(xa)) = x\pi_V(a) + (xa - x\pi_V(a))$$

with $\pi_V(xa), x\pi_V(a) \in V$ and $xa - \pi_V(xa), xa - x\pi_V(a) \in V^\perp$, it follows that $\pi_V(xa) = x\pi_V(a)$ and therefore $\pi_V \in M_n(N(G))$. The von Neumann dimension of V is now defined by

$$\dim_G V := \mathrm{trace}_G\, \pi_V \in \mathbb{R}.$$

Because π_V is a self-adjoint idempotent, we infer from Corollary 3.2.2 that $\dim_G V \geq 0$ and that $\dim_G V = 0$ implies $V = 0$.

For the general case we proceed as follows. Let M be an arbitrary Hilbert G-module and choose a G-equivariant isometric isomorphism

$$\alpha\colon M \xrightarrow{\cong} V \subset (\ell_2 G)^n.$$

Define the von Neumann dimension of M by

$$\dim_G M := \dim_G V.$$

We need to check that $\dim_G M$ does not depend on the choice of α. Suppose

$$\beta\colon M \xrightarrow{\cong} W \subset (\ell_2 G)^m$$

is another G-equivariant isometric isomorphism, with $m = n + k \geq n$. If V' denotes V considered as a subspace of $(\ell_2 G)^{n+k}$ using the inclusion

$$(\ell_2 G^n) \subset (\ell_2 G)^n \oplus (\ell_2 G)^k, \quad z \mapsto (z, 0),$$

one sees from the definition that $\dim_G V' = \dim_G V$. Therefore we may assume without loss of generality that $m = n$. Define then $h := \beta \circ \alpha^{-1}\colon V \to W$ and extend h to an operator $H \in M_n(N(G))$ by putting $H|V = h$ and $H|V^\perp = 0$, so H is a partial isometry $V \to W$. The composition H^*H is, by construction, the orthogonal projection onto V, and HH^* is the orthogonal projection onto W. It follows that

$$\dim_G V = \mathrm{trace}_G(H^*H) = \mathrm{trace}_G(HH^*) = \dim_G W,$$

showing that $\dim_G M$ is well-defined indeed.

An immediate consequence of the definition is that

- $\dim_G M \geq 0$, and $(\dim_G M = 0 \iff M = 0)$,
- $\dim_G M \oplus N = \dim_G M + \dim_G N$.

If $S < G$ has finite index m, then $G = \coprod_{i=1}^m Sx_i$, and $\ell_2 G$ decomposes as a Hilbert S-module into $\perp_{i=1}^m \ell_2 S \cdot x_i \cong (\ell_2 S)^m$. Thus if $F \in N(G)$ then

$$\mathrm{trace}_S F = \sum_{i=1}^m \langle F(x_i), x_i \rangle = \sum_{i=1}^m \langle F(1), 1 \rangle = m \cdot \mathrm{trace}_G F.$$

Similarly for $F \in M_n(N(G))$, so that for any Hilbert G-module, $\dim_S M = m \dim_G M$. The other properties stated in 3.1 are readily checked too and are left to the reader.

Remark 3.3.1: The general dimension theory in (complex, finite) von Neumann algebras goes back to the fundamental paper [20]. The dimension $\dim_G$ is closely related to the universal center-valued trace, and all our properties could be derived from it. The center-valued trace, however, is a rather difficult and deep concept while our treatment of $\dim_G$ is direct and elementary.

3.4.   In applications, we will be dealing with chain complexes of Hilbert G-modules.

Definition 3.4.1:   A chain complex

$$V_*: \cdots \to V_{i+1} \to V_i \to V_{i-1} \to \cdots$$

of Hilbert G-modules is called an ℓ_2G-**chain complex** if each $V_i \to V_{i-1}$ is a bounded G-equivariant operator.

Definition 3.4.2:   Let

$$V_*: \cdots \to V_{i+1} \to V_i \to V_{i-1} \cdots$$

be an ℓ_2G-chain complex. Then
- the **reduced homology modules** of V_* are the Hilbert G-modules

$$^{red}H_i(V_*) = \ker(V_i \to V_{i-1})/\overline{\mathrm{im}(V_{i+1} \to V_i)};$$

- the complex V_* is called **weak-exact**, if $^{red}H_i(V_*) = 0$ for all i.

Definition 3.4.3:   Let V_* and W_* be two ℓ_2G-chain complexes.
- A **morphism** $\phi_*: V_* \to W_*$ is an ordinary morphism of chain complexes consisting of bounded G-equivariant operators.
- Two morphisms $\phi_*, \psi_*: V_* \to W_*$ are ℓ_2G-homotopic if they are chain homotopic by a chain homotopy consisting of bounded G-equivariant operators.
- The complexes V_* and W_* are ℓ_2G-homotopy equivalent if there are morphisms $\phi_*: V_* \to W_*$ and $\psi_*: W_* \to V_*$ such that $\phi_* \circ \psi_*$ and $\psi_* \circ \phi_*$ are ℓ_2G-homotopic to the identity.

Clearly, a morphism $\phi_*: V_* \to W_*$ induces bounded G-equivariant operators $^{red}H_i(V_*) \to {}^{red}H_i(W_*)$, depending on the ℓ_2G-homotopy class of ϕ_* only. Indeed, the components of ϕ_* map cycles to cycles and, because of continuity, the closure of boundaries to the closure of boundaries; if ϕ_* is $\ell_2 G$-homotopic to $\tilde{\phi}_*$ then $\phi_i - \tilde{\phi}_i$ maps cycles to boundaries and thus they agree on reduced cohomology.

COROLLARY 3.4.4: *If the $\ell_2 G$-chain complexes V_* and W_* are $\ell_2 G$-homotopic, then the Hilbert G-modules $^{red}H_i(V_*)$ and $^{red}H_i(W_*)$ are isomorphic for all i.*

Definiton 3.4.5: A sequence $U \to V \to W$ of Hilbert G-modules is called **short weak-exact**, if

$$0 \to U \to V \to W \to 0$$

is a weak-exact $\ell_2 G$-chain complex.

Recall that for a G-equivariant bounded operator $\alpha : V \to W$ of Hilbert G-modules, the Hilbert G-modules $\overline{\alpha(V)}$ and $(\ker \alpha)^\perp \subset V$ are isomorphic, and the latter is isomorphic to $V/\ker \alpha$. It follows that

$$\dim_G V = \dim_G(\ker \alpha) + \dim_G(\overline{\alpha(V)}) = \dim_G(\ker \alpha) + \dim_G(V/\ker \alpha).$$

In particular, if $U \to V \to W$ is a short weak-exact sequence of Hilbert G-modules, then

$$\dim_G V = \dim_G U + \dim_G W.$$

COROLLARY 3.4.6: *Let*

$$V_*: 0 \to V_n \to V_{n-1} \to \cdots \to V_0 \to 0$$

be a chain complex of Hilbert G-Modules. Then

$$\sum_i (-1)^i \dim_G V_i = \sum_i (-1)^i \dim_G {}^{red}H_i(V_*).$$

Proof: Let $K_i = \ker(V_i \to V_{i-1})$ and $I_i = \overline{\text{im}(V_{i+1} \to V_i)}$. Then there are short weak-exact sequences of Hilbert G-modules

$$K_i \to V_i \to I_{i-1}, \quad I_i \to K_i \to {}^{red}H_i(V_*),$$

yielding the equations

$$\dim_G V_i = \dim_G K_i + \dim_G I_{i-1}, \quad \dim_G {}^{red}H_i = \dim_G K_i - \dim_G I_i.$$

The result now follows readily.   ∎

3.5.   Let Y be a free cocompact G-CW-complex with cellular chain complex $K_*(Y)$. Then $C_*(Y) = \ell_2 G \otimes_G K_*(Y)$ is an $\ell_2 G$-chain complex in the sense of Definition 3.4.1. The i-th ℓ_2-**Betti number** of Y (with respect to G) is defined by

$$\beta_i(Y;G) := \dim_G {}^{red}H_i(Y).$$

It is plain from the results proved earlier that the ℓ_2-Betti numbers enjoy the following properties:

- $\beta_i(Y;G)$ is a G-homotopy invariant of Y (and therefore a homotopy invariant of the orbit space Y/G; cf. 3.4.4);
- if $S < G$ is a subgroup of index m, then $\beta_i(Y;S) = m \cdot \beta_i(Y;G)$ (in this case Y/S is an m-sheeted finite covering space of Y/G);
- if G is finite, then $\beta_i(Y;G) = \frac{1}{|G|}b_i(Y)$, where $b_i(Y)$ stands for the ordinary ith Betti number of Y (in this case, $\dim_G {}^{red}H_i(Y) = \frac{1}{|G|}\dim_{\mathbb{R}} H_i(Y;\mathbb{R})$); in particular if Y is connected, $\beta_0(Y;G) = \frac{1}{|G|}$;
- if G is infinite and Y is connected, then $\beta_0(Y;G) = 0$ (cf. 2.7.2).

Definition 3.5.1:   Let X be a connected finite CW-complex. The ℓ_2-**Betti number** $\beta_i(X)$ of X is $\beta_i(\tilde{X};G)$, where $\tilde{X}$ denotes the universal covering space of X and $G = \pi_1(X)$.

Note that if α_i denotes the number of i-cells of X, the Hilbert G-module ${}^{red}H_i(\tilde{X})$ is isomorphic to a Hilbert submodule of $C_i(\tilde{X}) \cong (\ell_2 G)^{\alpha_i}$. Therefore the ℓ_2-Betti numbers satisfy

$$0 \le \beta_i(X) \le \alpha_i.$$

Furthermore, if $\bar{X}$ is a connected m-sheeted covering space of the finite complex X and $S < G$ denotes the fundamental group of $\bar{X}$, then

$$\beta_i(\bar{X}) = \dim_S {}^{red}H_i(\tilde{X}) = m \cdot \dim_G {}^{red}H_i(\tilde{X}) = m \cdot \beta_i(X),$$

which is quite different from the way the ordinary Betti numbers behave.

Example 3.5.2:   If $X = S^1$, then $\beta_i(X) = 0$ for all $i \ge 0$ (cf. 2.7.3). More generally, if X is a connected finite CW-complex which possesses a regular finite covering space $\bar{X} \to X$ of degree $m > 1$ with $\bar{X}$ homotopy equivalent to X, then the ℓ_2-Betti numbers of X all vanish, because in this case $\beta_i(X) = \beta_i(\bar{X}) = m \cdot \beta_i(X)$.

3.6.   Let X be a finite connected CW-complex with ordinary Betti numbers $b_i(X)$ and Euler characteristic $\chi(X) = \sum_i(-1)^i\alpha_i = \sum_i(-1)^ib_i(X)$, α_i the number of i-cells of X. It is an interesting fact that $\chi(X)$ can also be computed using the ℓ_2-Betti numbers.

THEOREM 3.6.1: *The Euler characteristic $\chi(X)$ of a finite connected CW-complex satisfies*

$$\chi(X) = \sum_i(-1)^i\beta_i(X).$$

Proof: We consider the ℓ_2-chain complex $C_*(\tilde{X}) = \ell_2G \otimes_G K_*(\tilde{X})$, where G denotes the fundamental group of X. Since $C_i(\tilde{X}) \cong (\ell_2G)^{\alpha_i}$ with α_i the number of i-cells of X,

$$\sum_i(-1)^i \dim_G C_i(\tilde{X}) = \sum_i(-1)^i\alpha_i = \chi(X).$$

On the other hand (cf. 3.4.6)

$$\sum_i(-1)^i \dim_G C_i(\tilde{X}) = \sum_i(-1)^i \dim_G {}^{red}H_i(\tilde{X}),$$

which proves the claim.   ∎

   The following is a slight generalization.

THEOREM 3.6.2: *Let X be a finite connected CW-complex and N a normal subgroup of π_1X with quotient group Q. Let X_N denote the covering space of X associated with N. Then*

$$\chi(X) = \sum_i(-1)^i\beta_i(X_N;Q).$$

Along the lines of 3.6.1 one can also establish the following **Morse Inequalities** (cf. 1.2.2).

COROLLARY 3.6.3: *Let X be a connected CW-complex with finite $(k+1)$-skeleton. Denote by α_i the number of i-cells and by β_i the ℓ_2-Betti numbers of X. Then*

$$\alpha_k - \alpha_{k-1} + \cdots + (-1)^k\alpha_0 \geq \beta_k - \beta_{k-1} + \cdots + (-1)^k\beta_0.$$

Remark 3.6.4:   We also like to mention (without proof) the following **Künneth Formula** for ℓ_2-Betti numbers. Let X be a free cocompact G-CW-complex, and Y a free cocompact H-CW-complex. Then, using the fact that $X \times Y$ is a free cocompact $(G \times H)$-CW-complex with $K_*(X \times Y)$ and $K(X) \otimes_{\mathbb{Z}} K_*(Y)$ isomorphic as $G \times H$-complexes, one can show that

$$\beta_j(X \times Y; G \times H) = \sum_{s+t=j} \beta_s(X; G) \cdot \beta_t(Y; H).$$

(*Question:* If $F \to E \to B$ is a fibration, under what condition is $\beta_j E = \sum_{s+t=j} \beta_s F \cdot \beta_t B$? For the case of $\beta_1 E$, see [18].)

3.7.   Let X be a connected CW-complex with fundamental group $\pi_1(X) =: G$. As earlier, we denote by $K_*(\tilde{X})$ the cellular chain complex of the universal cover $\tilde{X}$ of X. The associated n-**dual** ${}^nDK_*(\tilde{X})$ is defined by

$$\,^nDK_j(\tilde{X}) = \mathrm{Hom}_G(K_{n-j}(\tilde{X}), \mathbb{Z}[G])$$

which we consider as (left) G-modules via $(xf)(c) := f(c)x^{-1}$, where $x \in G$ and $f \in \mathrm{Hom}_G(K_{n-j}(\tilde{X}), \mathbb{Z}[G])$; the differential is the one induced from $K_*(\tilde{X})$.

Definiton 3.7.1:   We call a connected CW-complex X a **virtual** PD^n-**complex**, if there is a subgroup S of finite index in $\pi_1(X)$ such that $K_*(\tilde{X})$ is chain homotopy equivalent as a $\mathbb{Z}[S]$-complex to its n-dual ${}^nDK_*(\tilde{X})$. A group G is called a **virtual** PD^n-**group**, if $K(G, 1)$ is a virtual PD^n-complex.

For instance, a closed (not necessarily orientable) topological n-manifold is homotopy equivalent to a virtual PD^n-complex in the above sense. Also, the reader checks easily that a group G of type FP_∞ is a virtual PD^n-group in the sense of (3.7.1) if and only if it possesses a subgroup of finite index which is a PD^n-group in the "usual" sense (as defined for instance in [12]).

THEOREM 3.7.2:   *Let X be a finite virtual PD^n-complex. Then there exists a subgroup of finite index $S < \pi_1(X)$ such that the Hilbert S-modules ${}^{red}H_i(\tilde{X})$ and ${}^{red}H_{n-i}(\tilde{X})$ are isomorphic; in particular $\beta_i(X) = \beta_{n-i}(X)$ for all i and if $\pi_1(X)$ is infinite $\beta_n(X) = \beta_0(X) = 0$.*

Proof: Let $S < G$ be a subgroup of finite index such that the $\mathbb{Z}[S]$-complexes $K_*(\tilde{X})$ and ${}^nDK_*(\tilde{X})$ are chain homotopic. As in Section 2 we put $C_*(\tilde{X}) = \ell_2 S \otimes_S K_*(\tilde{X})$; define furthermore the $\ell_2 S$-chain complex ${}^nDC_*(\tilde{X})$ by

$$\,^nDC_j(\tilde{X}) = \mathrm{Hom}_S(K_{n-j}(\tilde{X}), \ell_2 S),$$

with S-action on $f \in \mathrm{Hom}_S(K_{n-j}(\tilde{X}, \ell_2 S)$ given by $(xf)(c) = f(c)x^{-1}$ for $x \in S$ and $c \in K_{n-j}(\tilde{X})$, and obvious differential. Since $K_*(\tilde{X})$ is chain homotopy equivalent to ${}^nDK_*(\tilde{X})$ as $\mathbb{Z}[S]$-complex, the $\ell_2 S$-chain complexes $C_*(\tilde{X})$ and

$^nDC_*(\tilde{X})$ are $\ell_2 S$-homotopic (cf. (3.4.3)). Thus by (2.6.2), the Hilbert S-modules $^{red}H_i(\tilde{X})$ and $^{red}H_i(^nDC_*(\tilde{X}))$ are isomorphic. But

$$^{red}H_i(^nDC_*(\tilde{X})) \cong {}^{red}H^{n-i}(\tilde{X}) \cong {}^{red}H_{n-i}(\tilde{X}),$$

with the first isomorphism following from the definition of the n-dual complex, and the second one was discussed at the end of (2.6). Since

$$\beta_j(X) = \dim_G {}^{red}H_j(\tilde{X}) = \frac{1}{[G:S]} \dim_S {}^{red}H_j(\tilde{X}),$$

the assertion concerning the ℓ_2-Betti numbers follows.   ∎

3.8.   If G is a group with a finite CW-model $K(G,1)$, we define its ℓ_2-Betti numbers by

$$\beta_i(G) = \beta_i(K(G,1)).$$

According to 2.6.3 we can extend this definition as follows. Suppose G has a CW-model with finite n-skeleton for some $n \geq 2$ (i.e., G is of type F_n). Then we put

$$\beta_i(G) = \beta_i(K(G,1)^n), \quad i < n.$$

In particular, $\beta_1(G)$ is defined for any finitely presented group G.

Example 3.8.1:   Let $X = \vee_k S^1$ be a wedge of k circles. Then $\chi(X) = k - 1$ and $X = K(*_k \mathbb{Z}, 1)$ so that

$$\beta_i(\vee_k S^1) = \beta_i(*_k \mathbb{Z}) = \begin{cases} k - 1, & \text{for } i = 1, \\ 0, & \text{else.} \end{cases}$$

In particular, $\beta_i(S^1) = \beta_i(\mathbb{Z}) = 0$ for all i. Using the Künneth formula we conclude that for any group G of type F_n

$$\beta_i(\mathbb{Z} \times G) = 0 \quad \text{for } i < n.$$

Example 3.8.2:   Let Σ_g be an orientable surface of genus $g \geq 0$, with fundamental group σ_g. Because $\Sigma_g = K(\sigma_g, 1)$ is a PD^2-complex of Euler characteristic $2 - 2g$,

$$\beta_i(\Sigma_g) = \beta_i(\sigma_g) = \begin{cases} 2g - 2, & \text{for } i = 1, \\ 0, & \text{else.} \end{cases}$$

Example 3.8.3:   Let X be a finite PD^2-complex with infinite fundamental group. Then $b_1(X) > \beta_1(X) \geq 0$. In particular, $b_1(X) > 0$. Indeed, one has

$$\chi(X) = 1 - b_1(X) + b_2(X) = -\beta_1(X) > -b_1(X).$$

(Actually, finite PD^2-complexes are homotopy equivalent to closed surfaces, see [12]; the result holds even for an arbitrary finitely dominated PD^2-complex, see [11].)

3.9.   We can extend the definition of our ℓ_2-Betti numbers as in (2.6.3). If Y is a G-CW-complex with cocompact j-skeleton, we put

$$\beta_i(Y;G) := \beta_i(Y^j;G), \quad i < j.$$

Similarly the ℓ_2-Betti numbers $\beta_i(X)$ are defined for $i < j$ if X^j is a finite connected complex. Note that if X^{j+1} is finite too, $\beta_j(X)$ is defined and it is obvious from the definition that it satisfies

$$\beta_j(X) \leq \beta_j(X^j).$$

COROLLARY 3.9.1: *Let X be a connected CW-complex with finite j-skeleton and $(j-1)$-connected universal cover. Then for all $i < j$*

$$\beta_i(X) = \beta_i(\pi_1(X)).$$

Proof: The assumptions on X imply that one can construct a model Y for $K(\pi_1(X), 1)$ by attaching cells of dimension $> j$ to X^j. Thus for all $i < j$

$$\beta_i(X) = \beta_i(X^j) = \beta_i(Y) = \beta_i(\pi_1(X)). \quad \blacksquare$$

In particular, for any connected space X with finite 2-skeleton one has

$$\beta_1(X) = \beta_1(\pi_1(X)).$$

Remark 3.9.2: Let $Y \to Y/G = X$ be a regular covering of a compact oriented Riemannian manifold X. The L^2-harmonic forms $^{dR}\mathcal{H}^p(Y)$ form a Hilbert G-submodule of the de Rham complex of L^2-forms on Y and integration of forms over cochains (with respect to a suitable triangulation of Y) defines a morphism of Hilbert G-modules

$$\int : {}^{dR}\mathcal{H}^p(Y) \to {}^{red}H^p(Y).$$

Dodziuk proved in [4] that this is an isomorphism of $\mathbb{R}$-vector spaces. In particular the Betti numbers $\beta_p(Y;G)$ agree with the corresponding de Rham ℓ_2-Betti numbers $\dim_G {}^{dR}\mathcal{H}^p(Y)$.

Atiyah asked in [1] whether these numbers are rational (resp. integers, in the case of a torsion-free group G). The question is related to the conjectures below and also to a question concerning zero divisors in $\mathbb{Q}[G]$.

3.10.   The following two statements are variations of what sometimes is referred to as **Atiyah's Conjecture.**

CONJECTURE A: *Let Y be a connected free cocompact G-CW-complex. Then all $\beta_i(Y; G)$ are rational numbers. If k is a positive integer such that the order of any finite subgroup of G divides k, then $k \cdot \beta_i(G)$ is an integer.*

Note that the group G in the conjecture is necessarily finitely generated, being a factor group of $\pi_1(Y/G)$.

CONJECTURE B: *Let $\phi: \mathbb{Z}[G]^m \to \mathbb{Z}[G]^n$ be a morphism of $\mathbb{Z}[G]$-modules, $\tilde{\phi}$ the induced bounded operator $\ell_2(G)^m \to \ell_2(G)^n$. Then $\dim_G \ker \tilde{\phi}$ is rational. If k is as above in $\mathbf{A}$ then $k \cdot \dim_G \ker \tilde{\phi}$ is an integer.*

Since the ϕ in the conjecture is induced by some $\bar{\phi}: \mathbb{Z}[\overline{G}]^m \to \mathbb{Z}[\overline{G}]^n$ for a suitable finitely generated subgroup $\overline{G} < G$, the conjecture holds if it holds for all finitely generated groups.

PROPOSITION 3.10.1: *For a finitely generated group G the two conjectures are equivalent.*

Proof: Assuming $\mathbf{A}$ and given ϕ as in $\mathbf{B}$, it is easy to construct a Y as in $\mathbf{A}$ such that $\ker \tilde{\phi}$ is isomorphic to $^{red}H_3(Y)$. (Choose a surjection $F \to G$ with F finitely presented and choose a $K(F, 1) = Z$ with finite 2-skeleton Z^2. Let $\vee_n S^2$ be a wedge of n two-spheres and define Y^2 to be the covering space of $Z^2 \vee (\vee_n S^2)$ associated with the kernel of $F \to G$. It is a free cocompact G-space; attach m free G-cells of dimension 3 to Y^2 to obtain Y with $\ker(K_3(Y) \to K_2(Y)) = \ker \phi$.) Conversely, assuming $\mathbf{B}$ one obtains $\mathbf{A}$ by observing that $^{red}H_i(Y) = \ker \tilde{\Delta}$, where the **combinatorial Laplacian** $\Delta: K_i(Y) \to K_i(Y)$ is defined in the obvious way (using the identification $K_j(Y) = \mathbb{Z}[G]^{\alpha_j}$, α_j the number of j-cells of Y/G).   ∎

The **Zero Divisor Conjecture** states that for a torsion-free group G the group ring $\mathbb{Q}[G]$ does not contain any zero divisors $\neq 0$. Clearly the conjecture holds, if it holds for finitely generated groups. It is known to hold for a large class of groups (cf. [17]).

THEOREM 3.10.2: *The conjectures $\mathbf{A}$ and $\mathbf{B}$ imply the Zero Divisor Conjecture.*

Proof: Let G be a finitely generated torsion-free group and let $a, b \in \mathbb{Q}[G]$ with $a \neq 0$ and $ab = 0$; we need to show that $b = 0$. Consider the bounded G-equivariant operator

$$L_b: \ell_2 G \to \ell_2 G, \quad z \mapsto zb$$

and write M for its kernel. Since $a \in M$, $M \neq 0$ and therefore $0 < \dim_G M \leq 1$. Replacing b by nb for some $n > 0$ if necessary, we may assume $b \in \mathbb{Z}[G]$ so that $L_b = \tilde{\phi}$ for $\phi \colon \mathbb{Z}[G] \to \mathbb{Z}[G]$ the right multiplication by b. Conjecture B now implies $\dim_G M = 1$, and therefore $M = \ell_2 G$ whence $b = 0$.   ∎

4. Applications (deficiency, amenable groups)

Suppose that the group G possesses a presentation with g generators and r relators. Then obviously

$$g - r \leq \text{rank}(G_{ab}) = b_1(G).$$

The maximal value $\text{def}(G)$ of the differences $g - r$ over all finite presentations of G is called the **deficiency** of G. For example, a finite group G has $\text{def}(G) \leq 0$. In the following we want to get some estimates for the deficiency of infinite groups.

If G is a finitely presented group with g generators and r relators we can construct a $K(G,1)$ with 2-skeleton $K(G,1)^2$ possessing 1 zero-cell, g one-cells and r two-cells. Taking Euler characteristics yields

$$r - g + 1 = b_2(K(G,1)^2) - b_1(K(G,1)^2) + 1.$$

But $b_1(K(G,1)^2) = b_1(G)$ and $b_2(K(G,1)^2) \geq b_2(G)$ so that in general

$$\text{def}(G) \leq b_1(G) - b_2(G).$$

For ℓ_2-Betti numbers we get

$$\text{def}(G) = 1 - \beta_0(G) + \beta_1(G) - \beta_2(K(G,1)^2).$$

Whence

THEOREM 4.1.1: *Let G be a finitely presented group. Then*

$$\text{def}(G) \leq 1 + \beta_1(G).$$

In particular $\beta_1(G) = 0$ implies $\text{def}(G) \leq 1$.

In case $K(G,1)$ has a finite 3-skeleton the Morse inequality (3.6.3, case $k = 2$) for the ℓ_2-Betti numbers of $K(G,1)^3$ yields

$$r - g + 1 \geq \beta_2(G) - \beta_1(G) + \beta_0(G).$$

Recall that a group G is of type F_n if and only if there is a $K(G,1)$ with finite n-skeleton.

THEOREM 4.1.2: *Let G be a group of type F_3. Then*

$$\det(G) \leq 1 + \beta_1(G) - \beta_2(G).$$

Example 4.1.3: If $G = *_n\mathbb{Z}$ is a free group of rank n, then using 3.8.1 we obtain

$$\det(*_n\mathbb{Z}) = n = 1 + \beta_1(*_n\mathbb{Z}).$$

Example 4.1.4: Let $G = \sigma_g$ be the fundamental group of an orientable surface of genus $g \geq 0$. The well known presentation for σ_g yields $\det(\sigma_g) \geq 2g - 1$ so that from 3.8.2 we obtain

$$\det(\sigma_g) = 2g - 1 = 1 + \beta_1(\sigma_g).$$

The next example is due to Lück [19].

Example 4.1.5: Let G be a finitely presented group possessing a finitely generated infinite normal subgroup N such that $\mathbb{Z} < G/N$. Then $\beta_1(G) = 0$ ([19], Theorem 0.7) and thus

$$\det(G) \leq 1.$$

The following variation of 4.1.2 is a consequence of 3.6.2.

THEOREM 4.1.6: *Let X be a connected finite CW-complex with fundamental group G, normal subgroup $N < G$ and $Q = G/N$. Let X_N^2 be the covering space of the 2-skeleton of X associated with N. Then*

$$\det(G) \leq 1 - \chi(X^2) \leq 1 + \beta_1(X_N^2; Q) - \beta_2(X_N^2; Q).$$

4.2.   Let G be a group and B the space of bounded $\mathbb{R}$-valued functions on G. We consider B as a G-module by putting $(xf)(y) := f(yx)$ for all $x, y \in G$ and $f \in B$. A **mean** on G is a linear map $M: B \to \mathbb{R}$ such that for all $x \in G$ and $f \in B$

- $M(1) = 1$ (1: $G \to \mathbb{R}$ the constant function 1),
- $M(xf) = M(f)$,
- $f \geq 0 \Longrightarrow M(f) \geq 0$.

The following notion goes back to von Neumann.

Definition 4.2.1: A group G is called **amenable** if it admits a mean.

Example 4.2.2: A finite group G is amenable: it has a unique mean, given by

$$M(f) = \frac{1}{|G|} \sum_{x \in G} f(x).$$

It is known that the infinite cyclic group $\mathbb{Z}$ is amenable and that the class of amenable groups is

(i) extension closed, closed with respect to passing to subgroups and factor groups,

(ii) closed with respect to taking directed unions.

In particular all abelian and all solvable groups are amenable. The smallest class of groups containing all finite and all abelian groups, and which satisfies the closure properties (i) and (ii), is the class of *elementary amenable* groups. There do exist examples (even finitely presented ones) of amenable groups which are not elementary amenable (cf. [15]). On the other hand, it is a classical result that a group which contains a non-abelian free group cannot be amenable. For more information on amenability the reader is referred to [22].

4.3.   Let Y be a free cocompact connected G-CW-complex. For results concerning the ℓ_2-Betti numbers $\beta_i(Y; G)$ for infinite amenable G, the following construction is most useful. Choose an (open) cell from each G-orbit of cells in Y and write $D \subset Y$ for their union and $\bar{D}$ for the closure of D in Y ($\bar{D}$ will not be a subcomplex of Y in general!). Since G is a factor group of $\pi_1(Y/G)$ it is countable: $G = \{g_\nu | \nu \in \mathbb{N}\}$. Construct an increasing family $\{Y_j\}_{j \in \mathbb{N}}$ of subspaces of Y as follows. Let $\{N_j\}$ be a strictly increasing sequence of natural numbers. Each Y_j is the union of N_j distinct translates $g_\nu D$, $\nu = 1, \ldots, N_j$, $g_\nu \in G$ and $Y = \bigcup Y_j$. Let $\dot{N}_j$ be the the number of translates of $\bar{D}$ which meet the topological boundary $\dot{Y}_j$ of $Y_j \subset Y$. Using the *combinatorial Følner criterion* for amenability of G [14] it follows that the sequences $\{N_j, Y_j\}$ can be chosen such that $\dot{N}_j/N_j \to 0$ for $j \to \infty$ (cf. [5, 8]); we will call such a family a **Følner exhaustion**.

Recall that there is a canonical map

$$\operatorname{can}^i : {}^{red}H^i(Y) \to H^i(Y; \mathbb{R})$$

induced by considering a harmonic ℓ_2-cocycle as an ordinary one. The following lemma is very useful.

LEMMA 4.3.1 (Cheeger–Gromov [3]): *Let Y be a connected free cocompact G-CW-complex, G an (infinite) amenable group. Then*

$$\operatorname{can}^i : {}^{red}H^i(Y) \to H^i(Y; \mathbb{R})$$

is injective for all $i \geq 0$.

Proof: Choose a Følner exhaustion $\{N_j, Y_j | j \in \mathbb{N}\}$ for Y. View the kernel $\mathcal{K}$ of ${}^{red}H^i(Y) \to H^i(Y; \mathbb{R})$ as a Hilbert G-submodule of $\mathcal{H}_i(Y) \subset C_i(Y) =: C_i$. Thus

$c \in \mathcal{K}$ is a harmonic chain $c = \sum c(\sigma)\sigma$ which, when considered as a cocycle $c(\ ): K_i(Y) \to \mathbb{R}$, is of the form $\delta^{i-1}b$ for some cochain $b: K_{i-1} \to \mathbb{R}$. Recall that C_i has a Hilbert basis which corresponds bijectively to the (open or closed) cells of Y and we will sometimes identify a cell of Y with the corresponding element in C_i. Let $P: C_i \to C_i$ stand for the orthogonal projection onto $\mathcal{K}$ and $\pi_{i,j} = \pi: C_i \to C_i$ the orthogonal projection onto the finite dimensional subspace spanned by the (open) i-cells which lie in Y_j. If R denotes the set of i-cells in D, then by definition

$$\dim_G \mathcal{K} = \sum_{\sigma \in R} \langle P(\sigma), \sigma \rangle.$$

Since $\pi \circ P: C_i \to C_i$ has a finite dimensional image, the ordinary trace

$$\operatorname{trace}_{\mathbb{R}} \pi \circ P = \sum_{\sigma \in R, x \in G} \langle ((\pi P)(x\sigma), x\sigma \rangle$$

is defined. Now $\langle \pi P(x\sigma), x\sigma) \rangle = 0$ for cells $x\sigma$ not in Y_j; for cells $x\sigma \subset Y_j$

$$\langle \pi P(x\sigma), x\sigma \rangle = \langle P(x\sigma), x\sigma \rangle = \langle P(\sigma), \sigma \rangle,$$

implying

$$\operatorname{trace}_{\mathbb{R}} \pi P = N_j \sum_{\sigma \in R} \langle P(\sigma), \sigma \rangle = N_j \dim_G \mathcal{K}.$$

Since for any $c \in C_i$, $\|\pi P(c)\| \leq \|c\|$,

$$\operatorname{trace}_{\mathbb{R}} \pi P \leq \dim_{\mathbb{R}} \operatorname{im} \pi P = \dim_{\mathbb{R}} \pi(\mathcal{K})$$

whence

$$\dim_G \mathcal{K} \leq \frac{1}{N_j} \operatorname{trace}_{\mathbb{R}} \pi P \leq \frac{1}{N_j} \dim_{\mathbb{R}} \pi(\mathcal{K}).$$

To complete the proof, we need an estimate on $\dim_{\mathbb{R}} \pi(\mathcal{K})$.

Let σ be a cell in Y whose closure does not meet $\dot{Y}_j$. Then the same holds for the cells in $d_i\sigma$ since they lie in the closure of σ. For such a σ one has either $\pi\sigma = \sigma$ (if σ is in Y_j) and $\pi d\sigma = d\pi\sigma$; or $\pi\sigma = 0$ (if σ is not in Y_j) and $\pi d\sigma = 0 = d\pi\sigma$. Writing C'_* for the subspace of C_* having as Hilbert basis all cells σ in Y whose closure does not meet $\dot{Y}_j$ it follows for $c \in C'_i$

$$d_i\pi c = \pi d_i c.$$

As earlier we identify ℓ_2-chains with ℓ_2-cochains, yielding inclusions

$$K_*(Y) \otimes \mathbb{R} \subset C_* = C^* \subset K^*(Y) \otimes \mathbb{R}.$$

For $a \in K_*(Y) \otimes \mathbb{R}$ and $b \in K^*(Y) \otimes \mathbb{R}$ we write $b(a) \in \mathbb{R}$ for the usual evaluation of the cochain b on the chain a; if b happens to lie in C^* and we consider a as an ℓ_2-chain, the inner product $\langle a, b \rangle = b(a)$.

Consider now $c \in \mathcal{K} \cap C'_*$. It satisfies $d_i c = 0$ and $c = \delta^{i-1} b$ for some $b \in K^{i-1}(Y) \otimes \mathbb{R}$, and $d_i \pi c = \pi d_i c = 0$, yielding

$$\|\pi(c)\|^2 = \langle \pi c, \pi c \rangle = \langle \pi c, c \rangle = \langle \pi c, \delta^{i-1} b \rangle$$
$$= (\delta^{i-1} b)(\pi c) = b(d_i \pi c) = b(\pi d_i c) = 0.$$

Whence

$$\mathcal{K} \cap C'_i \subset \ker\{\pi|_{\mathcal{K}}: \mathcal{K} \to \pi(\mathcal{K})\}$$

and therefore

$$\dim_{\mathbb{R}} \pi(\mathcal{K}) \leq \dim_{\mathbb{R}} \mathcal{K}/\mathcal{K} \cap C'_i = \dim_{\mathbb{R}} (\mathcal{K} + C'_i)/C'_i$$
$$\leq \dim_{\mathbb{R}} C_i/C'_i.$$

The orthogonal complement $\dot{C}_*$ of C'_* in C_* has as Hilbert basis all cells in Y whose closure meets $\dot{Y}_j$. Now

$$\dim_{\mathbb{R}} C_i/C'_i = \dim_{\mathbb{R}} \dot{C}_i \leq \dot{N}_j \alpha(D)$$

where $\alpha(D)$ denotes the number of cells in D. It follows that

$$\dim_G \mathcal{K} \leq \frac{\dot{N}_j \alpha(D)}{N_j} \to 0 \qquad \text{as } j \to \infty,$$

and we conclude $\mathcal{K} = 0$.   ∎

COROLLARY 4.3.2: *Let X be a finite connected CW-complex with amenable fundamental group G. Then $\beta_1(X) = \beta_1(G) = 0$.*

Proof: The universal cover Y of X satisfies $H^1(Y; \mathbb{R}) = 0$; thus the result follows from the Cheeger–Gromov Lemma.   ∎

The corollary also shows that —as remarked earlier— a group G which contains a non-abelian free group cannot be amenable, because $\beta_1(\mathbb{Z} * \mathbb{Z}) = 1 \neq 0$.

Applying the Cheeger–Gromov Lemma to the m-skeleton of the universal cover of a $K(G, 1)$ one obtains the following vanishing theorem for ℓ_2-Betti numbers.

THEOREM 4.3.3: *Let G be a finitely presented infinite amenable group. Then $\beta_1(G) = 0$; if G is of type F_m then $\beta_i(G) = 0$ for $i < m$.*

COROLLARY 4.3.4: *Suppose G is an infinite amenable group admitting a finite $K(G,1)$. Then $\chi(G) = 0$.*

For the next application we need a general fact on Hilbert modules which is a generalization of 2.7.1.

LEMMA 4.3.5: *Let G be an infinite group and W a Hilbert G-module of finite dimension as an $\mathbb{R}$-vector space. Then $W = 0$.*

Proof: We may assume $W \subset (\ell_2 G)^n$ and, by induction, that $n = 1$. Let $\Pi \in N(G)$ be the orthogonal projection onto W so that

$$\dim_G W = \langle \Pi(1), 1 \rangle = \langle \Pi(x), x \rangle \leq \|\Pi(x)\|, \quad \forall x \in G.$$

Choose an orthonormal basis $w_1, \ldots, w_n \in W$. Each w_i has the form $\sum_{g \in G} r_i(g) g$ with $\sum_{g \in G} r_i(g)^2 = 1$. Since G is infinite it is therefore possible to find for each $j \geq 0$ an element $x_j \in G$ such that $|r_i(x_j)| \leq 2^{-j}$, $i = 1, \ldots, n$. Note that

$$\langle \Pi(x_j), w_i \rangle = \langle x_j, \Pi(w_i) \rangle = \langle x_j, w_i \rangle = r_i(x_j).$$

With the x_j's above

$$\|\Pi(x_j)\|^2 = \sum_i r_i(x_j)^2 \leq n \cdot 2^{-2j} \longrightarrow 0 \quad \text{as } j \to \infty,$$

showing that $\dim_G W = 0$, whence $W = 0$. ∎

Combining the lemma with the Cheeger–Gromov Lemma yields

COROLLARY 4.3.6: *Let Y be a connected free cocompact G-CW-complex with (infinite) amenable G. Assume that the ordinary Betti number $b_i(Y) < \infty$ for some i. Then $\beta_i(Y; G) = 0$.*

COROLLARY 4.3.7: *Let G be a finitely presented quasi-amenable group (meaning that there exists a normal subgroup $N < G$ with $b_1(N) < \infty$ and G/N infinite amenable). Then $\mathrm{def}(G) \leq 1$.*

Proof: Choose a $K(G,1)$ with finite 2-skeleton X. Let Y be the covering space of X associated with N. From 4.1.6 we get

$$\mathrm{def}(G) \leq 1 + \beta_1(Y; G/N)$$

and the result follows since $\beta_1(Y; G/N) = 0$ for amenable G/N. ∎

For a free cocompact G-CW-complex Y there is a natural map

$$\mathrm{can}_i : H_i(Y;\mathbb{Z}) \to {}^{red}H_i(Y)$$

induced by considering an integral chain as an ℓ_2-chain; we can also view the map to be induced by the inclusion of chain complexes

$$K_i(Y) \to \ell_2(G) \otimes_{\mathbb{Z}[G]} K_i(Y) = C_i(Y).$$

LEMMA 4.3.8: *Let Y be an n-dimensional free cocompact G-CW-complex. Then*

$$\mathrm{can}_n : H_n(Y;\mathbb{Z}) \to {}^{red}H_n(Y)$$

is injective.

Proof: This follows immediately from the long exact homology sequence associated with

$$0 \to K_*(Y) \to C_*(Y) \to C_*(Y)/K_*(Y) \to 0,$$

and observing that because Y is n-dimensional, ${}^{red}H_n(Y) = H_n(C_*(Y))$. ∎

COROLLARY 4.3.9: *Let Y be a free cocompact $(n-1)$-connected G-CW-complex of dimension $n > 1$. Assume that one of the following conditions holds:*
- *the $\mathbb{R}$-vector space $H_n(Y;\mathbb{R})$ is finite dimensional,*
- *the ℓ_2-Betti number $\beta_n(Y;G) = 0$.*

Then Y is contractible.

Proof: Note that, because Y is n-dimensional, $H_n(Y;\mathbb{Z}) \subset H_n(Y;\mathbb{R})$. For the first case, use 4.3.5 and the previous lemma to conclude that $H_n(Y;\mathbb{Z}) = 0$. The Hurewicz Theorem then shows that Y is n-connected, thus contractible. Similarly for the second case. ∎

4.4.   In this section we present a few applications concerning the *partial* Euler characteristic $\mathfrak{q}_m(G)$ of (amenable) groups as well as the *Hausmann–Weinberger Invariant* $\mathfrak{q}(G)$. For more results along these lines the reader is referred to [8, 9, 10]. Suppose G admits a $K(G,1)$ with finite m-skeleton. Put $X = K(G,1)^m$ and consider

$$(-1)^m \chi(X) = \sum_{i=0}^{m-1} b_i(G) + b_m(X) \geq \sum_{i=0}^{m-1} b_i(G).$$

Define
$$\mathfrak{q}_m(G) = \min\{(-1)^m \chi(X)\},$$
the minimum being taken over all possible choices of finite X as above.

In particular
- $\mathfrak{q}_1(G) = \min\{\text{number of generators of } G\}$,
- $\mathfrak{q}_2(G) = 1 - \text{def}(G)$.

THEOREM 4.4.1: *Let G be an infinite amenable group of type F_m. Then $\mathfrak{q}_m(G) \geq 0$, and $\mathfrak{q}_m(G) = 0$ implies that the cohomology dimension of G over $\mathbb{Z}$ is $\leq m$.*

Proof: Let X be a finite model for $K(G,1)^m$. Since $\beta_i(X) = \beta_i(G) = 0$ for $i < m$ (cf. 4.3.3),
$$(-1)^m \chi(X) = \beta_m(X) \geq 0$$
and therefore $\mathfrak{q}_m(G) \geq 0$. If $\mathfrak{q}_m(G) = 0$ then we can choose $X = K(G,1)^m$ with $\beta_m(X) = 0$ and 4.3.9 applied to the universal cover of X implies that X is a $K(G,1)$ of dimension m (the case $m = 1$ cannot occur here, since for a non-trivial group $\mathfrak{q}_1 > 0$).   ∎

The following definition goes back to Hausmann–Weinberger [16]. Let M be a closed (smooth) oriented 4-manifold. Then, since $b_1(M) = b_3(M) = b_1(G)$ for G the fundamental group of M,
$$\chi(M) = 2 - 2b_1(G) + b_2(M) \geq 2(1 - b_1(G)).$$

Put
$$\mathfrak{q}(G) = \min\{\chi(M)\}$$
where M runs over all manifolds as above, with $\pi_1(M) = G$. In a similar way, various authors have defined (see [10])
$$\mathrm{P}(G) = \min\{\chi(M) + \sigma(M)\}$$
with M as before and $\sigma(M)$ the signature of M; the minimum exists because $|\sigma(M)| \leq b_2(M)$ so that
$$\chi(M) + \sigma(M) \geq 2(1 - b_1(G)).$$

Note also that $\mathrm{P}(G) \leq \mathfrak{q}(G)$ as $\sigma(-M) = -\sigma(M)$.

It is well known that there exists for any finitely presented group G a closed smooth oriented manifold M with fundamental group G. The invariants $\mathrm{P}(G)$ and $\mathfrak{q}(G)$ are therefore defined for any finitely presented group G.

Atiyah proved in [1] the ℓ_2-*Signature Theorem* which tells that ${}^{red}H^2(\tilde{M})$ splits into a Hilbert direct sum of two Hilbert G-modules with von Neumann dimensions $\beta_2^+(M)$ and $\beta_2^-(M)$ such that $\sigma(M) = \beta_2^+(M) - \beta_2^-(M)$. Expressing $\chi(M)$ in terms of ℓ_2-Betti numbers this leads in case of an infinite G to the formula

$$\chi(M) + \sigma(M) = -2\beta_1(G) + 2\beta_2^+(M).$$

Applying it to groups with vanishing first ℓ_2-Betti number we see that $\chi(M) + \sigma(M)$ is non-negative, which has interesting consequences, cf. [10]. In particular the case of an amenable G yields the following.

THEOREM 4.4.2: *Let G be a finitely presented amenable group. Then* $P(G)$ *and* $q(G)$ *are non-negative.*

References

[1] M. Atiyah, *Elliptic operators, discrete groups and von Neumann algebras*, Astérisque **32** (1976), 43–72.

[2] M. Burger and A. Valette, *Idempotents in complex group rings: theorems of Zalesskii and Bass revisited*, Journal of Lie Theory **8** (1998), 219–228.

[3] J. Cheeger and M. Gromov, *L^2-cohomology and group cohomology*, Topology **25** (1986), 189–215.

[4] J. Dodziuk, *De Rham–Hodge theory for L^2-cohomology of infinite coverings*, Topology **16** (1977), 157–165.

[5] J. Dodziuk and V. Mathai, *Approximating L^2 invariants of amenable covering spaces: a combinatorial approach*, Journal of Functional Analysis **154** (1998), 359–378.

[6] B. Eckmann, *Harmonische Funktionen und Randwertaufgaben in einem Komplex*, Commentarii Mathematici Helvetici **17** (1944/45), 240–255.

[7] B. Eckmann, *Coverings and Betti numbers*, Bulletin of the American Mathematical Society **55** (1949), 95–101.

[8] B. Eckmann, *Amenable groups and Euler characteristics*, Commentarii Mathematici Helvetici **67** (1992), 383–393.

[9] B. Eckmann, *Manifolds of even dimension with amenable fundamental group*, Commentarii Mathematici Helvetici **69** (1994), 501–511.

[10] B. Eckmann, *4-manifolds, group invariants, and ℓ_2-Betti numbers*, L'Enseignement Mathématiques **43** (1997), 271–279.

[11] B. Eckmann and P. Linnell, *Poincaré duality groups of dimension two, II*, Commentarii Mathematici Helvetici **58** (1983), 111–114.

[12] B. Eckmann and H. Müller, *Poincaré duality groups of dimension two*, Commentarii Mathematici Helvetici **55** (1980), 510–520.

[13] M. S. Farber, *Novikov-Shubin invariants and Morse inequalities*, Geometric and Functional Analysis **6** (1996), 628–665.

[14] E. Følner, *On groups with full Banach mean value*, Mathematica Scandinavica **3** (1955), 336–354.

[15] R. I. Grigorchuk, *An example of a finitely presented amenable group that does not belong to the class EG*, Matematicheskii Sbornik **189** (1998), no. 1, 79–100.

[16] J.-C. Hausmann and S. Weinberger, *Caractéristique d'Euler et groupes fondamentaux des variétés de dimension 4*, Commentarii Mathematici Helvetici **60** (1985), 139–144.

[17] P. Linnell, *Division rings and group von Neumann algebras*, Forum Mathematicum **5** (1993), 561–576.

[18] W. Lück, *L^2-Betti numbers of mapping tori and groups*, Topology **33** (1994), 203–214.

[19] W. Lück, *Hilbert modules over finite von Neumann algebras and applications to L^2-invariants*, Mathematische Annalen **309** (1997), 247–285.

[20] F. J. Murray and J. von Neumann, *On rings of operators*, Annals of Mathematics (2) **37** (1936), 116-229.

[21] S. P. Novikov and M. A. Shubin, *Morse inequalities and von Neumann invariants of nonsimply connected manifolds*, Uspekhi Matematicheskikh Nauk **41** (1986), 222–223.

[22] A. L. Paterson, *Amenability*, Mathematical Surveys and Monographs 29, American Mathematical Society, 1988.

[23] A. E. Zalesskii, *On a problem of Kaplansky*, Soviet Mathematics **13** (1972), 449–452.

Die Zukunft der Mathematik
Ein Rückblick auf Hilberts
programmatischen Vortrag vor 100 Jahren

Neue Zürcher Zeitung (Forschung und Technik) 11. 8. 2000, 46–51

Vor 100 Jahren, am 8. August 1900, hielt David Hilbert am Internationalen Mathematiker-Kongress in Paris einen Vortrag, in dem er 23 mathematische Probleme formulierte und ausführlich erklärte. Dieser programmatische Vortrag, der die weitere Entwicklung der Mathematik auf Jahrzehnte beeinflusst hat, ist in Fachkreisen bis heute berühmt geblieben.

Die Mathematik spielt eine dreifache Rolle: Sie gehorcht strenger Logik und setzt damit den Standard objektiver Wahrheit; wie die Künste ist sie eine freie Schöpfung unseres Geistes und damit Teil unserer kulturellen Tradition; und sie hat konkrete Anwendungen. Der Laie nimmt die Mathematik im Allgemeinen nur in ihrer praktischen Funktion wahr. Die Forschung in reiner Mathematik findet jedoch im Stillen statt.

Um das zu ändern, hat das Clay Mathematics Institute – eine Stiftung, die ihre Aufgabe erklärtermassen in der Mehrung und Verbreitung von mathematischem Wissen sieht – kürzlich ein Preisgeld von je einer Million Dollar für die Lösung von sieben «Millenniums-Probleme» ausgeschrieben. Bei den Problemen, die von Experten sorgfältig ausgewählt wurden, handelt es sich um schwierige Fragestellungen aus verschiedenen Gebieten der Mathematik. Zum Teil sind die Probleme neu, zum Teil harren sie seit über hundert Jahren einer Lösung.

Reminiszenz an Hilberts Rede

Die Millenniums-Probleme wurden im Mai an einer Konferenz in Paris der Öffentlichkeit vorgestellt. Dabei wurden ganz bewusst Erinnerungen an die berühmt gewordene Rede geweckt, die der deutsche Mathematiker David Hilbert vor 100 Jahren am zweiten Internationalen Mathematikerkongress in Paris gehalten hat. In dieser Rede stellte Hilbert 23 ältere und neuere mathematische Probleme vor, deren Lösung er zum Programm für das kommende Jahrhundert erhob. Während die Millenniums-Probleme nur Teilaspekte der heutigen Mathematik berühren, entfaltete Hilbert in seinem Vortrag eine Gesamtsicht der damaligen Mathematik. Das Hilbertsche Programm hat die Mathematik bis zur Hälfte des 20. Jahrhunderts und teilweise sogar darüber hinaus geprägt. Deshalb ist der Vortrag in Fachkreisen bis auf den heutigen Tag in lebendiger Erinnerung geblieben.

Die Ereignisse kurz vor und zu Beginn des Pariser Kongresses scheinen verwirrend gewesen zu sein. Der 38-jährige Hilbert war damals Professor in Göttingen, der Hochburg der Mathematik. Unter Kollegen galt Hilbert als bedeutendster deutscher Mathematiker. Anfang 1900 erhielt er die Einladung, am Mathematikerkongress in Paris den Eröffnungsvortrag zu halten. Hermann Minkowski, Professor an der ETH Zürich (damals Polytechnikum) und ein enger Freund Hilberts schon von der Studienzeit in Königsberg her, drängte ihn, «einen Blick in die Zukunft der Mathematik und ihre Probleme zu tun». Über einen solchen Gegenstand, so Minkowski, werde man noch Jahrzehnte später sprechen.

Minkowskis Brief blieb unbeantwortet. Erst in letzter Minute wurde der Vortragstext fertig und zur Durchsicht an Minkowski und Hilberts früheren Lehrer Adolf Hurwitz, der ebenfalls Professor an der ETH war, geschickt. Doch es war zu spät; der Vortrag konnte nicht mehr am Eröffnungstag gehalten werden. Er wurde deshalb auf den dritten Tag des Kongresses verschoben. Die Vortragszeit war kurz bemessen, und Hilbert musste sich nach einer ausführlichen Einführung auf die Darstellung von 10 der 23 Probleme beschränken. Aus der aufschlussreichen Einleitung seien an dieser Stelle nur zwei Aspekte hervorgehoben: das Verhältnis zwischen speziellem Problem und allgemeiner Theorie sowie die Notwendigkeit des «strengen Beweises».

Das Lösen eines speziellen Problems und das Bilden einer allgemeinen Theorie sollen laut Hilbert Hand in Hand gehen. Mehr noch, Hilbert betonte, dass die Bemühungen um ein schwieriges Problem für echte Fortschritte in der Mathematik nötig sind – selbst dann, wenn diese Bemühungen nicht zur Lösung des Problems führen. Er nannte zwei Parade-Beispiele: Die Fermatsche Vermutung und das Dreikörperproblem der Mechanik.

Die Fermatsche Vermutung besagt, dass die Gleichung $x^n + y^n = z^n$ für $n \geqq 3$ keine Lösung durch ganze Zahlen x, y und z hat. Diese Vermutung ist 1993 durch Andrew Wiles bewiesen worden und seitdem weiterum bekannt. Interessante Lösungsversuche hatte es schon im 19. Jahrhundert gegeben. Auch wenn diese Bemühungen nicht in der Lösung des Problems gipfelten, führten sie zur Algebraischen Zahlentheorie, einer neuen Disziplin mit einer Fülle von Begriffen, die sich später selbständig weiterentwickelten. Die Methoden von Wiles waren ganz anderer Art; sein Beweis baute auf einem Rahmen auf, in dem Überlegungen zur Geometrie von Kurven mit Symmetriebetrachtungen verknüpft waren.

Ähnliches trifft auf das Dreikörperproblem der Mechanik zu, das bis heute noch Fragen aufwirft. Das Problem inspirierte den Pariser Mathematiker Henri Poincaré (1854–1912) zu neuen Methoden der Himmelsmechanik, die heute zu einem unerlässlichen Hilfsmittel bei der Navigation von Satelliten geworden sind.

Hilbert und Poincaré kannten und bewunderten einander, aber ihre Auffassungen von Mathematik waren sehr verschieden. Für Poincaré spielten Intuition und physikalische Analogien eine grosse Rolle, für Hilbert aber der strenge logische Rahmen. Deshalb betonte Hilbert in seinem Vortrag die Notwendigkeit, dass jede Aussage, jede Lösung eines Problems bewiesen werden muss. Und wenn ein Problem keine Lösung hat wie etwa die Quadratur des Kreises durch Zirkel und Lineal, so muss auch das bewiesen werden. Dabei war Hilbert überzeugt, dass immer das eine oder das andere der Fall sei, ja er hätte gerne auch hiefür einen Beweis gehabt. Mit diesen Gedanken hat er manches ins Rollen gebracht.

Axiomatischer Aufbau der Mathematik

Unter Beweis verstand Hilbert eine endliche Folge von logischen Schlüssen. Gewiss gibt es andere Aspekte des Mathematisierens wie Plausibilitätsbetrachtungen, Analogien, physikalische Erfahrungen und anschauliche Intuition; aber all das darf den Beweis nie ersetzen. Es fällt auf, dass Hilbert strenge Beweise ausdrücklich nicht nur für Arithmetik und klassische Geometrie forderte, sondern auch für alle anderen Bereiche der Mathematik bis hin zur angewandten Mathematik. Offenbar war dies damals durchaus nicht so selbstverständlich wie heute. Die Forderung des strengen Beweisens ist nur im Rahmen des «axiomatischen Aufbaus» der Mathematik sinnvoll. Diese von Hilbert geprägte Auffassung wirft die Frage auf, ob es überhaupt möglich ist, die Mathematik auf eine Grundlage zu stellen, die in sich widerspruchsfrei ist.

Den Laien mag es erstaunen, dass überhaupt an den Grundfesten des mathematischen Gebäudes, an seiner Konsistenz, gezweifelt werden kann. Hier muss man ein wenig ausholen und daran erinnern, dass alle mathematischen Begriffe, die ständig um uns herum in Aktion sind, in unserer realen Beobachtungswelt gar nicht vorkommen. Es gibt weder ideale Punkte noch unendlich lange Geraden; sie sind Schöpfungen unseres Geistes so wie der Raum der Elementargeometrie oder die Räume höherer Dimension, die unendliche Reihe der ganzen Zahlen, von komplizierteren Begriffen ganz abgesehen.

Die bekannte philosophische Frage, ob es sich bei diesen Begriffen um Erfindungen oder Entdeckungen handelt, soll an dieser Stelle nicht weiter erörtert werden. Auf jeden Fall sind diese Begriffe nicht von ungefähr entstanden, sondern unserer Umwelt abgeschaut und dann sehr weitgehend verallgemeinert worden. Erst die völlige Loslösung vom Realen befähigt sie zur Beschreibung von Vorgängen, die wir noch nicht kennen, zum Voraussagen von Abläufen, die unserer Anschauung nicht zugänglich sind, und zu Berechnungen, die ein Formalismus für uns übernimmt.

Das mathematische Objekt ist zunächst einfach eine Menge von Elementen. Zwischen diesen sind Verknüpfungen definiert, zum Beispiel Addition, Multiplikation oder Nachbarschaft usw. Die Verknüpfungen müssen gewissen Axiomen genügen, die natürlich auch der Anschauung entnommen sind; es sind nicht evidente Tatsachen, sondern Spielregeln oder Rechenregeln der Theorie, und zum Beweisen dürfen nur sie verwendet werden und alles, was aus ihnen folgt. Das ist es, was Hilbert unter einem axiomatischen Aufbau der Mathematik verstand. Diese Auffassung Hilberts hat sich später unter dem Einfluss der Bourbaki-Gruppe mehr und mehr in der Mathematik durchgesetzt und entscheidend zu ihrem Fortschritt beigetragen.

Nachdem Hilbert eine axiomatische Grundlage für die Arithmetik und die klassische Geometrie entwickelt hatte, drängte sich ihm fast zwangsläufig die Frage nach der Widerspruchsfreiheit des Axiomensystems auf. Er selbst hegte in dieser Hinsicht keinen Zweifel. Aber Intuition allein genügte Hilbert nicht. Deshalb forderte er im zweiten seiner 23 Probleme einen Beweis dafür, dass es nicht möglich ist, im Axiomensystem der Arithmetik durch logische Schlüsse eine Aussage und ihr Gegenteil zu beweisen.

Dieser Beweis wurde nicht erbracht. Im Gegenteil: 30 Jahre nach Hilberts Vortrag, also zu einer Zeit, als Hilbert noch aktiv war, zeigte der Logiker Kurt Gödel, dass es einen mathematischen Beweis für die Widerspruchsfreiheit der Arithmetik gar nicht geben kann. Streng genommen war damit das Hilbertsche Programm gescheitert, ist doch die Widerspruchsfreiheit eines Axiomensystems die Grundlage für einen axiomatischen Aufbau der Mathematik.

Nicht nur das zweite, auch das erste der von Hilbert aufgeworfenen Probleme erhielt eine ganz andere Antwort als von ihm erwartet. In dieser «Kontinuumshypothese» geht es darum, wie sich eine Teilmenge eines Kontinuums zum Kontinuum als Ganzem verhält. Dabei versteht man unter einem Kontinuum die Menge der reellen Zahlen oder gleichbedeutend: alle Punkte einer Geraden. Die Behauptung lautet: Jede Teilmenge des Kontinuums ist entweder abzählbar (also durch ganze Zahlen nummerierbar) oder gleich mächtig wie das ganze Kontinuum. Im Jahr 1963 zeigte jedoch der Mathematiker Paul Cohen etwas Überraschendes. Sowohl die Behauptung als auch ihr Gegenteil – dass es zwischen dem Abzählbaren und dem Kontinuum noch andere Mächtigkeiten gibt – ist mit den Axiomen verträglich. Man kann also die eine oder die andere Behauptung als neues Axiom hinzunehmen.

Auch die oben schon erwähnte Überzeugung Hilberts, dass man jedes Problem lösen oder aber seine Unlösbarkeit beweisen kann, stimmt nicht. Im Laufe der Jahre haben die Logiker solche Fragen in sehr subtiler Weise bearbeitet.

Für die historische Beurteilung von Hilberts Rede ist es unwichtig, dass Antworten auf fundamentale Fragen radikal anders ausfielen, als Hilbert es vorausgesehen hatte. Entscheidend war, dass er die Fragen so klar und eindeutig formuliert hat, dass sich später mutige Forscher der Herausforderung stellen konnten. Vor allem aber – dies sei betont – hat die Tatsache, dass die Widerspruchsfreiheit der Axiome nicht bewiesen werden kann, die Mathematik samt ihren Anwendungen wenig gestört. Man kann den allgemeinen Standpunkt so formulieren: Die Mathematik ist vermutlich widerspruchsfrei; man kann es nur nicht in finitistischer Weise beweisen. Die unübersehbare Bewährung, die die Mathematik seit Jahrhunderten erfährt, hat ganz einfach eine Atmosphäre des Vertrauens geschaffen.

Einige der speziellen Probleme

Die genannten Grundlagenfragen bildeten nur einen kleinen Teilaspekt von Hilberts Vortrag. Daneben formulierte er eine ganze Reihe von spezifischen Problemen. Der Bogen ist weit gespannt und umfasst «alle Teile des mathematischen Denkens»; so heisst es in der Laudatio des Ehrendoktorats, das die ETH Hilbert 1922 für sein Gesamtwerk verliehen hat. An dieser Stelle können nur wenige Probleme erwähnt werden. – Von den spezifischen Problemen sind heute alle bis auf drei gelöst. Eines dieser drei ist die «Riemannsche Vermutung», die auf den Göttinger Mathematiker Bernhard Riemann (1826–66) zurückgeht. Es gilt heute als schwierigstes und wichtigstes der ungelösten Probleme und figuriert deshalb auch unter den Millenniums-Problemen. Die Riemannsche Vermutung handelt von den Nullstellen einer mit den Primzahlen zusammenhängenden Funktion und ist mit fast allen Disziplinen der Mathematik verknüpft. Die Vermutung widerstand bisher allen Angriffen. Umfangreiche Computer-Berechnungen machen sie zwar plausibel – aber das genügt eben nicht.

Ein weiteres Problem betrifft die Diophantischen Gleichungen. Dabei handelt es sich um Polynomgleichungen mit ganzen Zahlen als Koeffizienten. Im Fermat-Problem geht es um eine äusserst spezielle Gleichung dieser Form. Hilbert suchte ein allgemeines Verfahren, mit dem sich feststellen lässt, ob diese Gleichungen ganzzahlige Lösungen besitzen. 1970 wurde gezeigt, dass es kein solches Verfahren geben kann. Neben Fragen der Algebraischen Zahlentheorie, der Funktionentheorie und der Algebra berührt Hilberts Vortrag auch Fragen, die scheinbar elementargeometrischer Art sind. Letztere haben ihre Aktualität bis heute nicht verloren. Hierher gehören zum Beispiel die bekannten Penrose-Pflasterungen.

Im sechsten der 23 Probleme setzte sich Hilbert mit den axiomatischen Grundlagen der Physik auseinander. So forderte er zum Beispiel eine strengere Begründung des Wahrscheinlichkeitsbegriffs, der damals in der kinetischen Gastheorie eine wichtige Rolle spielte. In der Quantentheorie sollte diesem Begriff später eine zentrale Bedeutung zukommen. Die Axiomatik der Wahrscheinlichkeit liess allerdings noch lange auf sich warten. Heute gehört dieses sehr wichtige und aktive Gebiet ganz in die Mathematik, und es hat auch die Finanzwelt erobert.

Mathematik gestern und heute

Die Spannweite von Hilberts Vortrag reichte also von fundamentalen Existenzfragen über abstrakte und konkrete Fragen der Mathematik bis hin zu Anwendungen. Wegen dieses Weitblicks wird der Vortrag bis heute als bedeutendstes historisches Ereignis der neueren Mathematik betrachtet.

Warum sollte man verschweigen, dass es nicht nur Lob, sondern auch leise Kritik an «Hilbert» gibt? Kritik nicht an ihm und seinem Werk, auch nicht an den Problemen selbst, sondern an der Bedeutung, die man ihnen zumisst. Die allgemeinen Fragestellungen seien zu allgemein; darüber kann man streiten. Die spezielleren zu speziell, so dass sie nur in Sackgassen führen. Das trifft bei den gelösten Problemen nicht zu; fast immer haben sich aus der Beschäftigung mit einem Problem ganze Theorien entwickelt wie beim Fermat-Problem. Und wieder andere reagieren auf «Hilbert» mit einem nachsichtigen Lächeln. «One squints at old daguerreotypes, charming mementos of the period» (Barry Mazur).

Ja, man muss zugeben, dass in der zweiten Hälfte des Jahrhunderts Theorien von anderer Grössenordnung entstanden sind. Hierzu gehören die aus neuerer Zeit stammenden Millenniums-Probleme und viele andere, deren Bedeutung heute noch gar nicht abschbar ist. Hinzu kommen anwendungsnahe Gebiete der Mathematik wie astrophysikalische Mathematik, theoretische Biologie, Codierung und Kryptographie, neuronale Netze und vieles mehr.

Allerdings haben sich auch die Bedingungen, unter denen mathematische Forschung stattfindet, in den letzten 100 Jahren verändert. Heute dürfte die Zahl der forschenden Mathematiker über 50 000 betragen. Dank Information und Kommunikation bilden sie so etwas wie einen Organismus, in dem jeder Teil sehr rasch auf jeden anderen reagiert. Um 1900 war die Zahl der forschenden Mathematiker hingegen sehr klein. Die Kommunikation war umständlich, ebenso wie das Reisen. Hilbert hat – von Briefkontakten mit Minkowski und Hurwitz abgesehen – im Alleingang einen Überblick über fast alle Gebiete der Mathematik gegeben. Was er in Bewegung gesetzt hat, inspirierte Generationen von Mathematikern, und manches steckt in neuesten Gedankengängen. Vergleicht man die heutige weltweite Mathematik mit dem Dreieck Göttingen–Paris–Zürich um 1900, so muss man mit Bewunderung feststellen, dass von so wenigen so vieles vollbracht wurde.

Kolmogorov and contemporary mathematics

Newsletter of the European Mathematical Society 50 (2003), 13

It is a great honor for me to address the opening ceremony of this wonderful event celebrating the 100th anniversary of Andrei Kolmogorov, and I express my sincere thanks to the organizing committee, in particular to Albert Shiryaev. My words will be of a very personal nature.

I have heard the name Kolmogorov for the first time when I was a beginning mathematics student, around 1936. Here is the story. In a course by *George Polya* ("Aufgaben und Lehrsätze") there were nice and amusing applications of probability, including his favorite topic Random Walks. But we were unhappy with the so-called definitions of the probability concept. For Polya, intuition and application were more important.

Then there happened to be a Colloquium lecture in Zurich by *Richard von Mises* – reducing the concept to something we couldn't understand at all (maybe independence, or was it a probabilistic definition of probability?). As you can imagine, our confusion was absolutely complete!

Michel Plancherel ("Plancherel formula") also one of our professors, told us to consult the 1933 Ergebnisse book by a certain Russian named Kolmogorov, where the concept of probability was put on a final rigorous axiomatic basis. I still remember like today: it was sensational, dramatic, enlightening. I am not certain that Plancherel really liked the new type of abstraction (he actually was more fascinated by a very early construction by Kolmogorov, in 1923, of a Lebesgue-Fourier series diverging almost everywhere; he did not tell us that the author was not even 20 when he wrote that paper!). But we, the young ones were happy, the 1933 logical foundation of probability was crystal-clear, abstract and at the same time relating everything to the standard terminology and to concrete application. We just liked the new trend of axiomatic abstraction. As for the name Kolmogorov, it also was quite abstract for us – just a name.

But in 1939 I got in real contact with Kolmogorov when I began my PhD work in topology under *Heinz Hopf* ("Hopf algebras", "Hopf fiberings"). Of course not personal contact, but Hopf told me about the famous 1935 topology conference in Moscow. At that conference Kolmogorov had introduced a very important new concept (independently of J. W. Alexander): namely cohomology. Homology is due to Poincaré (around 1900 already), and cohomology is its dual; what topologists had not expected was the product structure for arbitrary complexes. Well, cohomology

---

Lecture delivered at the opening of the centenary commemoration, Moscow 2003.

in very different appearances has been taken over by so many people, more and more in close connection with algebra, functional analysis, measure theory and so on that it is not well known that the concept goes back to Kolmogorov! Although my thesis was about homotopy, cohomology became one of my favorite research topics until today.

Could this be the same person as the probabilist? Believe it or not, Hopf said, it is! Not only that; it is in every respect a remarkable young man, of unusual physical strength, climbing mountains of over 4000 meter altitude, skiing enormous distances, swimming in ice-cold water – almost like his very close friend Paul Alexandroff. Alexandroff in turn was a very good friend of Heinz Hopf; their joint book on topology is well-known.

During these early years Kolmogorov made many other important contributions to algebraic topology, for various types of spaces. At that time not many could anticipate the importance topological ideas and all their ramifications would gain during the century – but Kolmogorov did! Thus we must admire the really independent mind of this young man – independent of fashion. Fashion was in that part of the century to work on the Hilbert problems. Surely he did that too, in the axiomatization of the applied field probability.

Let me remark that in the famous Hilbert 1900 list algebraic topology (homology, analysis situs) is not mentioned with one single word! It was not considered to be a respectable field for a long time. I heard this much later from *Hermann Weyl*; he even told me that around 1925 he had published two papers on algebraic topology in Spanish in a South-American Journal, because he did not want his colleagues to read them!

Clearly I very much wanted to meet that man Kolmogorov. But there was world war II and then the cold war. Communication by letter was slow if not impossible, contacts were limited. Finally in 1954 I could meet Kolmogorov, he gave the famous plenary lecture at the International Congress in Amsterdam – about Dynamical Systems, the beginning of what would later become the KAM theory. As for myself, surely enough I was lecturing about cohomology of groups.

Now things go on in the same way: At the next International Congress 1958 in Edinburgh I heard Kolmogorov lecturing, this time, about Functional Analysis (just a short communication), at the 1962 Congress in Stockholm about Optimal Approximation of Functions and so on and so forth. No need to continue, you all know what I want to say: That I was lucky to know Kolmogorov and to see him develop into one of the truly universal mathematicians of our time, covering all fields (with the exception of number theory, as far as I know), original, creative, deep and broad – whatever you want and whatever will be said in this conference.

Contacts with Russia became easier during the years 1952 to -60 when I was Secretary of the International Mathematical Union. It was the time when we succeeded in making the Union truly international, in spite of great political difficulties. All the countries of Eastern Europe, and China became members. As a new member of the Executive Committee of IMU Paul Alexandroff was appointed – for me it should have been Kolmogorov, but I understood that he was unpolitical in every respect. Of course we liked Paul Alexandroff. He told us many stories about his friend Kolmogorov. For example about their long voyage of several months on the

Wolga and beyond in 1929 – what impressed me tremendously was the fact that Andrei had with him on that long trip, apart from mathematics books, the Odyssey! And the better I knew Kolmogorov the more I realised that his cultural universality went much beyond mathematics, into logic and foundations, into arts and poetry and history and education. His human and humanistic universality enabled him to be an extraordinary teacher.

His students are here, they can tell about that better than I. He inspired them to do mathematics according to its true nature and unity: abstract, valid within its strict context, universal and precisely for that reason eminently practical.

With the passing away of each human being a mystery disappears from the world, a mystery that nobody else will be able to rediscover (Friedrich Hebbel). Words, and certainly my words, are inadequate to describe the mystery of Andrei Kolmogorov. But with regard to our common profession we can learn from him that mathematics is a manifestation of the free creative power of the human mind and the organ for world understanding through theoretical construction. And that it is part of the cultural tradition we have to transmit to the next generation.

To achieve more we dare not hope, to achieve less we must not try. We are grateful to Andrei Kolmogorov for his outstanding contribution towards that goal.

Heinz Hopf

Hermann Weyl

Is algebraic topology a respectable field?

Preliminary remark

This is the text of a lecture delivered shortly before the 40 years's celebration
of FIM, as the last lecture of the *Zürich Graduate Colloquium 2003/04*, which
took place in the Hermann Weyl Zimmer of the FIM. I had been asked to recall
some memories of my long life in mathematics. Without revealing the topic,
I suggested the title "Some Old Time Mathematics: 40 Years and Beyond".
The topic was only formulated after I had mentioned my personal contacts with
Hermann Weyl.

Is Algebraic Topology a respectable field? Of course it is. Even more
than that: it is commonplace that today Algebraic Topology is a general name
for various more or less different branches, like differential topology, manifold
theory, combinatorial methods, ℓ_2-cohomology, general homology and K-theory,
homological algebra – each of them interesting in itself but also for applications
in many other fields of mathematics. But this was not always the case. After
the discovery – or invention? – of Algebraic Topology (called Analysis Situs) by
Poincaré in 1895 it took many decades for this field to be recognized generally
as a "respectable" field of mathematics.

What follows is not meant to be a historical survey of that long development. There exist many very detailed writings about it, and comparing them
closely one realizes that the history was quite complicated indeed. I just want
to describe, mostly from my own personal experience in that field, some of the
facts which support the claims formulated above, tell how gradually the field
became respectable and fully accepted in the family of mathematicians. Thus
there is no claim of completeness; to the contrary, what follows is just a number
of specific items chosen from a personal viewpoint.

1. Hermann Weyl, 1923/24

*"Why did you publish your two 1923/1924 papers on Algebraic Topology ("Analy-
sis Situs Combinatorio") in Spanish in the Revista Matematica Hispano-Ameri-
cana, a periodical which was not well-known and not easily accessible at that
time?"*

After his retirement from the Institute for Advanced Study, Hermann Weyl spent most of his time in Zurich. I had known him before in Princeton and our contacts continued in Zurich. I asked him the above question in 1954 when he was just preparing the laudatio for the Fields Medals to be awarded to J-P. Serre and to K. Kodaira at the International Congress in Amsterdam. Hermann Weyl answered that he simply did not want to draw attention to those two publications [484]*, the colleagues should not read them! The field was not considered to be serious mathematics like the classical fields of Analysis, Algebra, Geometry. In the spirit of the modern term political correctness it was at that early time not "mathematically correct" to work in such a field. But one has to recall that the medal was awarded to Serre for his famous thesis work in Algebraic Topology (homotopy groups of spheres) [429]. So in the meanwhile things must have changed considerably.

The two articles by Weyl give an elegant, very detailed and largely algebraic presentation of Combinatorial Topology as described by Poincaré in the *Compléments* (see below).

Before going further into the development of "mathematical correctness" of Algebraic Topology one has to take a short look at the early history from the very beginning. This of course took place long before I was involved in mathematics and topology. I say what I can find in the original papers.

2. Poincaré, 1895–1904

The birth of Algebraic Topology can be fixed historically in a very precise way: the papers of Henri Poincaré from 1895 to 1904 [369] began with "Analysis Situs" and were continued in a series of "Compléments". They clearly do not look like Algebraic Topology in a modern book. But everything connected with homology of spaces and homological algebra can be traced back to these old papers. This applies in particular to the multiple applications in Complex Analysis, in Algebraic Geometry, in Algebra and Group Theory, and in Theoretical Physics.

Thus not only the vast fields of the various aspects of modern topology, but many concepts used in mathematics today go back to one person, Henri Poincaré. His Analysis Situs was inspired by earlier ideas of Riemann and Betti, but these could not really be called a theory.

In Poincaré we find the concepts of cell complex, the cells being portions of bounded manifolds; incidence numbers describing the boundary of a cell, i.e. the way boundary cells of the next-lower dimension lie on a cell; cycles and homology; Betti numbers β_i and Euler characteristic $\chi = \sum(-1)^i \alpha_i$ where α_i is the number of cells of dimension i; the Euler–Poincaré formula

$$\chi = \sum(-1)^i \alpha_i = \sum(-1)^i \beta_i.$$

---

* Our references in [ ] refer to the bibliography of the monumental work by Dieudonné "A History of Algebraic and Differential Topology 1900–1960"

and Poincaré duality for a closed manifold of dimension n

$$\beta_i = \beta_{n-i}.$$

In the beginning everything was topologically invariant, at least in the differentiable sense, not really rigorous by today's standards. Then Poincaré turned to the rigorous concept of simplicial complex with invariance of homology under subdivision. But there topological invariance got lost – this is something we all know from our own work: you gain something, but you have to pay for it! The idea of simplicial approximation was already in the air; it later became one of the most important tools.

3. Hilbert, 1900

Many of us have reread, in 2000, Hilbert's famous address at the 1900 International Congress of Mathematicians, when the Millennium mathematical problems of the Clay Institute were formulated. Hilbert had established a program for the development of mathematics in the century to come (from letters addressed to his friends one knows that the original title was "the future of mathematics"). Partly he formulated explicit problems and partly he asked, in a more general way, for certain fields to be investigated and developed. Everywhere he insisted on rigor in the sense of axioms and proofs. One knows to what extent that lecture influenced mathematical research at least for the first half of the century, and in certain fields up to now.

But – not a word about Analysis Situs, not a word of the tremendous effort of Poincaré to establish this entirely new field! Was it on purpose, or a Freudian slip? One must admit that Hilbert simply did not realize that here was something to become more and more important throughout the century. This is in strong contrast to his remarkable anticipation of things to come in practically all other fields.

It is interesting to note that the papers by Hermann Weyl mentioned above are presented in a rigorous axiomatic way, in contrast to Poincaré's highly intuitive approach. Maybe this would have been more to Hilbert's taste.

4. After Poincaré

So it is a fact, mentioned explicitly by Hadamard in [217], that at the beginning of the twentieth century only a few mathematicians were interested in Analysis Situs. On the other hand those who were made very remarkable contributions; we mention some of them. Brouwer [89] proved in 1911 topological invariance of the dimension of $\mathbb{R}^n$; he solved a problem which had intrigued analysts since Cantor's (not continuous) bijective map of the real interval onto higher dimensional cubes, and Peano's continuous not bijective map of the interval onto the square. Very important for the future development was Brouwer's method of simplicial approximation and the concept of degree for mappings of manifolds. It is not clear whether even in the small family of topologists all this was really known.

In 1915 the topological invariance of the Betti numbers, and thus of the Euler characteristic, of a cell complex, was proved by Alexander [9]. In 1922 Alexander [11] found another interesting result: his duality theorem generalizing the classical Jordan curve theorem to all higher dimensions.

As for the topological invariance proof simplicial maps and simplicial approximation played an important role, combined with the concept of homotopy (making precise the earlier rather vague idea of deformation). Much later the invariance proofs became very simple thanks to the concept of homotopy equivalence and its algebraic counterpart.

All such results were considered as ingenious but somewhat exotic achievements, and it seems that not many mathematicians really knew exactly about them.

5. Heinz Hopf

With the appearance of Heinz Hopf's thesis and with his papers and lectures immediately afterwards [238] things seem to have changed considerably. Topology - that was now the standard name – was somehow accepted, though still considered a strange field. This change, what was the reason? Was it the fact that Hopf's work was intimately linked to easily accessible problems in differential geometry (Clifford-Klein problem, Curvatura Integra)? Was it his style, clear and rigorous, his inventing methods and solving "concrete" problems at the same time? Or his wonderful personality? Or his collaboration with Paul Alexandroff, beginning in Göttingen 1926 and lasting for many years? Hopf used to say later that his main merit was to have read, understood and made accessible the difficult work of Brouwer. According to Alexandroff and Hopf they both had largely been inspired by wonderful lectures of Erhardt Schmidt, Hopf's thesis adviser, on some of Brouwer's papers. In any case, certain papers of Hopf had a decisive influence on the later place of topology within mathematics, and we list them in more detail.

5.1 Hopf, 1925

In close connection with his work relating topological arguments to global differential geometry Hopf [240] proved for arbitrary dimension the famous theorem on tangent non-zero vector fields on a closed manifold (extending Poincaré's result for surfaces): if the field has isolated singularities (or zeros) then the sum of their indices is equal to the Euler characteristic of the manifold – whence a topological invariant. The index is an integer, defined as a mapping degree, which is zero if and only if the field can be modified in the neighborhood of the singularity so that the singularity disappears.

It follows, in particular, that a sphere of even dimension cannot admit tangent vector fields without singularities, while on an odd-dimensional sphere such fields exist (and can easily be described).

5.2 Hopf, 1928

On the other hand the influence of Emmy Noether on Hopf must have played a decisive role. In a 1928 paper by Hopf [241] algebraic concepts such as groups and homomorphisms were used for the first time to describe "combinatorial" aspects of (finite) cell complexes and homology. Instead of the matrices of incidence numbers, the free Abelian groups C_i generated by the i-dimensional cells of a complex were considered. The boundary ∂ becomes a homomorphism $C_i \longrightarrow C_{i-1}$, its kernel is the cycle group Z_i and $Z_i/\partial C_{i-1}$ is the Homology group H_i of the cell complex; the Betti number β_i is its $\mathbb{Q}$-rank. The sequence

$$C_n \longrightarrow ... C_{i+1} \longrightarrow C_i \longrightarrow C_{i-1} \longrightarrow ... C_0 \longrightarrow \mathbb{Z} \longrightarrow 0$$

was later called the chain complex of the cellular space: the boundary of a 0-cell, a vertex, is by definition $= 1 \in \mathbb{Z}$. That chain complex is exact (kernel=image) if and only if all homology groups with $i \geq 1$ are 0.

Very soon algebraization took over; this may also be one of the reasons why, after the first papers of Hopf, some more people got interested in what could now truly be called Algebraic Topology. The term Analysis Situs disappeared, the name Topology seems to be old – after Poincaré both terms had been used for some time. In the thirties the field was pretty well established. Several books appeared and special meetings were organized.

5.3 Hopf, 1931 and 1935

In 1931 Hopf [243] showed that there are (infinitely many) maps $S^3 \longrightarrow S^2$ which are contractible i.e. not homotopic to the constant map. This fact, quite unexpected from the viewpoint of homology, was not recognized as being important – for example topologists like Lefschetz did not find it interesting. It turned out later to be the starting point of a new branch of topology, homotopy theory.

In 1935 Hopf [245] extended that result to maps $S^{4k-1} \longrightarrow S^{2k}$ for all $k \geq 1$. In an appendix special such maps are constructed with the help of a simple geometrical idea, namely "fibrations". Later these again turned out to be the root of a very vast and important theory.

The fibrations considered were essentially the following

(1) $S^{2k+1} \longrightarrow \mathbb{C}P^k$, with fiber $S1$, $k \geq 1$
(2) $S^{4k+3} \longrightarrow \mathbb{H}P^k$, with fiber $S3$, $k \geq 1$
(3) $S^{8k+7} \longrightarrow \mathbb{O}P^k$, with fiber $S8$, $k = 1$ only

The spheres on the left are the unit spheres in complex (or quaternionic, or octonionic respectively) number space of dimension $k + 1$. The arrows denote the passage to homogeneous coordinates and thus are (continuous) maps onto the respective projective spaces. Since the octonions are not associative, the procedure is possible in (3) for $k = 1$ only. The fibers, the inverse images of the points of these projective spaces, are easily seen to be the respective spheres.

Since the projective lines ($k = 1$) are the spheres $S2$, $S4$, and $S8$ respectively one gets maps

(1') $S3 \longrightarrow S2$
(2') $S7 \longrightarrow S4$
(3') $S^{15} \longrightarrow S8$

which according to Hopf's method are non-contractible.

Before telling about the generalization of the Hopf fiberings (fiber spaces) and further results of Hopf we turn to another important event in Algebraic Topology:

6. Hurewicz, 1935/36

The four Dutch Academy Notes by Witold Hurewicz [256] on the "Theory of Deformations" had a great impact on the whole further development, although in the beginning they remained almost unnoticed. There are two aspects:

6.1 Homotopy groups

A few words about the definition of the homotopy groups $\pi_i(X)$ of a path-connected space X with base-point, $i \geq 1$. Its elements are the homotopy classes of based maps $S^i \longrightarrow X$, thus for $i = 1$ the homotopy classes of loops, and the group operation is a natural generalisation of the composition of loops. The structure of the group $\pi_i(X)$ is independent of the base-points. For $i \geq 2$ these groups are Abelian. They had been proposed, in 1932 already, by Cech: but then topologists did not consider them as important because of the commutativity – Hurewicz however put them to work. For any covering $\overline{X}$ of X the homotopy groups $\pi_i(\overline{X})$ and $\pi_i(X)$ are isomorphic for $i \geq 2$. X is called aspherical if all $\pi_i(X)$, $i \geq 2$ are 0.

6.2 Homotopy equivalence

A most important concept introduced by Hurewicz is homotopy equivalence, generalizing homeomorphism. A map $f : X \longrightarrow Y$ is called a homotopy equivalence if there is a map $g : Y \longrightarrow X$ such that the two compositions gf and fg are homotopic to the respective identities. The spaces X and Y are then called homotopy equivalent. Their homotopy groups and their homology groups are isomorphic.

Hurewicz proved, in particular, that two aspherical spaces X and Y with isomorphic fundamental groups are homotopy equivalent; any isomorphism between their fundamental groups is induced by a homotopy equivalence. Thus, in particular, an aspherical space with vanishing fundamental group is homotopy equivalent to the trivial space consisting of a single point (contractible).

7.

We approach the time when my own research began [148, 149, 151]. In 1939 Hopf asked me to study the papers of Hurewicz mentioned above. Some of my other Professors said that with Hopf I could certainly not go wrong, although Topology was not a well-known field. But something exotic like homotopy groups? Who might be interested?

Well, I was impressed by what I read and very soon noticed two extraordinary things – miracles.

7.1 Miracle one.

The degree of a map $S^n \longrightarrow S^n$ could easily be seen to be a homomorphism $\pi_n(S^n) \longrightarrow \mathbb{Z}$, and by simplicial approximation one realized that $\pi_n(S^n)$ is generated by the identity (degree $= 1$). Thus

$$\pi_n(S^n) = \mathbb{Z},$$

i.e. one recovers by this simple argument Hopf's Theorem that the homotopy classes of maps $S^n \longrightarrow S^n$ are characterized by the degree.

7.2

A concept which proved to be very suitable in connection with homotopy groups was that of fiber spaces (or fibrations) generalizing the Hopf fibrations (see 5.2). A fiber space is in the simplest case a map of spaces $p : E \longrightarrow B$ such that the fibers F, i.e. the inverse images of the points of B are all homeomorphic among themselves and constitute locally a topological product. The map p is called projection, the space B the base space of the fibration. In the context of homotopy groups, E and B are path-connected and have base-points (respected by maps and homotopies), and F is the inverse image of the base-point of B.

I noticed that a fibration gives rise to an exact sequence

$$...\pi_i(F) \longrightarrow \pi_i(E) \longrightarrow \pi_i(B) \longrightarrow \pi_{i-1}(F) \longrightarrow...$$

(The lowest dimensions require some changes which we do not mention here.) The first homomorphism is induced by the imbedding of F into E, the second by the map p. To define the third homomorphism and to prove exactness an additional property is required, the *homotopy lifting*. It tells that if f is a map $f = pg : X \longrightarrow B$ via E then any homotopy of f is also obtained via E by a homotopy of g. This "axiom" for fibrations (there were later many variants of it) was easily verified in all geometrical examples I was dealing with. Then the third map in the sequence is constructed as follows: one represents an element of $\pi_i(B)$ by a map of the i-ball into B with boundary sphere S^{i-1} mapped to the base-point and lifts it up to a map into E with S^{i-1} mapped into F.

7.3 Miracle two.

We apply the sequence to the Hopf fibration $S^3 \longrightarrow S^2$ above and get

$$...\pi_3(S^1) \longrightarrow \pi_3(S^3) \longrightarrow \pi_3(S^2) \longrightarrow \pi_2(S^1) \longrightarrow...$$

But $\pi_i(S^1) = 0$ for $i \geq 2$ since the universal covering is contractible. Thus

$$\pi_3(S^2) = \pi_3(S^3) = \mathbb{Z}$$

and we get (even in a more precise way) Hopf's result about non-contractible maps $S^3 \longrightarrow S^2$.

7.4 Using homotopy groups, the homotopy lifting, and exactness, various problems of geometrical nature could be solved but many questions remained open. We mention here only the vector field problem.

On a sphere S^n of odd dimension n there exist tangent unit vector fields without singularities. Do there exist two or more (or even the maximum possible number n) of such fields which are linearly independent at each point of S^n? I proved that for $n = 4k+1$ there cannot exist two independent such fields. Later, with the development of algebraic topology, more and more results of this kind were obtained: Kervaire [272] and Milnor showed that only the spheres S^n with $n = 1$, 3, 7 admit the maximum number n of independent fields (parallelizability). This problem is related to (actually a special case of) the existence of a continuous multiplication in $\mathbb{R}^{n+1}$ with two-sided unit and with norm-product rule. Adams [2] showed in 1960 that this is possible for $n + 1 = 1$, 2, 4, 8 only; in these cases bilinear multiplications of the required type were known already before 1900.

8. Hopf, 1944

According to Hurewicz (see **6.2**) aspherical spaces X and Y with isomorphic fundamental group G are homotopy equivalent and thus have isomorphic homology groups. Thus these homology groups are determined by G. A natural problem came up: to express them in a purely algebraic way from the group G.

Hopf [249] solved this problem by constructing a free resolution of a module M over the group algebra $\mathbb{Z}G$ of G (actually over any ring). This was a fundamental concept in the development of the algebraic field which later was called Homological Algebra. A free resolution of M is an exact sequence

$$\ldots \longrightarrow C_i \longrightarrow C_{i-1} \longrightarrow \ldots C_1 \longrightarrow C_0 \longrightarrow M \longrightarrow 0$$

where all C_i are free $\mathbb{Z}G$-modules. It can easily be constructed since any module is the quotient of a free module.

This was, of course, patterned after the methods of Hurewicz. If X is an aspherical (cellular) space then its universal covering $\widetilde{X}$ is contractible and has vanishing integral homology groups $H_i(\widetilde{X})$ for $i \geq 1$ and $\mathbb{Z}$ for $i = 0$. The fundamental group G acts freely on $\widetilde{X}$ and the chain groups are free $\mathbb{Z}G$-modules. Thus the chain complex of $\widetilde{X}$ is precisely a free resolution of $\mathbb{Z}$ over $\mathbb{Z}G$. The homotopy equivalence of all aspherical spaces with the same G was imitated in an algebraic way by Hopf; thus all free resolutions of $\mathbb{Z}$ yield the same homology groups with various coefficients, in particular those with coefficients $\mathbb{Z}$ (trivial action of $\mathbb{Z}G$); these yield in the case of $\widetilde{X}$ the homology of X.

9. The exact sequence

Here comes a correction: All the sequences, exact or not, mentioned in our text so far were NOT at all expressed with arrows. The arrow notation for maps $A \longrightarrow B$ with domain A and range B did not yet exist. Maps were just described by words. Arrows occurred together with a certain sequence for the first time in 1941 in a short announcement by Hurewicz [257] which seems to have remained unnoticed. Even in a note by Hurewicz and Steenrod (1941) [260] where the exact homotopy sequence appears implicitly no arrows nor sequences occur. As late as 1947 the importance of arrows and sequences was emphasized by Kelley and Pitcher [271]; they invented the name "exact" and showed that exact sequences play an important role in Algebraic Topology. Immediately this was taken up by topologists and algebraists. The simplification in notation and in concepts was so evident that Henri Cartan said in an Oberwolfach-meeting 1952:

S'il est vrai que la mathématique est la reine des sciences, qui est la reine de la mathématique? La suite exacte!

This plaisanterie was not meant too seriously. But it showed that here was a real improvement, in notation, concept and intuition. Not only sequences, but large diagrams of sequences were used very soon (Eilenberg-Steenrod, Foundations of Algebraic Topology). To express more complicated statements (and to prove them!) without that new notation was almost impossible.

In the *pre*-arrow and *pre*-exact sequences time we (Hopf, the author, and everybody else) used lengthy descriptions of the maps and of the fact that an image was equal to the kernel of another map – or not. It is today, for the authors themselves, but even more so for younger mathematicians, difficult to read the "old" papers.

10. After World War II

During the War a great deal of work was done independently on both sides of the Atlantic. Communication was almost impossible. After the War people got together and were happy to compare results. In the meanwhile Algebraic Topology had become a respectable field, recognized world-wide.

Not only that; the interest in this field seemed to grow every day. People learned about various applications and wanted to understand the techniques, which were more and more simplified and elegant, and useful here and there.

Most famous was certainly Hopf's Theorem [246] on the Betti numbers of compact Lie groups, as follows.

10.1 Hopf algebras

This had occurred in 1939 already. The paper was submitted to Compositio, but that periodical stopped publication. The manuscript found its way to the U.S. and was published in 1941 in Annals of Mathematics [246]. It became really known after the war only. It was a real surprise: the results of Elie Cartan (1936) on the topology of certain compact Lie groups turned out to be a corollary of a topological theorem. It was about closed manifolds provided with a multiplication with unit; the results were valid for all compact Lie groups without using their deep Lie structure. This was exactly what Elie Cartan had asked for, namely to find a general reason for the special topology of compact Lie groups.

The multiplication was used by Hopf to give the cohomology ring of the manifold (modern terminology) a second structure, a co multiplication. Such a superposition was called later a Hopf Algebra; it turned out to be one of the most important concepts, until today, in many fields beyond topology (e.g. theoretical physics).

11. A list of highlights

There was, in the years following 1946, a real explosion of interesting applications of Algebraic Topology to various fields, due to a continuous development of the techniques. We mention only some spectacular ones, with very few explanations.

11.1 Serre 1953

In his Ph.D thesis [429], Serre obtained a wealth of results on the homotopy groups of spheres; before, only very little was known. Serre used the Hopf algebra structure of the cohomology of loop spaces and other recent techniques.

11.2 Cartan-Serre

In the 1953 paper "Variétés analytiques complexes et et cohomologie" [105] cohomology with sheaf coefficients was applied to the Cousin problem in the theory of functions of several complex variables. They consider a complex manifold X and the sheaves Ω and $\mathcal{M}$ of germs of local holomorphic, and meromorphic respectively, functions. Since Ω is contained in $\mathcal{M}$ one has an exact coefficient cohomology sequence

$$... \longrightarrow H^i(X; \Omega) \longrightarrow H^i(X; \mathcal{M}) \longrightarrow H^i(X; \mathcal{M}/\Omega) \longrightarrow H^{i+1}(X; \Omega) \longrightarrow ...$$

where the quotient sheaf is the sheaf of germs of locally given principal parts. $H0(\mathcal{M})$ is the group of global meromorphic functions, and $H0(X; \mathcal{M}/\Omega)$ of global principal parts on X. The existence of a meromorphic function on X with given principal part (additive Cousin problem) is thus guaranteed if $H1(X; \Omega) = 0$. This is proved for Stein manifolds X (complex manifolds with enough holomorphic functions).

11.3 Hirzebruch, 1953/54

The Hirzebruch–Riemann–Roch Theorem for algebraic manifolds [234, 235] expressed, in its simplest form the holomorphic Euler–Poincar'e characteristic in terms of topological invariants (Chern classes). It was based on many topological theories established before (Thom cobordism theory, Steenrod operations, sheaf theory etc). There were later many generalizations, in particular Atiyah-Hirzebruch, "Differentiable Riemann-Roch and K-Theory".

11.3 Bott, 1956

It was known in the thesis of the author already (1942) [148] that the homotopy groups $\pi_i(U(n))$ of the unitary groups $U(n)$ are constant for $n \geq 1/2(i+2)$ for even i and $n \geq 1/2(i+1)$ for odd i: these "stable" groups were known to be $= 0$ for $i = 0, 2, 4$ and $= \mathbb{Z}$ for $i = 1, 3, 5$. Bott [77] proved by very elaborate combination of Morse theory and differential geometry that the stable group is $= 0$ for all even i and $= \mathbb{Z}$ for all odd i (periodicity modulo 2; similar result for the orthogonal groups with periodicity modulo 8). There were later many different and more transparent proofs. Bott's theorem stimulated other developments: topological K-theory, general cohomological functors.

11.4 Adams, 1960 and 1962

In 1960 appeared Adams' theorem [2] about continuous multiplications in $\mathbb{R}^n$ with unit and norm product rule: they exist for $n = 1$, 2, 4, 8 only, with many interesting corollaries (parallelizability of spheres, bilinear division algebras etc). The proof was a real tour de force using the whole range of cohomological techniques developed before. Later the proof could be simplified thanks to topological K-theory and the Atiyah-Hirzebruch integrality results.

In 1962 Adams [Ann.of Math 75] solved completely the vector field problem for spheres: the maximum number of independent tangent vector fields on S^n is exactly the same as the corresponding number for vector fields which are linear with respect to the coordinates of S^n in $\mathbb{R}^{n+1}$ – known long ago. Here no simplification of the proof seems to be known.

12. The climax

12.1 ICM Stockholm, 1962

The International Congress Stockholm witnessed the triumph of Algebraic Topology (after that things calmed down). But there everything was topology even if the field was very different; some connections could always be established. The enthusiasm went very far. A joke went around, even quoted by the Congress president L. Garding at the official dinner: *All the different sections of the Congress should be named "Topology" with some attribute, Algebraic. Differential, Manifold-, Combinatorial, Geometrical, Analytical, Arithmetical, Nu-*

merical, Computational, etc etc, and finally there should even be a Section on Topological Topology!

12.2 Topology and Differential Geometry Zurich, 1960. FIM, 1964

The Swiss Mathematical Society organized in 1960 an international meeting devoted mainly to topology and global geometry. There was great general interest for this "new" field of mathematics. An article for a general public appeared in the Neue Zürcher Zeitung on the front page.

12.3. FIM, 1964

After Zurich 1960 and Stockholm 1962 I felt, and so did many others, that the rapid development in all fields of mathematics – algebraic topology was just a striking example – required much more and different contacts between mathematicians. The idea was that there should be at the Department of the ETH Zurich an institution for inviting people from all over the world, involved in newest research for extended stays in Zurich. Thus professors and students could learn from them and exchange views and problems, and collaboration would be stimulated. The system should be as flexible as possible and provide all necessary facilities for the visitors.

I approached President H. Pallmann of the ETH Zurich. I went to see him and explained the idea, really quite new at that time. After thinking for a few moments he said: "We have no funds, no rooms, no infrastructure for this, nothing. But we will get it. You have the idea, just go ahead".

Before any formal decision, we were allowed to start the Forschungsinstitut für Mathematik on January 1, 1964, with distinguished visitors, among them K. Chandrasekharan and L. Bers.

Social choice and topology.
A case of pure and applied mathematics

Expo. Math. 22 (2004), 385–393

Abstract: The existence of a social choice model on a preference space P is a topological, even homotopical problem. It has been solved 50 years ago, under different terminology, by the author and, a little later, jointly with T. Ganea and P. J. Hilton. P must be an H-space and either contractible or homotopy equivalent to a product of Eilenberg-MacLane spaces over the rationals.

Introduction

This text is largely a historical survey. It is about a case of quite recent unexpected application of algebraic topology to a different field of intellectual enterprise; namely a "Social Choice" model in Economics (see [C]) where results are based on topological arguments. It has turned out that the main topological theorem appears already in a 50 year old paper of mine, of course under different terminology ("Spaces with Means"); there the existence problem is reduced, via the homotopy groups to "Groups with Means". A summary of the arguments is given, together with personal reminiscences of the historical context. In a later publication (1962, jointly with Tudor Ganea and P.J.Hilton) further properties of means were established. Consequences of these may in turn present new aspects of the social choice models.

---

* Lecture delivered at the ETH Zurich, October 24, 2003

1. The new social choice model

1.1. The term social choice appears in models of mathematical economy for a long time already (cf. Remark 1.3.). Here we refer to a new version, a few years old. It is a model for decision making in economic, social, political, etc contexts. It leads to topological, even homotopical problems.

One considers a set P of elements called preferences. It has the structure of a topological space (normally given by a metric). This is natural, because the preferences are numbers, real or complex, point sets in number spaces, vectors, configurations etc. One further considers a society of n agents numbered $1, 2, ..., n$. If $p_j \in P$ is the preference of the agent j the element $(p_1, ..., p_n)$ of the topological product P^n of n copies of P is the "profile" of a society. The *Social Choice* is a function

$$F : P^n \longrightarrow P$$

which associates to each profile $\in P^n$ a social preference $\in P$. It has to fulfill the following properties

(a) Continuity
(b) Unanimity
(c) Anonymity

(a) is a natural requirement: small changes of a profile should lead to small changes of the outcome.

(b) means
$$F(p, p, ..., p) = p$$
for all $p \in P$. In other words, if Δ is the map given by $\Delta(p) = (p, p, ..., p)$ then
$$P \longrightarrow P^n \longrightarrow P$$
is the identity of P. The subspace $\Delta(P)$ of P^n is called the diagonal.

(c) means that F is invariant under all permutations of the indices $1, 2, ..., n$, i.e. under the symmetric group Σ_n (all agents are equal).

1.2. The existence of such a map F according to properties of the space P is clearly a topological problem. It is even a homotopy problem: One actually is looking for a map of the symmetric product P^n/Σ_n (one identifies points of P^n equivalent under all permutations of the indices) to P. On the diagonal, which survives in P^n/Σ_n, the map is given and it has to be extended to all of P^n/Σ_n. But the extension property of a continuous map only depends on the homotopy type of the spaces involved – we avoid pathologies and assume throughout that P is a "polyhedron", i.e. a cell complex; and that it is connected.

The existence problem of such a function F from P^n to P with properties (a), (b), (c) has been treated 50 years ago in one of my papers [E], of course under different terminology. I did not know the term social choice. On the other hand it is clear that the economists' group investigating the new social choice concept, G.Chichilnisky and her collaborators (cf [C]) did not know of my old paper. I was informed of this new work through a paper by Ch.Horvath [H] published in 2001; there it is mentioned that "the fundamental result ... was established by B.Eckmann in 1954".

1.3. Remark. In the famous work of Arrow (see e.g. [K-S]) the term social choice is used in a different sense (this is why I say "the new social choice model"). There the social choice function σ from P^n to P must fulfill (b), among other things, but (c) is not required. From Arrow's axioms, it follows that σ is necessarily a projection onto one copy, say number k, in P^n, i.e $\sigma(p_1,...,p_n) = p_k$ for all elements of P^n: agent number k is a dictator. This cannot happen in the "new " model because of (c). Moreover in [K-S] the preference set is not a topological space, but consists of preorders of a certain set X.

2. Spaces with means

2.1. The paper I published in 1954 is on "Spaces with means" (Räume mit Mittelbildungen). A mean, or an n-mean, or a generalized mean, in a connected space P is exactly the same as a social choice function F above.

Elementary examples of n-means are the arithmetic mean of n numbers in an interval of the real line, and the geometric mean.

Generalized means have a long history. They started in a paper by Kolmogorov in 1930 [K] on quasi-arithmetic means and were continued by various authors in connection with analytic or arithmetic questions. An important step was the thesis of G.Aumann (1933, under Caratheodory). In a further paper (1935) Aumann formulated the conditions (a), (b), (c) above, mainly for subsets of number spaces. In 1943 he states that the existence is a topological problem for arbitrary spaces; he solves it, however, for a few very special cases only. Thus the problem was open and well-known at the time I decided to look at it from the viewpoint of algebraic topology, with emphasis on algebraic.

2.2. At this point a historical reminiscence may be of interest. Let me recall that the period 1950 to 1954 was very significant not only for algebraic topology and algebra, but also for other fields of mathematics. It was the time when categories and functors became mathematical tools; they had been, in the original version of Eilenberg and MacLane, rather a language to formulate naturality. But now functors from modules to modules or Abelian groups were used by Cartan and Eilenberg (their book appeared in 1956, more general categories are mentioned in an appendix by D.Buchsbaum). The book "Foundations of Algebraic Topology" by Eilenberg and Steenrod (1952) is about functors from spaces to Abelian groups, but categories and functors appear in Chap.IV.

2.3. In order to transform the n-mean in a space into an n-mean in a group I applied functors from spaces to groups (continuous maps to homomorphisms). This clearly works if the functor preserves products so that the power P^n of n copies of P becomes the direct power of n copies of the corresponding group.

Each homotopy group π_i is such a functor. We first recall some of the properties of $\pi_i(P)$. The elements of $\pi_i(P)$ are homotopy classes of maps of the i-sphere S^i into P. (Actually a base-point in S^i and in P has to be chosen and respected in all maps and homotopies. However the structure of $\pi_i(P)$, for arc-connected spaces, is independent of the choice of base-points.) The group operation is similar to that of the fundamental group (case $i = 1$) where the group need not be abelian; however for $i \geq 2$ it is abelian. Composition with a map $P \longrightarrow P'$ yields an induced map $\pi_i(P) \longrightarrow \pi_i(P')$ which is a homomorphism, compatible with further compositions of maps and such that the identity map induces the identity isomorphism – i.e. π_i is indeed a functor. Moreover maps of S^i into a topological product are determined by the maps into the factors, i.e. π_i preserves products.

Let us also mention in that context that a space where all π_i, $i \geq 2$, vanish is called aspherical. If it is, moreover, simply connected then a famous theorem of J.H.C. Whitehead tells that the space is contractible (can be deformed in itself into a point); this is true for spaces which are connected polyhedra as considered here.

2.4. Thus the passage from spaces P with an n-mean to their homotopy groups $\pi_i(P)$ yields groups with n-means as discussed in the next section. That passage can be done in one stroke using the functor concept. In my old paper I preferred, however, to establish everything by explicit computation – it seems that the time was not ripe to use the functorial concept in a publication!

Proposition 1. An n-mean in the space P induces a homomorphic n-mean in each homotopy group $\pi_i(P)$, $i = 1, 2, ...$

3. Groups with means

3.1. We consider an arbitrary group G, Abelian or not, but use additive notation, so that the neutral element is 0. An n-mean in G is a function

$$f : G^n \longrightarrow G$$

with properties (a) homomorphism, (b) and (c) as before, for some $n \geq 2$. It implies strong restrictions on G.

First we note that for any $x \in G$

$$f(x, 0, ..., 0) = f(0, x, 0, ..., 0) = .. = f(0, ..., 0, x)$$

written $g(x)$. Now $g(x + y) = g(x) + g(y)$ and

$$g(x) + g(y) = f(x, 0, 0, ..., 0) + f(0, y, ..., 0) = f(x, y, 0, ..., 0)$$

whence $g(x) + g(y) = g(y) + g(x)$. Further

$$ng(x) = g(nx) = f(x, x, ..., x) = x.$$

Thus $x + y = y + x$, i.e. G is Abelian. Multiplication in G by n is therefore an endomorphism of G; since by the above g is its inverse it follows that multiplication by n is an automorphism of G.

Theorem 2. If the group G admits an n-mean for some $n \geq 2$ then it is Abelian and multiplication by n is an automorphism of G, and G is uniquely divisible by n.

3.2. We can write n^{-1} for the automorphism g above. Since

$$nf(x_1, ...x_n) = ng(x_1 + ... + x_n) = x_1 + ... + x_n$$

it follows that

$$f(x_1, ..., x_n) = n^{-1}(x_1 + ... + x_n.$$

Thus there is only one n-mean on G, the "arithmetic mean".

Conversely, if the Abelian group G is divisible by n then it admits an n-mean (and only one).

3.3. From Theorem 2 it follows that if G is finitely generated and admits an n-mean then the order of any element must be prime to n; in particular there cannot be elements of infinite order. If G is finitely generated and admits n-means for all n then G is $= 0$.

4. Application to spaces

4.1. Applying the results on groups to spaces with means it follows that a space P with an n-mean for $n \geq 2$ has Abelian fundamental group and all its homotopy groups $\pi_i(P)$ are uniquely divisible by n. Example. $P = S^k$, $k \geq 1$. Since $\pi_k(S^k) = \mathbb{Z}$, the sphere S^k does not admit any n-mean, $n \geq 2$.

4.2. Let P be a polyhedron (cell-complex) with finitely generated integral homology groups. We assume that it admits n-means for all n. Since $\pi_1(P)$ is Abelian, it is isomorphic to $H_1(P)$, the first homology group of P with integer coefficients, whence finitely generated and divisible by all n, and thus $= 0$. By the Hurewicz theorem $\pi_2 = H_2$, finitely generated and divisible by all n, whence $= 0$, etc. Thus all $\pi_i(P)$ are zero, and the theorem of J.H.C.Whitehead tells us that P is contractible.

Theorem 3. If a polyhedron with finitely generated homology groups admits n-means for all n then it is contractible.

This applies in particular to a finite polyhedron (consisting of finitely many cells).

Conversely any contractible space P admits n-means for all n, since P is homotopy equivalent to a point.

4.3. Serre's generalization of the Hurewicz theorem modulo a class $\mathcal{C}$ of abelian groups tells, in our case, that all homotopy groups being uniquely divisible by n is equivalent to this holding for all homology groups (with integer coefficients). I knew already Serre's theorem but proved directly that homology groups are divisible by n; I probably thought again that using new generalizations would make the paper too difficult.

Thus, for example, a closed orientable manifold cannot admit any n-mean for $n \geq 2$ since in the top dimension the homology group is infinite cyclic.

5. Non-contractible spaces with means, H-spaces

5.1. The discussion of groups with means in Section 3 yields further results. They are mostly contained in a later paper (1962) by Ganea, Hilton and myself [EGII].

A non-contractible (infinite) polyhedron admitting n-means for all n can be constructed as follows. We consider $P = K(\mathbb{Q}, k)$, the Eilenberg-MacLane space with $\pi_i(P) = 0$ for $i \neq k$ and $= \mathbb{Q}$ for $i = k$ where k is an arbitrary integer ≥ 1. Such a space can be constructed by means of cells (starting with a k-sphere and using mapping cylinders); it is finite dimensional if k is odd. Since the homotopy group of a topological product is the direct product of the homotopy groups of the factors we have $P^n = K(\mathbb{Q}^n, k)$. For any two spaces $K(G_1, k)$ and $K(G_2, k)$ the homotopy classes of maps are in bijective correspondence with the homomorphisms $G_1 \longrightarrow G_2$. The group $\mathbb{Q}$ is divisible by all n. Thus the n-mean $\mathbb{Q}^n \longrightarrow \mathbb{Q}$ defines a homotopy class of maps $P^n \longrightarrow P$. We can take a map which is the identity on the diagonal $\Delta(P)$; it is homotopically Σ_n-invariant. By a lemma of Grothendieck on cohomology with rational coefficients (the homotopy classes of maps from a space to $K(\mathbb{Q}, k)$ constitute such a cohomology group) we can obtain a map which is strictly Σ_n-invariant.

Theorem 4. $K(\mathbb{Q}, k)$ admits n-means for all n.

5.2. This example turns out to be fundamental for any (non-contractible) space with n-means for all n. We first show ([EGH]) that such a space is an H-space; it suffices to assume that there is an n-mean for a single $n \geq 2$.

We recall that an H-space is a space X admitting a continuous multiplication $\mu : X \times X \longrightarrow X$ with two-sided unit $e \in X$ up to homotopy. This means that the two maps $X \longrightarrow X$ defined by $\mu(p, e)$ and $\mu(e, p)$, $p \in X$ are homotopic to the identity.

Let $F : P^n \longrightarrow P$ be an n-mean for some $n \geq 2$, and e a point $\in P$. The map $\phi : P \longrightarrow P$ defined by $\phi(p) = F(p, e, ..., e)$ for all $p \in P$ induces in each homotopy group of P precisely the homomorphism called g in 3.1 which is an automorphism. Thus ϕ induces isomorphisms in all homotopy groups and is therefore, by the theorem of J.H.C.Whitehead, a homotopy equivalence. Let ψ be its inverse so that $\psi\phi$ and $\phi\psi$ are both homotopic to the identity of P.

We now put for $p, q \in P$

$$\mu(p, q) = \psi F(p, q, e, ..., e).$$

Then $\mu(p, e) = \psi F(p, e, ..., e) = \psi\phi(p)$ for all $p \in P$. This map is homotopic to the identity of P, and the same argument works for $\mu(e, p)$.

Theorem 5. If P admits an n-mean for some $n \geq 2$ then it is an H-space.

5.3. Using a theorem of W.Browder [B] we can draw a very strong conclusion from that result. If P is a finite polyhedron (even a little more generally) and an H-space then Browder's theorem tells that either P is contractible or fulfills Poincaré duality like an orientable manifold. Thus some homology group $H_k(P)$, $k \geq 1$ is infinite cyclic so that no n-mean, $n \geq 2$ can exist.

Theorem 6. If a finite polyhedron admits an n-mean for some $n \geq 2$ then it is contractible – and thus all n-means exist.

Note that this is much stronger than the version of Theorem 3 for finite polyhedra.

5.4. For (infinite) non-contractible polyhedra P admitting n-means for all n the H-space poperty yields a complete description. Using more elaborate techniques due essentially to Hopf and Serre, and the fact that all homotopy groups of P are $\mathbb{Q}$-vector spaces one can prove:

Theorem 7. If P (non-contractible) admits n-means for all n then it is of the homotopy type of a topological product of Eilenberg-MacLane spaces $K(\mathbb{Q}, k_\nu)$.

Here interesting consequences for the social choice model must be mentioned. Take for the preference space P a product as in Theorem 7 with social choice function F and an arbitrary number n of agents. Then for any constant $d > 0$ there are profiles $(p_1, ..., p_n)$ such that the distance between $F(p_1, ..., p_n)$ and p_j is $> d$ for all j. [We assume that the topology is given by a suitable metric.] This means that no agent gets approximately what he wants, to the contrary! Would the agents agree to a solution by mediation?

The fact above is due to the construction of P which starts with spheres, and these do not admit any n-mean with $n \geq 2$. There are more precise ways to express the result. I just wanted to show that except for the contractible case either no social choice function can exist on P, or if it exists for all n then unexpected properties turn up.

5.5. Remark. I would like to mention a recent paper by Shmuel Weinberger [W] on the topological aspects of the social choice model. Weinberger did not know of our papers [E] and [EGH] on generalized means; these contain all topological results of [W] – except Theorem 7 above: we knew it but did not consider it important enough! In [W] its significance is explained very nicely and there are some further interesting observations concerning various aspects of the social choice model.

References

[B]   W.Browder, On torsion in H-spaces, Ann.Math.(2) 74 (1961), 24-51

[C]   G.Chichilnisky, Intersecting families of sets and the topology of cones in economics, Bull.AMS 29(1993), 189-207

[E]   B.Eckmann, Räume mit Mittelbildungen, Comment.Math.Helv. 28(1954), 329-340

[EGH] B.Eckmann, T.Ganea, P.J.Hilton, Generalized means, Studies in Mathematical Analysis, Stanford University Press(1962), 82-92

[H]   C.D.Horvath, On the topological social choice problem, Soc.Choice Welfare 18(2001), 227-250

[K-S] Alan P.Kirman and Dieter Sondermann, Arrow's Theorem, many agents, and invisible dictators, CORE Discussion Paper 7142 (1971/72)

[K]   A.Kolmogorov, Sur la notion de moyenne, Rendiconti Acad.dei Lincei 12(1930), 388-391

[W]   S.Weinberger, On the topological social choice model, preprint 2003, to appear in Soc.Choice Welfare

Acknowledgements

The author and Springer-Verlag thank the original publishers for permission to publish the articles in this volume.

American Mathematical Society: 3, 12, 16
Elsevier: 8, 10, 13, 21
Institut Mittag-Leffler: 6
Israel Journal of Mathematics: 17
Istituto Nazionale di Alta Matematica: 7
L'Enseignement Mathématique: 14
Mathematical Association of America: 11
Naturforschende Gesellschaft in Zürich: 2
Neue Zürcher Zeitung: 18
Newsletter of the European Mathematical Society: 19
Princeton University Press: 9
Société Mathématique de France: 5